OXFORD STATISTICAL SCIENCE SERIES

# OXFORD STATISTICAL SCIENCE SERIES

A. C. Atkinson: *Plots, Transformations, and Regression*
M. Stone: *Coordinate-Free Multivariable Statistics*

# Plots, Transformations, and Regression

An introduction to graphical methods of diagnostic regression analysis

A. C. ATKINSON
*Imperial College, London*

CLARENDON PRESS · OXFORD

*Oxford University Press, Walton Street, Oxford OX2 6DP*
*Oxford New York Toronto*
*Delhi Bombay Calcutta Madras Karachi*
*Petaling Jaya Singapore Hong Kong Tokyo*
*Nairobi Dar es Salaam Cape Town*
*Melbourne Auckland*
*and associated companies in*
*Beirut Berlin Ibadan Nicosia*

*Oxford is a trade mark of Oxford University Press*

*Published in the United States*
*by Oxford University Press, New York*



*First published 1985*
*Reprinted (with corrections) 1987*
*First published in paperback (with corrections) 1987*



*British Library Cataloguing in Publication Data*
*Atkinson, A. C.*
*Plots, transformations, and regression: an introduction to graphical methods of diagnostic regression analysis.—(Oxford statistical science series)*
*1. Regression analysis—Data processing*
*I. Title*
*519.5'36'02854 QA278.2*
*ISBN 0-19-853359-4*
*ISBN 0–19–8533713 (Pbk)*

*Library of Congress Cataloging in Publication Data*
*Atkinson, A. C. (Anthony Curtis)*
*Plots, transformations, and regression.*
*(Oxford statistical science series)*
*Bibliography: p.*
*Includes index.*
*1. Regression analysis. I. Title. II. Series.*
*QA278.2.A85 1985 519.5'36 85-5140*
*ISBN 0-19-853359-4*
*ISBN 0–19–8533713 (Pbk)*

*Printed in Great Britain by St Edmundsbury Press,*
*Bury St Edmunds, Suffolk*

For
Alison and Rachel,
who insisted

# Preface

This book is about methods of checking statistical models and the data to which they are fitted. The main interest is in multiple regression models and least squares. The essential feature of the procedures which are described is that they are suitable for routine use. Thus they require little extra computation above that required to fit the model. The results are often best presented as plots which have a more immediate impact than lists of numbers and so are more suitable for routine checking. A selection of the methods is available in several statistical packages under the name regression diagnostics. I hope the book will provide a clear introduction to these methods and suggest useful extensions to them.

The material of the book falls into three main parts. Chapters 1–5 describe methods of checking regression models for inadequacies both of the model and of the data. Inadequacies of the data include the presence of outliers and of individual observations which strongly influence the conclusions drawn from the data. Chapters 6–9 are concerned with the transformation of variables in an equation, especially the response. Problems of influence are again important. The remaining three chapters are concerned with more advanced and specialized topics including generalized linear models and robust estimation.

Because this is a book about statistical methods, examples are important. The theoretical discussion is accordingly accompanied by the analysis of 14 sets of data, many of which make several appearances during the course of the book. The analyses are illustrated by over 150 plots.

In writing the book I have had three groups of readers in mind. One group consists of scientists and engineers with some experience in analysing their own data. These readers will probably find the first six chapters of greatest interest. Another group consists of undergraduate and postgraduate students in statistics. I trust both groups will find the book of assistance in interpreting diagnostic output from computer regression packages. The mathematical level of the book is low—an introductory statistics course including some exposure to regression analysis is all that is needed. The third group, consisting of academic statisticians, will find that some of the discussions in later chapters indicate the need for further research. I hope they will feel tempted to undertake it.

My interest in this area of statistics began with a talk to a joint university–

industry symposium at the University of Wales, Aberystwyth in the autumn of 1979. Subsequently I have benefited greatly from the helpful comments and interest of many statisticians including my colleagues at Imperial College, especially David Cox, Peter McCullagh and Richard Smith and my ex-colleague Brian Ripley. My first attempt to give large-scale order to my ideas on the subject was prompted by the invitation from Philip Dawid to give a course of postgraduate lectures on regression diagnostics. Subsequently Professor Dawid read the manuscript of the book and made many helpful comments, for which I am grateful. The idea that I should write a book on this subject came originally from Maurits Dekker. At the moment of finishing the two years' labour that writing the book has entailed, it is not clear to me how grateful I should be to him.

*October 1984* A. C. A.
London

# Contents

# List of examples

# 1
# Introduction

## 1.1 Strange observations

In its simplest form, regression analysis is concerned with the fitting to data of models in which there is a single continuous response. It is hoped that the response depends upon the values of several explanatory variables. The response variable is observed with error, whereas the explanatory variables are taken to be known without error. The relationship between the two sets of variables depends upon the values of a set of unknown parameters which are estimated by least squares. That is, perhaps after a transformation of the response, those values of the parameters are found which minimize the sum of squared deviations between the observations and the predictions from the model. In linear regression the model depends linearly on the vector of unknown parameters. Two more general classes of models consist of those which are nonlinear in the parameters and generalized linear models in which the response need not be continuous and in which the dependence is on a function of a linear model. Although both classes of model are mentioned in this book, the principal concern is with linear multiple regression, in which the response may depend upon several explanatory variables.

The mathematics of multiple regression has long been understood. As a tool for the statistical analysis of data the method has been widely used since the advent of programmable electronic computers. An influential statistics book from the beginning of this period is Brownlee (1960), which describes multiple regression in Chapter 13. The emphasis is on the derivation of the normal equations yielding the least squares estimates of the parameters and on numerical methods for solving the equations, suitable both for electronic computers and for desk calculators. There is one famous example on the oxidation of ammonia which we consider in Chapter 6. The analysis given by Brownlee consists of fitting two models and of testing hypotheses about the values of the parameters. Graphical techniques are not used. The data are not plotted, nor are the residual differences between the observations and the fitted model examined.

Since Brownlee's book was published there has been an appreciable change in emphasis in statistical analysis. With the wide availability of graphics terminals and of high-quality graphics for microprocessors, it is now easy to explore the properties of many models rather than difficult to fit only a few. For each of the fitted models a wide variety of quantities can then be plotted.

One of the main purposes of the present book is to provide copious examples of the use of many kinds of plots in the construction and assessment of regression models.

A particular use of these plots, to which the name regression diagnostics has become attached, is to determine whether there is anything strange about one or more of the observations. Data which are in some way strange can arise in several ways. Some possibilities are as follows.

1. There may be gross errors in either response or explanatory variables. These could arise from incorrect or faulty measurements, from transcription errors or from errors in key punching and data entry.
2. The linear model may be inadequate to describe the systematic structure of the data.
3. It may be that the data would be better analysed in another scale, for example after a logarithmic or other transformation of the response.
4. The systematic part of the model and the scale may both be correct, but the error distribution of the response is appreciably longer tailed than the normal distribution. As a result least squares may be far from the best method of fitting the model.

One potent source of gross errors is omission of the decimal point in data which is being read in using the F format in Fortran. Depending upon the circumstances, the number entered can be multiplied or divided by a power of ten. For this reason free format input, in which the position of the decimal point is not pre-specified, is safer.

The particular discrepancy between the data and the model will determine whether the departure is easy to detect. One difficulty is that the fitted model may try to accommodate points which are in some way 'strange', leading to residual differences between the data and the fitted model which are not sufficiently large to be noteworthy. In fact, in some cases, a wrong value of one or more explanatory variables can lead to a very small residual for the observation in question.

The material presented in this book provides a series of methods which are sensitive to the first three kinds of discrepancy in the list above. The fourth departure in the list, that of the distribution of the response, is more amenable to treatment by the methods of robust regression described in Chapter 12. In Chapters 1–5 interest is in obtaining diagnostic information on response and explanatory variables. In the second part of the book, Chapters 6–9, the concern is with information for a transformation of the response. The effect of individual observations, or of groups of observations, on inferences drawn from the data, is determined by examining the effect of deletion of subsets of the data. Although this might seem a formidable computational undertaking, the deletion formulae derived in Section 2.2 mean that such information can be obtained from standard regression calculations. The diagnostic checks on data and models that are derived in the succeeding chapters should require little extra computational effort above that needed to fit the regression equation.

Of course, identification of some observations as strange is not an end in itself. The observations may be informative, for example about improvements to the model, or they may simply be wrong. If the discrepancy remains after the data have been checked and all plausible models exhausted, further observations may have to be taken. Or it may be that the observations diagnosed as strange have no influence on the conclusions drawn from the data and so can be ignored with impunity. One of the main methods used in this book is the identification of observations which are influential, in the sense that they are crucial to the inferences drawn from the data.

## 1.2 Residual plots

Several plots have long been used to detect certain departures from the fitted model. The popularity of the plots goes back at least as far as the first edition of Draper and Smith (1966) who give some examples. The most useful plots seem to be:

1. A scatter plot of the response $y$ against each of the explanatory variables $x_j$.

2. A plot of the residuals, that is of the differences between observed and predicted responses, against each explanatory variable in the model. The presence of a curvilinear relationship suggests that a higher-order term, perhaps a quadratic in the explanatory variable, should be added to the model.

3. A plot of the residuals against explanatory variables not in the model. A relationship would suggest that the variable should be included in the model.

4. A plot of the residuals against predicted values from the fitted model. If the variance of the residuals seems to increase with predicted value, a transformation of the response may be in order.

5. A normal plot of the residuals. After all the systematic variation has been removed from the data, the residuals should look pretty much like a sample from the normal distribution. A plot of the ordered residuals against the expected order statistics from a normal distribution, that is the normal scores, provides a visual check of this assumption.

6. Data are often collected in time order. Even if time is not one of the explanatory variables included in the model, plots of the response and residuals should, if possible, be made against time. These plots sometimes lead to the detection of unsuspected patterns due either to time itself or to other variables closely correlated with time. One example in an experiment with which I was associated was the change in ambient temperature inside an

inadequately insulated building during a two-month period in the spring. Such unsuspected variables are sometimes called 'lurking' variables. Examples of plots exhibiting these unsuspected effects are given by Joiner (1981).

Part of the argument of this book is that these plots are sometimes not sufficiently powerful and can, in fact, be misleading. Preferable alternatives for many of these purposes are derived in the next two chapters, with examples. But for a first example we consider a set of data in which the residual plots do exhibit the main shortcoming of the observations.

**Example 1 Forbes' data 1** The data in Table 1.1, described in detail by Weisberg (1980, pp. 2–4), are used by him to introduce the ideas of regression analysis. There are 17 observations on the boiling point of water in °F at different pressures, obtained from measurements at a variety of elevations in the Alps. The purpose of the original experiment was to allow prediction of pressure from boiling point, which is easily measured, and so to provide an estimate of altitude. Weisberg gives values of both pressure and $100 \times \log$ (pressure) as possible responses. For the moment we follow him and use the latter as the response. The variables are therefore:

$y$: $100 \times \log$ (pressure, inches of mercury)
$x$: boiling point, °F.

**Table 1.1** Example 1: Forbes' boiling point data

| Observation | Boiling point (°F) | Pressure (in. Hg) | 100 × log (pressure) |
|---|---|---|---|
| 1 | 194.5 | 20.79 | 131.79 |
| 2 | 194.3 | 20.79 | 131.79 |
| 3 | 197.9 | 22.40 | 135.02 |
| 4 | 198.4 | 22.67 | 135.55 |
| 5 | 199.4 | 23.15 | 136.46 |
| 6 | 199.9 | 23.35 | 136.83 |
| 7 | 200.9 | 23.89 | 137.82 |
| 8 | 201.1 | 23.99 | 138.00 |
| 9 | 201.4 | 24.02 | 138.06 |
| 10 | 201.3 | 24.01 | 138.05 |
| 11 | 203.6 | 25.14 | 140.04 |
| 12 | 204.6 | 26.57 | 142.44 |
| 13 | 209.5 | 28.49 | 145.47 |
| 14 | 208.6 | 27.76 | 144.34 |
| 15 | 210.7 | 29.04 | 146.30 |
| 16 | 211.9 | 29.88 | 147.54 |
| 17 | 212.2 | 30.06 | 147.80 |

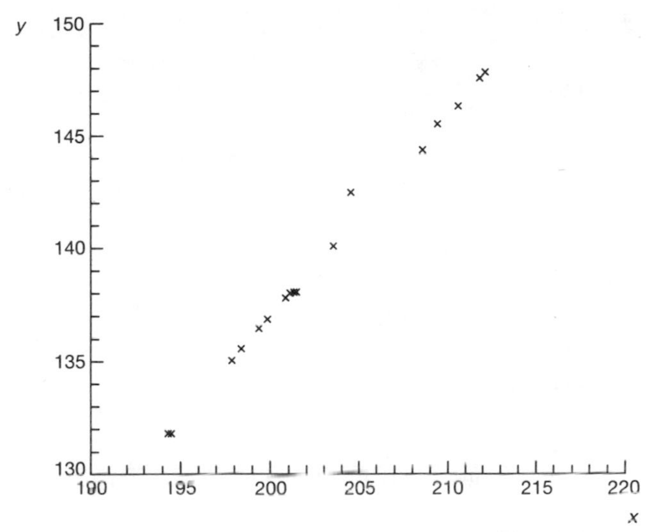


**Fig. 1.1** Example 1: Forbes' data: scatter plot of $y$ against $x$

Figure 1.1 shows a scatter plot of $100 \times \log$ (pressure) against boiling point. There is clearly a strong linear relationship between the two variables, although one point seems not to lie quite on the line. Linear regression of $y$ on $x$ yields a residual sum of squares of 2.15 on 15 degrees of freedom, since there are 17 observations and 2 parameters have been estimated. The $t$ value of 54.45 for the test of the significance of the regression confirms what is abundantly obvious from the plot, that there is a highly significant relationship between the two variables. However, the plot of residuals against $x$ in Fig. 1.2 shows a series of small residuals, mostly negative, with a large positive residual for observation 12. Because of the strong relationship between $y$ and $x$, the plot of residuals against predicted values is virtually indistinguishable from Fig. 1.2 and so is not shown. The normal plot of the residuals, Fig. 1.3, consists of a slightly sloping straight line from which the residual for observation 12 is appreciably distanced.

The strong evidence from Figs. 1.2 and 1.3 supports the suggestion from the scatter plot of Fig. 1.1 that there is something strange about observation 12. It could be that either the $x$ value or the $y$ value is wrong. In the absence of any formal approach at this stage, we merely delete observation 12 and refit the model. The effect is dramatic. The residual sum of squares decreases from 2.15 to 0.18 and the $t$ statistic for the regression increases from 54.45 to 180.73. The residual plots also show no evidence of any further departures. The plot of residuals against $x$, Fig. 1.4, shows a scatter of positive and negative values with no observable trend with $x$. The normal plot of the residuals, Fig. 1.5, is, if not straight, at least a decided improvement on Fig. 1.3.

This example provides a clear illustration of the use of plots of residuals to

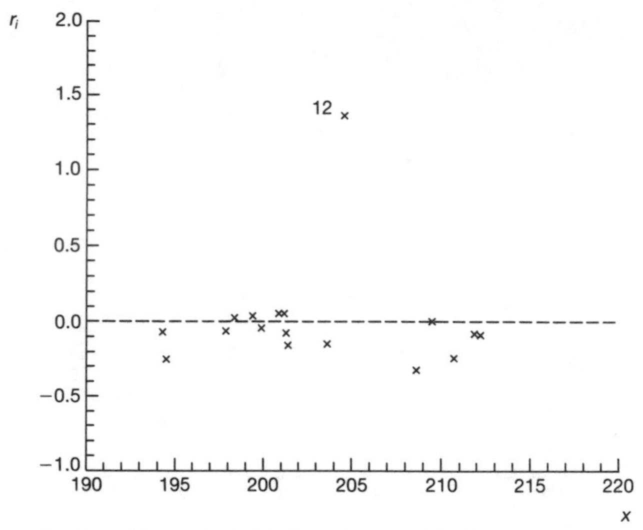


**Fig. 1.2** Example 1: Forbes' data: residuals $r_i$ against $x$

identify an outlying observation which does not agree with the rest of the data. It remains to discuss the choice of the scale of $y$ in which the analysis was conducted. We chose to work with log (pressure). An alternative is to work with pressure itself, rather than its logarithm, as the response. The difference this choice of response makes to the analysis of the data is described in Chapter 6, where methods for the analysis of transformations are developed. □

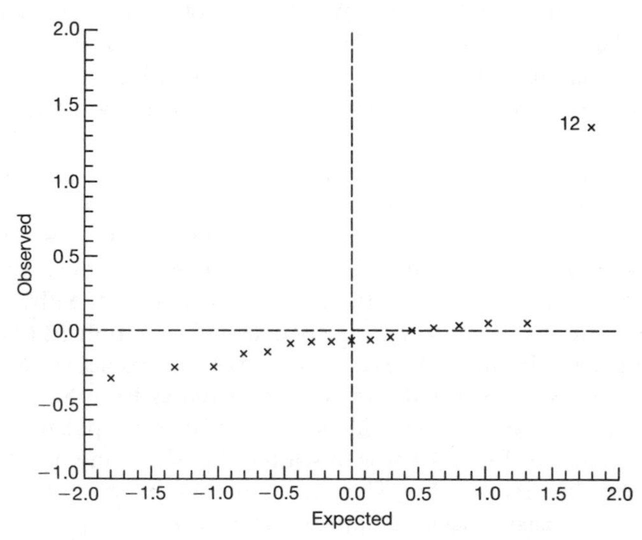


**Fig. 1.3** Example 1: Forbes' data: normal plot of residuals

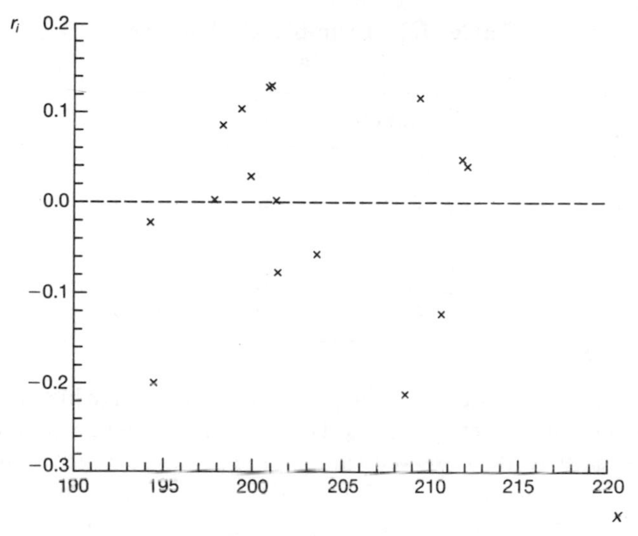


**Fig. 1.4** Example 1: Forbes' data with observation 12 deleted: residuals $r_i$ against $x$

## 1.3 Leverage

In Example 1, Forbes' data, it was easy to see that there was something strange about observation 12 because the explanatory variable fell in the middle of the range of $x$ values. It is however much less easy to see if there is something strange about an observation which has an extreme value of one

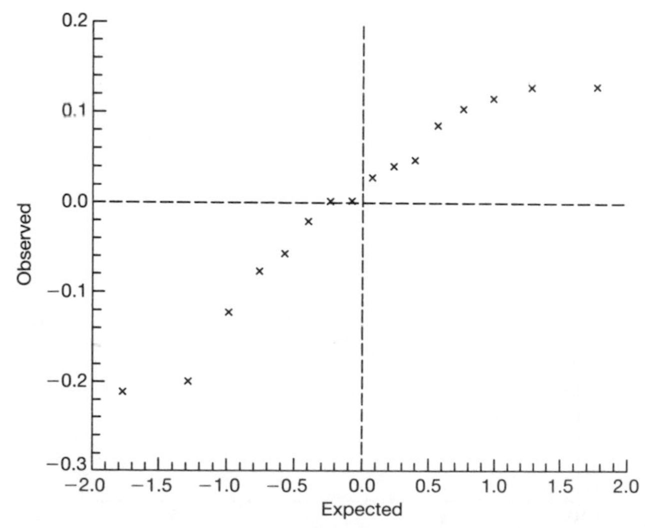


**Fig. 1.5** Example 1: Forbes' data with observation 12 deleted: normal plot of residuals

**Table 1.2** Example 2: Huber's 'data'

| Observation | $x$ | $y$ |
|---|---|---|
| 1 | −4 | 2.48 |
| 2 | −3 | 0.73 |
| 3 | −2 | −0.04 |
| 4 | −1 | −1.44 |
| 5 | 0 | −1.32 |
| 6 | 10 | 0.00 |

or more explanatory variables. Such points are said to have high leverage and are sometimes called leverage points. The effect of high leverage is to force the fitted model close to the observed value of the response. One result is a small residual. To exhibit this effect, and some of the difficulties to which it can give rise, consider the following small synthetic example used by Huber (1981, Chapter 7) to introduce his discussion of robust regression.

**Example 2 Huber's 'data' 1** There are only six pairs of $x$ and $y$ values, which are given in Table 1.2. The scatter plot of $y$ against $x$, Fig. 1.6, suggests that one possible model would be a straight line through five points, with observation 6 disregarded. As an alternative suppose a straight line were fitted to all the data. A plot of the resultant residuals against $x$ is shown in Fig. 1.7. The largest residual belongs not to observation 6 but to observation

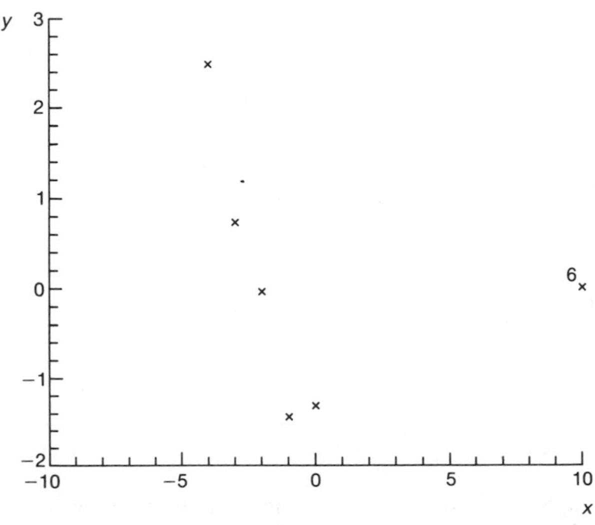


**Fig. 1.6** Example 2: Huber's 'data': scatter plot of $y$ against $x$

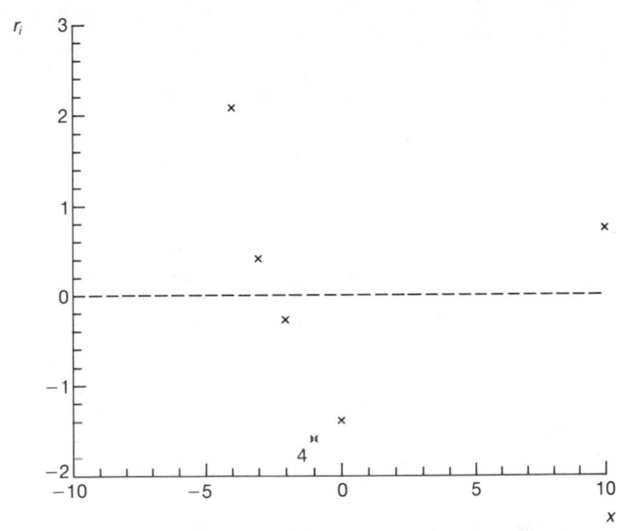


**Fig. 1.7** Example 2: Huber's 'data': residuals $r_i$ against $x$

4. Inspection of Fig. 1.7 might lead to the conclusion that a quadratic should be fitted in $x$. The $t$ value for the quadratic term in such a model is 7.37 which, even with only three degrees of freedom for error, is significant at the 1 per cent level. A plot of the residuals from this quadratic fit against $x$, Fig. 1.8, shows no peculiarities, although it is noteworthy that observation 6 has a very small residual. The normal plot of the residuals, Fig. 1.9, falls on a straight line; the quadratic fit is obviously satisfactory. □

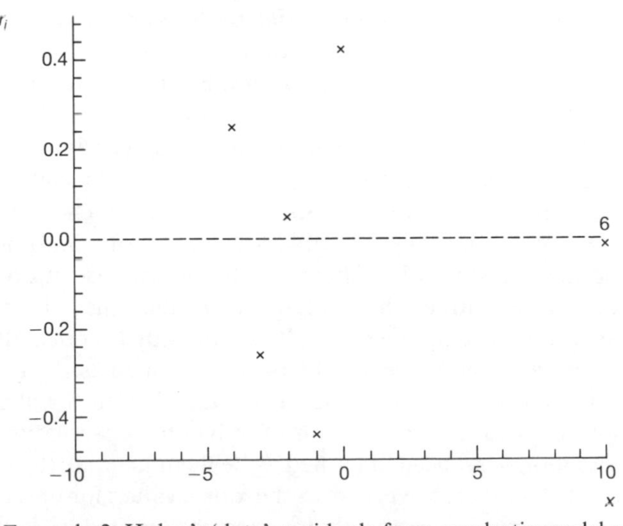


**Fig. 1.8** Example 2: Huber's 'data': residuals from quadratic model against $x$

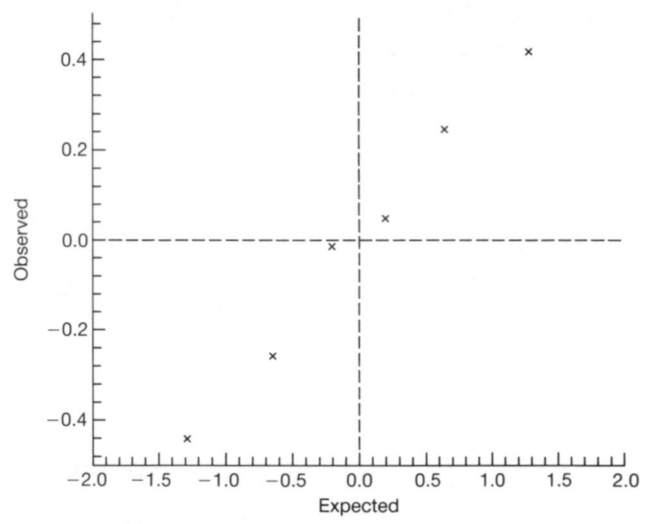


**Fig. 1.9** Example 2: Huber's 'data': normal plot of residuals from quadratic model

Although this small synthetic example is of no scientific interest and our analysis, which continues in later chapters, is over elaborate, it does illustrate many of the difficulties which have given rise to the diagnostic methods described in the later chapters of this book. The most important is that because observation 6 has high leverage, the residual associated with it is small. The contribution that observation 6 is making is therefore not detected by plots of residuals, in the way that observation 12 was detected in Example 1. Standard regression procedures in Example 2 led to fitting a quadratic model. Checking of the regression model by looking at residuals fails to reveal anything strange about observation 6.

One way in which the effect of observation 6 can be detected is to fit the model with and without it. Omission in turn of observations 1–5 in Example 2 would have little effect on the fitted model, whereas omission of observation 6 would cause the fitted model to change dramatically through exclusion of the quadratic term. Because observation 6 has such an effect on the fitted model it is said to be an influential observation. An observation with high leverage need not necessarily be influential. If the response at observation 6 had been such as to continue the straight line of the other observations, its omission would not have appreciably altered the fitted model, although the variance of the estimated slope would have been increased. In succeeding chapters the technique of omitting each observation in turn will be the chief method employed for the identification of influential observations.

Once observation 6 has been identified as being influential, the next stage is to determine whether there are errors in the $x$ or $y$ values for observation 6 or whether it is correctly indicating that a straight-line model is inadequate.

This requires recourse to sources of information beyond the bare data in Table 1.2, which we forego until the conclusion of the analysis of this example in Section 4.2. One purpose of the example is to suggest how the diagnostic technique of omitting one observation at a time can call attention to important observations. For outlying response values in densely sampled regions, as in Example 1, it is easy enough, by use of standard residual plots, to detect that something strange or discordant is happening. But, as was shown in Example 2, leverage points can give rise to patterns of residuals from which they are not easily detected. In the case of Example 2 the configuration was easily identifiable from the scatter plot of $y$ against $x$ (Fig. 1.6). But with two or more explanatory variables there can be influential observations at points of high leverage which are not detectable by marginal plots of observations or residuals against individual explanatory variables. Plots of pairs of explanatory variables may reveal information about leverage, but can tell us nothing about influence. The techniques of diagnostic regression analysis provide a way of identifying such influential observations.

## 1.4 Computing

Many of the diagnostic procedures described in this book are available in several statistical packages. For example, at the time of writing in 1984, BMDP can provide an almost overwhelming choice of diagnostic quantities derived from residuals. Several, but not all, versions of Minitab also provide diagnostic information, although this is not at present described in the *Minitab Student Handbook* (Ryan *et al.*, 1976). Because diagnostic methods are easily programmed, the situation can be expected to change radically in the lifetime of this book. It is therefore not appropriate to give a detailed description of what is currently available. Rather I hope this book may serve as a useful complement to a wide variety of packages and the related manuals.

## 1.5 Further reading

An excellent introduction to regression analysis, including some of the diagnostic material to be described in the next few chapters, is given by Weisberg (1980). An introduction at appreciably greater length (709 pages as opposed to 283 for Weisberg) is the second edition of Draper and Smith (1981). Residual analysis is described in their Chapter 3 together with an introduction to deletion diagnostics. Rao (1973) and Seber (1977) provide a more theoretical introduction to regression analysis. Chapters 1 and 2 of Barnett and Lewis (1978) contain a general discussion of outlying observations and their treatment. Diagnostic procedures for regression models are the main subject of Belsley *et al.* (1980). Cook and Weisberg (1982) give an encyclopaedic presentation of material on diagnostic regression analysis as it had developed to the end of 1981. The book provides further information on many of the topics to be covered in the next few chapters and will be frequently referenced. Cook and Weisberg, unlike the other authors, describe some of the recently developed extensions of diagnostic techniques to the analysis of transformations. This is the main subject of the second part of the present book, beginning with Chapter 6.

A major theme of this book is the importance of graphical methods in the analysis of data. Tufte (1983), in a book which is as physically gratifying as it is intellectually challenging, describes and illustrates many methods for the visual presentation of data. His book could, with advantage, be on the coffee tables not only of statisticians, but of all those, from newspaper proprietors to government clerks, who are involved in the communication of numerical information. Many examples of graphical procedures in the service of statistical methods are given by Chambers, J. M. *et al.* (1983).

# 2
# The algebra of deletion

## 2.1 Least squares

Two ideas that emerged informally in the first chapter were those of leverage and of influence. In the first section of this chapter the nomenclature and algebra for least squares are developed and the idea of leverage is quantified. The example due to Huber from the first chapter is used to exemplify the calculation of leverage. The investigation of influential behaviour begins in the second section of the chapter with calculation of the effect on parameter estimates and the residual sum of squares of the deletion of one or more observations. These results are used in Chapter 3 in the description of various measures of influence. But first we consider multiple regression.

For the linear regression model with $n$ observations on a response variable $y$, the expected value of the response, $E(Y)$, is related to the values of $p$ known constants by the relationship

$$E(Y) = X\beta . \tag{2.1.1}$$

In this representation $Y$ is the $n \times 1$ vector of responses, $E(\;)$ denotes expected value, $X$ is the $n \times p$ matrix of known constants and $\beta$ is a vector of $p$ unknown parameters. Capital letters will customarily be used to denote both matrices and random variables. As in (2.1.1), the context will establish the meaning. In general, bold-faced symbols are not employed. The matrix $X$ will be called the matrix of carriers which are assumed non-random. Time series models are therefore excluded. If some columns of $X$ are produced by a random mechanism, we argue conditionally on the observed values and so ignore the random element. It will be assumed that the model is of full rank $p$. There are two reasons for this assumption. One is that models that are not of full rank can always be rewritten in a full rank form, although often with a lack of symmetry and a consequent loss of ease of interpretation. But, more importantly, models which are not full rank arise naturally in the analysis of balanced experiments where there is usually little or no variation in leverage. For such data the difficulties of leverage and influence do not arise in a form to necessitate methods other than those based on residuals. The $p$ carriers are functions of $k$ explanatory variables, the values of which are known. For Forbes' data, Example 1 in Chapter 1, there is one explanatory variable, boiling point, and two carriers: the constant 1 and $x$. For the quadratic fit to Huber's 'data', Example 2, there is one explanatory variable and three carriers.

The $i$th of the $n$ observations is

$$y_i = x_i^{\mathrm{T}}\beta + \varepsilon_i$$

where, under second order assumptions, the errors $\varepsilon_i$ have zero expectation, constant variance $\sigma^2$ and are uncorrelated. That is

$$E(\varepsilon_i) = 0$$

$$E(\varepsilon_i \varepsilon_j) = \begin{cases} \sigma^2 & i = j \\ 0 & i \neq j \end{cases}. \tag{2.1.2}$$

Unless otherwise stated it will also be assumed that the errors are normally distributed.

The least squares estimates $\hat{\beta}$ of the parameters $\beta$ minimize the sum of squares

$$R(\beta) = (y - X\beta)^{\mathrm{T}}(y - X\beta)$$

and satisfy the relationship

$$X^{\mathrm{T}}(y - X\hat{\beta}) = 0. \tag{2.1.3}$$

This form, relevant in the discussion of robust regression, Section 12.1, is usually rewritten to yield the normal equations

$$X^{\mathrm{T}}X\hat{\beta} = X^{\mathrm{T}}y. \tag{2.1.4}$$

The required least squares estimates are therefore

$$\hat{\beta} = (X^{\mathrm{T}}X)^{-1}X^{\mathrm{T}}y. \tag{2.1.5}$$

For this vector of estimates, the minimized value of the sum of squares is

$$\begin{aligned} R(\hat{\beta}) &= (y - X\hat{\beta})^{\mathrm{T}}(y - X\hat{\beta}) \\ &= y^{\mathrm{T}}y - y^{\mathrm{T}}X(X^{\mathrm{T}}X)^{-1}X^{\mathrm{T}}y \\ &= y^{\mathrm{T}}\{I - X(X^{\mathrm{T}}X)^{-1}X^{\mathrm{T}}\}y. \end{aligned} \tag{2.1.6}$$

An alternative representation which emphasizes that the residual sum of squares depends solely on the parameter estimates and on the sufficient statistics $y^{\mathrm{T}}y$ and $X^{\mathrm{T}}y$ is

$$R(\hat{\beta}) = y^{\mathrm{T}}y - \hat{\beta}^{\mathrm{T}}X^{\mathrm{T}}y. \tag{2.1.7}$$

Calculation of the residual sum of squares thus does not require calculation of the individual residuals.

The vector of $n$ predictions from the fitted model is

$$\hat{y} = X\hat{\beta}.$$

From the expression for $\hat{\beta}$ (2.1.5) this can be written as

$$\hat{y} = X(X^{\mathrm{T}}X)^{-1}X^{\mathrm{T}}y = Hy, \tag{2.1.8}$$

where $H$ is the 'hat' matrix, so called because it 'puts the hats on $y$'.

The hat matrix recurs repeatedly in the development of regression diagnostics. One feature of great importance is that $H$ provides a measure of leverage. The $i$th diagonal element of $H$ is $h_{ii}$, which will be abbreviated to $h_i$ so that

$$h_i = x_i^{\mathrm{T}}(X^{\mathrm{T}}X)^{-1}x_i \qquad (i = 1, \ldots, n), \tag{2.1.9}$$

which is a measure of the remoteness of the $i$th observation from the remaining $n - 1$ observations in the space of the carriers $X$. For an observation with high leverage the value of $h_i$ is near one and the prediction $\hat{y}_i$ is virtually determined by the value of $y_i$.

**Example 2 (continued) Huber's 'data' 2** The use of $H$ in assessing leverage is conveniently illustrated by the fit of the first-order model to Huber's 'data'. For the two-carrier model

$$E(Y_i) = \beta_0 + \beta_1 x_{1i},$$

it is easy to show that the definition of $h_i$ (2.1.9) yields

$$h_i = 1/n + (x_i - \bar{x})^2 / \sum (x_i - \bar{x})^2 \tag{2.1.10}$$

where $\bar{x} = \sum x_i/n$ is the arithmetic mean of the $x_i$. If $x_i = \bar{x}$, $h_i$ has the minimum value, $1/n$. As the distance of $x_i$ from $\bar{x}$ increases, so does $h_i$.

It is clear from Figs. 1.6–1.8 that observation 6 is remote from the rest of the data. The $6 \times 6$ hat matrix for the example is given in Table 2.1. The diagonal elements of the matrix yield the values of $h_i$. Since $x_5 = \bar{x}$, observation 5 has the smallest value of $h_i$, namely $\frac{1}{6}$ as there are six observations. For observations 4, 3, 2 and 1 which are increasingly distanced from $\bar{x} = 0$, the values of $h_i$ rise steadily to 0.290. But the value of $h_6$ is 0.936,

**Table 2.1** Example 2: Huber's 'data'. Hat matrix $H = X(X^{\mathrm{T}}X)^{-1}X^{\mathrm{T}}$ for first-order model.

| Observation | $j = 1$ | 2 | 3 | 4 | 5 | 6 |
|---|---|---|---|---|---|---|
| $i = 1$ | 0.290 | 0.259 | 0.228 | 0.197 | 0.167 | −0.141 |
| 2 | 0.259 | 0.236 | 0.213 | 0.190 | 0.167 | −0.064 |
| 3 | 0.228 | 0.213 | 0.197 | 0.182 | 0.167 | 0.013 |
| 4 | 0.197 | 0.190 | 0.182 | 0.174 | 0.167 | 0.090 |
| 5 | 0.167 | 0.167 | 0.167 | 0.167 | 0.167 | 0.167 |
| 6 | −0.141 | −0.064 | 0.013 | 0.090 | 0.167 | 0.936 |

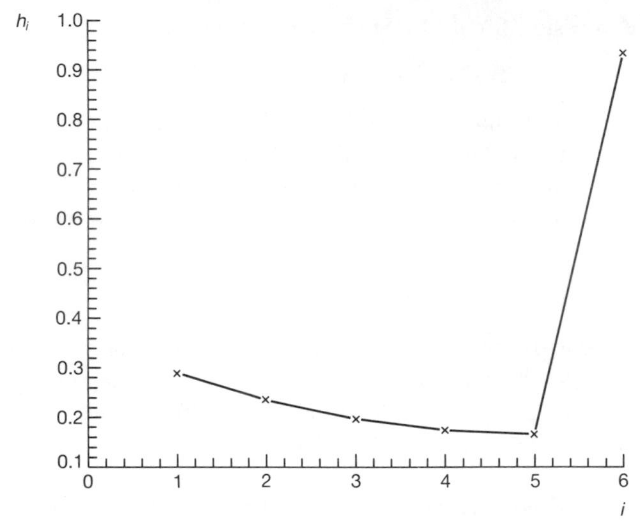


**Fig. 2.1** Example 2: Huber's 'data': index plot of leverage measure $h_i$

confirming that observation 6 has appreciably higher leverage than any of the other observations.

For this small and straightforward example, the structure of the $h_i$ can be understood from a list and summarized verbally. For larger and more complicated problems it is often easier to appreciate the structure by looking at an index plot of the $h_i$, that is a plot of $h_i$ against observation number. Figure 2.1 shows, for this example, a frequently used form of index plot. For large problems, the line joining the points in Fig. 2.1 may become a hindrance and an alternative form, such as Fig. 2.2, may be preferable. Both plots clearly show the high leverage associated with observation 6. In data where the largest values of $h_i$ are less extreme than they are here, an index plot more sensitive to large values is provided by plotting the leverage measure $h_i/(1 - h_i)$ which is unbounded as $h_i$ approaches its upper limit of one.

The effect of leverage on the predictions can be shown by consideration of two fitted values

$$\hat{y}_1 = 0.290y_1 + 0.259y_2 + 0.228y_3 + 0.197y_4 + 0.167y_5 - 0.141y_6$$

and

$$\hat{y}_6 = -0.141y_1 - 0.064y_2 + 0.013y_3 + 0.090y_4 + 0.167y_5 + 0.936y_6.$$

The predicted value $\hat{y}_1$ thus depends appreciably on all six observations, although to a lesser degree on observation 6 than on any other. But $\hat{y}_6$ is determined almost exclusively by the value of $y_6$.

The hat matrix is also of value in understanding the properties of residuals.

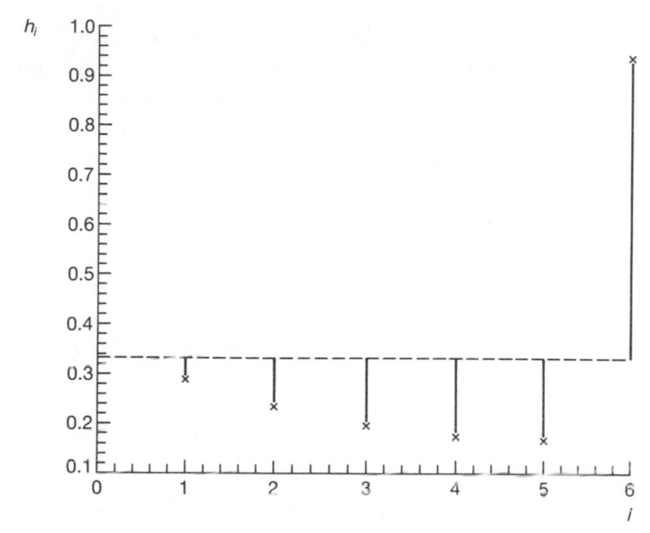


**Fig. 2.2** Example 2: Huber's 'data': alternative form for index plot of leverage measure $h_i$

Let $r_i = y_i - \hat{y}_i$, so that the vector of residuals

$$r = y - \hat{y} = y - X\hat{\beta} = (I - H)y \tag{2.1.11}$$

where, as in (2.1.6), $I$ is the $n \times n$ identity matrix. To see the effect of leverage on the residuals consider again observations 1 and 6 in Huber's example. The residual at the first observation

$$r_1 = 0.710y_1 - 0.259y_2 - 0.228y_3 - 0.197y_4 - 0.167y_5 + 0.141y_6,$$

which depends on the values of all the $y_i$, but principally on the value of $y_1$. However, for the sixth observation,

$$r_6 = 0.141y_1 + 0.064y_2 - 0.013y_3 - 0.090y_4 - 0.167y_5 + 0.064y_6.$$

This residual is almost independent of the value of $y_6$ and does not greatly depend on the values of the other five observations. Whatever the value of $y_6$, within reason, the residual $r_6$ will be small. This reflects the position of the sixth observation in $X$ space. Because observation 6 is so far from the other observations, the fitted line is forced near the observed value and the residual will be small. From the values of the elements of $H$ in Table 2.1 the largest effect of a wrong value of $y_6$ is on the residual for the fifth observation. □

The discussion of the example centres on the idea of leverage for a single observation which is measured by the value of $h_i$. The extension of the concept to a group of two or more observations is discussed in Section 10.1. It is clear from the expression for $h_i$ for the two-carrier linear model (2.1.10) that the minimum value of $h_i$ is $1/n$. This minimum value holds for all models which include a constant. But for regression through the origin the minimum value is 0 for an observation at $x = 0$. The maximum value of $h_i$ is 1, occurring when the fitted model is irrelevant to the prediction at $x_i$ and the residual is identically 0. The other bound is on the total value of the $h_i$. Since the trace

$$\text{tr}(H) = \text{tr}\{X(X^TX)^{-1}X^T\} = \text{tr}\{(X^TX)^{-1}X^TX\} = \text{tr}(I_p) = p,$$

it follows that

$$\sum_{i=1}^{n} h_i = \text{tr}(H) = p$$

and that the average value of $h_i$ is $p/n$.

For a balanced experimental design the values of all $h_i$ will be near the average with equality for a D-optimum design. (Silvey, 1980, gives a succinct account of optimum design theory). Values of $h_i > 2p/n$ are taken by Belsley *et al.* (1980, p. 17) to indicate observations with sufficiently high leverage to require further investigation. We have already seen in the index plots of Figs. 2.1 and 2.2 another method of inspecting leverage. These and other methods, particularly graphical, of calling attention to observations with appreciable leverage or influence are the subject of Chapter 4. The remainder of the present section is concerned with the effect of leverage on the variance of the residuals.

The residuals $r$ defined in (2.1.11) do not all have the same variance. To find the variance we note that $I - H$ is idempotent, that is that $(I - H)(I - H) = (I - H)$. It therefore follows that

$$\text{var}(r) = \text{var}\{(I - H)y\} = (I - H)\sigma^2$$

and, in particular,

$$\text{var}\, r_i = \sigma^2(1 - h_i). \tag{2.1.12}$$

Therefore a residual at a point with high leverage has smaller variance than those at points of lower leverage. This can be seen by comparing the expressions for $r_1$ and $r_6$ given above for Example 2.

To find residuals with constant variance (2.1.12) is divided by its estimated standard error. Let $s^2$ be the residual mean square estimate of $\sigma^2$, that is

$$s^2 = R(\hat{\beta})/(n - p). \tag{2.1.13}$$

The *standardized residual* $r'_i$ is then defined as

$$r'_i = \frac{r_i}{s\sqrt{(1 - h_i)}} = \frac{y_i - \hat{y}_i}{s\sqrt{(1 - h_i)}}. \tag{2.1.14}$$

The advantage of the standardized residuals $r'_i$ is that if the model (2.1.1) is correct they all have the same variance. They are therefore particularly appropriate for normal probability plots or for plotting to check for homogeneity of variance. A slight disadvantage is that nomenclature for these and other kinds of residuals is not standardized. Thus Hoaglin and Welsch (1978) and Ryan *et al.* (1976) call (2.1.14) a standardized residual. Because the dependence of the residual on $\sigma^2$ has been removed by use of the estimate $s^2$ it is also correct, but potentially confusing, to refer, as do Cook (1977) and Cook and Weisberg (1982, Section 2.2.1) to $r_i'$ as a studentized residual. Since Hoaglin and Welsch use the name studentized residual for the rather different deletion residual to be introduced in Section 3.1, it seems safer to avoid the name altogether and to call (2.1.14) a standardized residual.

## 2.2 Deletion of observations

### *2.2.1 General*

The basic components of a least squares analysis have now been derived. These are the parameter estimates $\hat{\beta}$, the vector of residuals $r$ and the residual sum of squares $R(\hat{\beta})$ which furnishes $s^2$ as the residual mean square estimate of the error variance $\sigma^2$. It is now time to consider the algebraic effect on these quantities of the deletion of one or more observations. The use of the quantities so calculated in the analysis of data, in particular in the detection of influential observations, is described and illustrated in Chapters 3 and 4.

If one observation, the $i$th, is to be deleted, the associated carriers are given by the $i$th row of $X$, that is by the row vector $x_i^{\mathrm{T}}$. If the number of observations for deletion is $m > 1$, belonging to a set $I$, the rows of $X$ for deletion form an $m \times p$ matrix, which will be denoted by $X_I$. The effect of deleting the $i$th observation on the estimate $\hat{\beta}$ is to yield the estimate $\hat{\beta}_{(i)}$, where the subscripted $i$ in parentheses is to be read as 'with observation $i$ deleted'. Similarly the effect on $\hat{\beta}$ of deleting the $m$ observations indexed by $I$ is to yield the estimate $\hat{\beta}_{(I)}$. The calculation of these quantities begins with the effect of deletion on $(X^{\mathrm{T}}X)^{-1}$.

Let $A$ be a square $p \times p$ matrix and let $U$ and $V$ be matrices of dimension $p \times m$. Then it is easy to verify that

$$(A - UV^{\mathrm{T}})^{-1} = A^{-1} + A^{-1}U(I - V^{\mathrm{T}}A^{-1}U)^{-1}V^{\mathrm{T}}A^{-1}, \qquad (2.2.1)$$

where $I$ is the $m \times m$ identity matrix and it is assumed that all necessary inverses exist. For regression let

$$A = X^{\mathrm{T}}X \qquad \text{and} \qquad U^{\mathrm{T}} = V^{T} = X_I.$$

Then

$$X_{(I)}^{\mathrm{T}}X_{(I)} = (X^{\mathrm{T}}X - X_I^{\mathrm{T}}X_I)$$

and, from (2.2.1),

$$(X_{(I)}^{\mathrm{T}}X_{(I)})^{-1} = (X^{\mathrm{T}}X)^{-1} + (X^{\mathrm{T}}X)^{-1}X_I^{\mathrm{T}}\{I - X_I(X^{\mathrm{T}}X)^{-1}X_I^{\mathrm{T}}\}^{-1}X_I(X^{\mathrm{T}}X)^{-1}.$$

In (2.1.9) $h_i$, the $i$th diagonal element of the hat matrix, was defined as $h_i = x_i^{\mathrm{T}}(X^{\mathrm{T}}X)^{-1}x_i$. The analogous submatrix of $H$ for observations belonging to the set $I$ is given by

$$H_I = X_I(X^{\mathrm{T}}X)^{-1}X_I^{\mathrm{T}}, \tag{2.2.2}$$

which yields

$$(X_{(I)}^{\mathrm{T}}X_{(I)})^{-1} = (X^{\mathrm{T}}X)^{-1} + (X^{\mathrm{T}}X)^{-1}X_I^{\mathrm{T}}(I - H_I)^{-1}X_I(X^{\mathrm{T}}X)^{-1}. \tag{2.2.3}$$

This important formula gives the inverse for the reduced data resulting from deletion in terms of the inverse for all $n$ observations and quantities readily calculated from the hat matrix.

The purpose in deriving (2.2.3) is not to provide a numerical method for calculation of the reduced inverse, but as an intermediate stage in the evaluation of the effect of deletion. For (2.2.3) to be valid it is necessary that deletion of the subset of $m$ observations does not lead to a singular model, in which case it would not be possible to invert $X_{(I)}^{\mathrm{T}}X_{(I)}$.

Two further definitions are required which relate to the subset $I$ of the observations. The predicted values of these $m$ observations are

$$\hat{y}_I = X_I\hat{\beta} = X_I(X^{\mathrm{T}}X)^{-1}X^{\mathrm{T}}y. \tag{2.2.4}$$

The associated vector of residuals is denoted by

$$r_I = y_I - \hat{y}_I,$$

where $y_I$ is the vector of observed values of the response for the set $I$.

### 2.2.2 *The parameter estimate* $\hat{\beta}_{(I)}$

The parameter estimate $\hat{\beta}$ was defined in (2.1.5) for all $n$ observations. For the $n - m$ observations remaining after deletion of the set $I$ it follows that

$$\begin{aligned}\hat{\beta}_{(I)} &= (X_{(I)}^{\mathrm{T}}X_{(I)})^{-1}X_{(I)}^{\mathrm{T}}y_{(I)} \\ &= (X_{(I)}^{\mathrm{T}}X_{(I)})^{-1}(X^{\mathrm{T}}y - X_I^{\mathrm{T}}y_I).\end{aligned}$$

Use of the expression for $(X_{(I)}^{\mathrm{T}}X_{(I)})^{-1}$ given by (2.2.3) yields

$$\begin{aligned}\hat{\beta}_{(I)} = \hat{\beta} &- (X^{\mathrm{T}}X)^{-1}X_I^{\mathrm{T}}y_I + (X^{\mathrm{T}}X)^{-1}X_I^{\mathrm{T}}(I - H_I)^{-1}\hat{y}_I \\ &- (X^{\mathrm{T}}X)^{-1}X_I^{\mathrm{T}}(I - H_I)^{-1}H_Iy_I\end{aligned}$$

which simplifies to

$$\hat{\beta}_{(I)} - \hat{\beta} = -(X^{\mathrm{T}}X)^{-1}X_I^{\mathrm{T}}(I - H_I)^{-1}r_I. \tag{2.2.5}$$

This expression for the change in the parameter estimate following deletion of the set $I$ will be repeatedly used in the calculation of the effect of deletion on a variety of diagnostic statistics.

### 2.2.3 *Effect on the residual sum of squares*

To calculate the effect of deletion on the residual sum of squares $R(\hat{\beta})$ it is most convenient to work in terms of the residual mean square estimate of error $s^2$ defined by (2.1.13). From (2.1.7) this definition can be rewritten as

$$(n-p)s^2 = y^{\mathrm{T}}y - \hat{\beta}^{\mathrm{T}}X^{\mathrm{T}}y. \tag{2.2.6}$$

After deletion of $m$ observations the corresponding expression is

$$(n-p-m)s_{(I)}^2 = y^{\mathrm{T}}y - y_I^{\mathrm{T}}y_I - \hat{\beta}_{(I)}^{\mathrm{T}}(X^{\mathrm{T}}y - X_I^{\mathrm{T}}y_I).$$

The combination of these two expressions with the value of $\hat{\beta}_{(I)}$ given by (2.2.5) leads to

$$\begin{aligned}(n-p-m)s_{(I)}^2 &= (n-p)s^2 + \hat{y}_I^{\mathrm{T}}(I-H_I)^{-1}r_I \\ &\quad - y_I^{\mathrm{T}}r_I - y_I^{\mathrm{T}}H_I(I-H_I)^{-1}r_I \\ &= (n-p)s^2 - r_I^{\mathrm{T}}(I-H_I)^{-1}r_I. \end{aligned} \tag{2.2.7}$$

The deletion of $m$ observations therefore causes the residual sum of squares to be reduced by an amount which depends on $r_I$, the vector of residuals of the $m$ observations, and on the inverse of a matrix containing elements of the hat matrix $H$. These are all quantities which can be calculated from the fit for all $n$ observations.

### 2.2.4 *Deletion of one observation*

The chief application, in succeeding chapters, of the results derived above is to the deletion of a single observation. For later reference the two main results of this chapter are repeated here in the slightly simpler form which results when $m=1$. For the deletion of observation $i$,

$$\hat{\beta}_{(i)} - \hat{\beta} = -(X^{\mathrm{T}}X)^{-1}x_i r_i/(1-h_i) \tag{2.2.8}$$

and

$$(n-p-1)s_{(i)}^2 = (n-p)s^2 - r_i^2/(1-h_i). \tag{2.2.9}$$

# 3
# Diagnostic quantities

## 3.1 Deletion residuals

The formulae derived in Section 2.2 give the changes in the parameter estimates and in the residual sum of squares when $m$ observations are deleted. A difficulty with the deletion of more than one observation at a time is the combinatorial explosion of the number of cases to be considered. For example, if the number of observations $n = 30$ and all triplets of observations are considered for deletion, that is $m = 3$, then there are $30.29.28/1.2.3 = 4060$ possibilities. Both 3 and 30 are quite moderate numbers in this context, so it is clear that unthinking application of deletion methods can rapidly turn a simple analysis of data into a computational nightmare. The discussion and partial solution of the problems of multiple deletion is deferred until Chapter 10. In this chapter interest is in statistically useful quantities derived from deletion of just the $i$th observation.

The first statistically interesting quantity comes from considering whether deletion of the $i$th observation has a marked effect on prediction. In particular, does the observed $y_i$ agree with the prediction $\hat{y}_{(i)}$ which arises when the $i$th observation is not used in fitting? Since $y_i$ and $\hat{y}_{(i)}$ are independent the difference

$$y_i - \hat{y}_{(i)} = y_i - x_i^{\mathrm{T}}\hat{\beta}_{(i)}$$

has variance

$$\sigma^2\{1 + x_i^{\mathrm{T}}(X_{(i)}^{\mathrm{T}}X_{(i)})^{-1}x_i\}.$$

To estimate $\sigma^2$ we use the deletion estimate $s_{(i)}^2$ which is also independent of $y_i$. The test for agreement of the predicted and observed values,

$$r_i^* = \frac{y_i - x_i^{\mathrm{T}}\hat{\beta}_{(i)}}{s_{(i)}\sqrt{\{1 + x_i^{\mathrm{T}}(X_{(i)}^{\mathrm{T}}X_{(i)})^{-1}x_i\}}}, \tag{3.1.1}$$

accordingly has a $t$ distribution on $n - p - 1$ degrees of freedom. This expression simplifies appreciably. From (2.2.3)

$$x_i^{\mathrm{T}}(X_{(i)}X_{(i)})^{-1}x_i = x_i^{\mathrm{T}}(X^{\mathrm{T}}X)^{-1}x_i + \frac{x_i^{\mathrm{T}}(X^{\mathrm{T}}X)^{-1}x_ix_i^{\mathrm{T}}(X^{\mathrm{T}}X)^{-1}x_i}{1 - h_i}$$

$$= h_i/(1 - h_i).$$

Also, from (2.2.8),

$$x_i^{\mathrm{T}}\hat{\beta}_{(i)} = x_i^{\mathrm{T}}\hat{\beta} - \frac{x_i^{\mathrm{T}}(X^{\mathrm{T}}X)^{-1}x_i r_i}{1-h_i}$$

$$= \hat{y}_i - \frac{h_i r_i}{1-h_i}.$$

It therefore follows that

$$r_i^* = \frac{r_i}{s_{(i)}\sqrt{(1-h_i)}} = \frac{s r_i'}{s_{(i)}}. \tag{3.1.2}$$

There are many names for $r_i^*$, some of which reflect the $t$ distribution of the statistic. Belsley *et al.* (1980, p. 20) use the name *RSTUDENT*. Comparison of (3.1.2) with the definition of the standardized residual (2.1.14) shows that the two differ only in the estimate of $\sigma$ employed. Since the estimate $s_{(i)}$ does not depend on $r_i$, Cook and Weisberg (1982, p. 20) call $r_i^*$ an externally studentized residual. Atkinson (1981), to stress that $r_i^*$ is a residual which results from deletion, suggests either cross-validatory or jack-knife residual. A simpler alternative adopted here is to call $r_i^*$ the deletion residual.

An alternative form for (3.1.2) can readily be obtained and is informative about the relationships between the various kinds of residuals. Substitution for $s_{(i)}$ from (2.2.9) in (3.1.2) yields

$$r_i^* = \frac{r_i'}{\left(\dfrac{n-p-r_i'^2}{n-p-1}\right)^{1/2}}. \tag{3.1.3}$$

The deletion residuals are thus a monotone, but nonlinear, function of the standardized residuals. From (3.1.3), or equivalently from the expression for $s_{(i)}^2$ (2.2.9), it follows that the maximum value of $r_i'^2$ is $n-p$. For this value $r_i^*$ is infinite. An interpretation in terms of a single outlier is that if the $i$th observation has an amount $\sigma\Delta$ added to it, the residual $r_i$, from (2.1.11), has mean $\sigma\Delta(1-h_i)$. It then follows from (3.1.2) that the deletion residual $r_i^*$ has a non-central $t$ distribution with mean $\Delta\sqrt{(1-h_i)}$. For $j \neq i$ all $r_j^*$ have a shrunken distribution due to over-estimation of $\sigma$. Likewise, all standardized residuals $r_j'$, but now including $r_i'$, are shrunken due to over-estimation of $\sigma$. It follows that detection of an outlying observation is made easier by consideration of all values of $r_i^*$, not just the largest value. This point is considered further in the analysis of the examples of Sections 4.2 and 4.3.

Provided the linear model holds and the errors are normally distributed the distribution of $r_i^*$ is Student's $t$ on $n-p-1$ degrees of freedom. It has already been argued that the distribution of the standardized residual $r_i'$ must be such that $r_i'^2$ lies between 0 and $n-p$. That the distribution of $r_i'^2/(n-p)$ is beta with parameters $\frac{1}{2}$ and $(n-p)/2$ follows from interpreting (2.2.7) divided by $\sigma^2$ as an equality in chi-squared random variables. An outline of an algebraic derivation is given by Cook and Weisberg (1982, p. 19). The result serves as a warning against the facile assumption that the distribution of the residuals from a fitted model is necessarily close to that of the errors.

## 3.2 Cook's statistic

The deletion residuals $r_i^*$ derived in Section 3.1 lead to detection of significant differences between the predicted and observed values of the response for each observation. But these quantities do not provide information about changes in the parameter estimates due to the deletion of observations. To answer questions about the influence of individual observations on inferences drawn from the data it is necessary to derive some further, related, quantities.

One way of determining the influence of individual observations is to look at the changes, as each observation is deleted in turn, in the estimates of the $p$ parameters given by (2.2.8). For comparisons between the parameters these differences need to be scaled by their standard errors. This leads to the inspection of $np$ quantities. If some particular feature of the model is of interest, the $p$ quantities for each observation can be reduced to a single measure of influence. In this section we look at one measure based on the confidence region for all the parameters in the model. Extensions for other measures concerned, for example, with a subset of the parameters are described in Section 10.2. A more general discussion of measures of influence is given in Chapter 3 of Cook and Weisberg (1982).

A confidence region at level $100(1-\alpha)\%$ for the parameter vector $\beta$ is given by those values of the parameters for which

$$(\beta - \hat{\beta})^{\mathrm{T}} X^{\mathrm{T}} X (\beta - \hat{\beta}) \leqslant p s^2 F_{p, \nu, \alpha}, \tag{3.2.1}$$

where $s^2$ is an estimate of $\sigma^2$ on $\nu$ degrees of freedom and $F$ is the $100\alpha\%$ point of the $F$ distribution on $p$ and $\nu$ degrees of freedom. For differing values of $\alpha$ the regions are a series of ellipsoids centred on the least squares estimate $\hat{\beta}$. These regions provide a metric for determining whether $\hat{\beta}$ is near $\hat{\beta}_{(i)}$. For a model with only two parameters a graphical display of influence can be achieved by plotting all $n$ values of $\hat{\beta}_{(i)}$ with selected confidence regions from (3.2.1). An example is given in Fig. 3.5.1 of Cook and Weisberg (1982). This plot is not available if the number of parameters is greater than two, except for the clumsy device of plotting for selected pairs of parameters. For a general number of parameters Cook (1977) proposed the statistic

$$D_i = \frac{(\hat{\beta}_{(i)} - \hat{\beta})^{\mathrm{T}} X^{\mathrm{T}} X (\hat{\beta}_{(i)} - \hat{\beta})}{p s^2} \tag{3.2.2}$$

for detecting influential observations. Large values of $D_i$ indicate observations which are influential on joint inferences about all the parameters in the model. An alternative form for $D_i$ is obtained by writing

$$D_i = \{X(\hat{\beta}_{(i)} - \hat{\beta})\}^{\mathrm{T}}\{X(\hat{\beta}_{(i)} - \hat{\beta})\}/ps^2$$
$$= (\hat{Y}_{(i)} - \hat{Y})^{\mathrm{T}}(\hat{Y}_{(i)} - \hat{Y})/ps^2 \qquad (3.2.3)$$

where the vector of predictions $\hat{Y}_{(i)} = X\hat{\beta}_{(i)}$. Amongst other interpretations, the quantity $D_i$ thus measures the change in the vector of all $n$ predicted values when observation $i$ is not used in estimating $\beta$.

A more convenient form for $D_i$ follows from substitution for $\hat{\beta}_{(i)} - \hat{\beta}$ from (2.2.8) which yields

$$D_i = \frac{r_i^2 h_i}{ps^2(1-h_i)^2} = \frac{r_i'^2 h_i}{p(1-h_i)}, \qquad (3.2.4)$$

a form which stresses the relationship between $D_i$ and the standardized residuals $r_i'$ defined by (2.1.14).

A list, or an index plot, of the values of $D_i$ can be used to identify influential observations. As the basis of graphical procedures there are advantages in using the square root of $D_i$, which gives the standardized residual weighted by the square root of the leverage measure $h_i/(1-h_i)$. Atkinson (1981) suggests graphical use of the modified Cook statistic

$$C_i = \left\{\frac{n-p}{p} \cdot \frac{h_i}{1-h_i}\right\}^{1/2} |r_i^*|. \qquad (3.2.5)$$

In addition to taking the square root of $D_i$, the modifications leading to $C_i$ are:

1. Scaling by the factor $(n-p)/p$. As was mentioned in Section 2.1, for a $D$-optimum design, which is the most balanced case possible, all $h_i$ are equal and take the value $p/n$. The effect of the scaling is to make $C_i = |r_i^*|$ for this case of equal leverage.
2. Use of $s_{(i)}^2$, rather than $s^2$, as an estimate of $\sigma^2$. The advantages are similar to those from use of the deletion residual $r_i^*$ instead of the standardized residual $r_i'$.

Apart from the absolute value and the factor in $n$ and $p$, the modified Cook statistic is the quantity which Belsley *et al.* (1980, p. 15) call $DFFITS_i$. Use of $DFFITS_i$ is clearly equivalent to use of $C_i$. Because of the estimate of $\sigma^2$ which is used, both of these measures give greater weight to outlying observations than does $D_i$.

The modified Cook statistic $C_i$ (3.2.5) and the deletion residual $r_i^*$ (3.1.2) are only two of several related quantities which have been suggested as diagnostic for leverage and influence. Obenchain (1977) recommends separate use of the components of $C_i$, namely $r_i^*$ and $h_i/(1-h_i)$ as do Hoaglin and Welsch (1978). Another possibility is to use values of $r_i^*$ and those of $h_i$, noting values of this latter quantity which are larger than $2p/n$. An advantage of $C_i$, or equivalently of $D_i$, is that they call attention to leverage points where the values of $y$ and $x$ do not agree. A frequent cause of such points is that one or more of the explanatory variables is in error. Use of the value of $h_i$ alone does not help to discriminate between observations of high leverage which are influential and those which are not.

**Example 1 (continued) Forbes' data 2** In the analysis of these data in Chapter 1 it was shown that there was a strong linear relationship between the response and the explanatory variable. Observation 12 was suspected of being an outlier and the data were re-analysed with this observation omitted.

Table 3.1 gives values of the deletion residuals $r_i^*$ and the modified Cook statistic for all 17 observations, together with values of the leverage measures $h_i$ and $h_i/(1 - h_i)$. For the original 17 observations the values of both diagnostic quantities are a maximum for observation 12 with $r_{12}^* = 12.40$ and $C_{12} = 8.88$. Because the observations are, with negligible exceptions, ordered in the single explanatory variable, the leverage measures $h_i$ and $h_i/(1 - h_i)$ decrease and then increase in a roughly parabolic fashion. For observation 12, which is near the centre of the range of $x$, $h_{12} = 0.0639$ and $h_{12}/(1 - h_{12})$

**Table 3.1** Example 1: Forbes' data. Diagnostic quantities for all 17 observations and for 16 observations with observation 12 deleted

| Observation | All 17 observations | | | | Observation 12 deleted | |
|---|---|---|---|---|---|---|
| $i$ | $r_i^*$ | $C_i$ | $h_i$ | $h_i/(1-h_i)$ | $r_i^*$ | $C_i$ |
| 1 | −0.71 | 0.96 | 0.1934 | 0.2398 | −2.22 | 2.88 |
| 2 | −0.19 | 0.26 | 0.1999 | 0.2498 | −0.20 | 0.27 |
| 3 | −0.17 | 0.16 | 0.1069 | 0.1197 | 0.01 | 0.01 |
| 4 | 0.06 | 0.05 | 0.0979 | 0.1085 | 0.79 | 0.70 |
| 5 | 0.09 | 0.08 | 0.0826 | 0.0900 | 0.97 | 0.78 |
| 6 | −0.11 | 0.09 | 0.0764 | 0.0827 | 0.26 | 0.20 |
| 7 | 0.14 | 0.10 | 0.0668 | 0.0715 | 1.20 | 0.87 |
| 8 | 0.14 | 0.10 | 0.0653 | 0.0699 | 1.21 | 0.87 |
| 9 | −0.41 | 0.29 | 0.0634 | 0.0677 | −0.69 | 0.49 |
| 10 | −0.20 | 0.14 | 0.0640 | 0.0683 | 0.02 | 0.01 |
| 11 | −0.38 | 0.26 | 0.0596 | 0.0634 | −0.51 | 0.35 |
| 12 | 12.40 | 8.88 | 0.0639 | 0.0683 | — | — |
| 13 | 0.00 | 0.00 | 0.1396 | 0.1622 | 1.12 | 1.23 |
| 14 | −0.90 | 0.91 | 0.1189 | 0.1349 | −2.28 | 2.28 |
| 15 | −0.69 | 0.86 | 0.1719 | 0.2076 | −1.22 | 1.50 |
| 16 | −0.22 | 0.31 | 0.2096 | 0.2652 | 0.47 | 0.65 |
| 17 | −0.25 | 0.36 | 0.2199 | 0.2819 | 0.40 | 0.57 |

**Table 3.2** Example 2: Huber's 'data'. Diagnostic quantities for first-order model

| Observation $i$ | $r_i^*$ | $C_i$ | $h_i$ | $h_i/(1-h_i)$ |
|---|---|---|---|---|
| 1 | 2.30 | 2.07 | 0.2897 | 0.4079 |
| 2 | 0.27 | 0.21 | 0.2359 | 0.3087 |
| 3 | −0.17 | 0.12 | 0.1974 | 0.2460 |
| 4 | −1.18 | 0.77 | 0.1744 | 0.2112 |
| 5 | −0.98 | 0.62 | 0.1667 | 0.2 |
| 6 | 5.32 | 28.73 | 0.9359 | 14.6000 |

= 0.0683. Since $p/n = 2/17 = 0.1176$, the leverage of observation 12 is less than the average value and, as has been observed, $C_{12} < |r_{12}^*|$.

Table 3.1 also gives the values of $r_i^*$ and $C_i$ when the model is refitted after observation 12 has been deleted. It is clear that the largest values of $r_i^*$ and $C_i$ are now much nearer the remaining values, with the maximum value of $C_i = 2.88$. It is not clear from inspection of these lists whether this largest value is significantly large. We leave until Chapter 4 discussion of the ways in which the diagnostic quantities can be used either as tests or to yield plots. It will then be possible to assess the value 2.88. □

**Example 2 (continued) Huber's 'data' 3** An important feature of Example 2 is that the value of the explanatory variable for the sixth observation is far from the $x$ values for the other observations. This was reflected in Table 2.1 in the value of 0.936 for $h_6$. Analysis of the data clearly shows the effect of leverage on the diagnostic quantities. In this respect Example 2 differs importantly from Example 1 in which the information yielded by $r_i^*$ and $C_i$ was not markedly different.

Table 3.2 gives values of $r_i^*$, $C_i$, $h_i$ and $h_i/(1-h_i)$ for the six observations of Huber's synthetic example when a first-order model is fitted. The deletion residual $r_6^*$ is appreciable at 5.32 and so is the leverage measure $h_6 = 0.9359$. The resulting modified Cook statistic $C_6^*$ is 28.73, clearly indicating what we already know to be a highly influential observation.

If, as in Section 1.3, a quadratic model is fitted, the value of $r_6^*$, which is given in Table 3.3, is reduced to −1.53. This is further confirmation of the earlier result that a quadratic model provides an adequate explanation of the data. But the value of $C_6$ is now 58.53. The value for $h_6$ of 0.9993 indicates that the fit at this point is virtually independent of the other observations and that the ordinary residual $r_6$ will be close to zero, whatever the value of the response at this point. The analysis of this example is concluded in Section 4.2 with a graphical assessment of the values of $r_i^*$ and $C_i$. □

**Table 3.3** Example 2: Huber's 'data'. Diagnostic quantities for quadratic model

| Observation $i$ | $r_i^*$ | $C_i$ | $h_i$ | $h_i/(1-h_i)$ |
|---|---|---|---|---|
| 1 | 1.06 | 1.47 | 0.6610 | 1.9502 |
| 2 | −0.68 | 0.43 | 0.2861 | 0.4008 |
| 3 | 0.11 | 0.06 | 0.2086 | 0.2636 |
| 4 | −1.63 | 1.12 | 0.3189 | 0.4683 |
| 5 | 2.40 | 2.52 | 0.5260 | 1.1095 |
| 6 | −1.53 | 58.53 | 0.9993 | 1454.8857 |

With the derivation in this chapter of the deletion residuals $r_i^*$ and of the modified Cook statistic $C_i$, we have the basic tools that will be used for the analysis of data in succeeding chapters. In Chapter 10 these methods are developed in three ways. The first extension is to the effect of the deletion of several observations at once. The second is to the effect of deletion on other aspects of the model, for example on the estimates of a subset of the parameters. Chapter 10 concludes with an outline of diagnostic methods for more general models not necessarily fitted by least squares.

# 4 The presentation of results

## 4.1 Lists, tests, and plots

In this chapter we describe and exemplify uses of the deletion residuals $r_i^*$ and the modified Cook statistics $C_i$, resulting from the deletion of each observation in turn, which were derived in Sections 3.1 and 3.2. Even with this restricted set of diagnostic measures, each sample of $n$ observations gives rise to $2n$ quantities. There is therefore ample scope for over-interpretation of the data, with potential for the discovery of numerous outliers and influential observations.

How then should the $2n$ diagnostic quantities be presented? Belsley *et al.* (1980, Chapter 2) give lists of diagnostic quantities, similar to those in Tables 3.1–3.3 but with values greater than certain thresholds indicated by asterisks. This is the approach followed by several computer packages, where an option is sometimes only to print values which lie above the threshold. Two disadvantages of such lists are that they are time consuming to print and that the information contained in the lists is often not easy to assimilate. For these reasons plots are usually preferable to lists.

A difficulty with thresholds for lists of diagnostic quantities is to establish appropriate values, even when these are regarded as the percentage points of statistical tests. Mention was made in Section 2.1 of the arbitrary threshold of $2p/n$ which Belsley *et al.* suggest for values of $h_i$ to call attention to points of high leverage. This is derived on the assumption that the explanatory variables are not arbitrary fixed constants, but arise as a sample from a multivariate normal distribution. The derivation is intended only to provide a 'rough cutoff'. In passing, it is interesting to note that this threshold fails to detect observation 6 when a quadratic model is fitted to Huber's 'data'. For, with six observations and three parameters, the threshold has the value one. But these parameter values are outside the range for which the threshold is intended to be useful.

Given the values of the explanatory variables, the leverage measures $h_i$ are known constants, the values of which do not have distributional properties. But the deletion residuals $r_i^*$ are random variables which, under the assumption of normal errors, have a Student's $t$ distribution. For the individual observed values the percentage points of the distribution can therefore be used to provide a test for a single outlier. But, if the largest of the $n$ absolute values of $r_i^*$ is selected for the outlier test, allowance has to be made for the selection in the assessment of the level of the test. A conservative procedure for an $\alpha$-level significance test for the presence of a single outlier is to perform the test for the maximum value at the $\alpha/n$ level. A thorough discussion of outlier tests is given by Hawkins (1980).

In Section 3.1 brief mention was made of the single outlier model in which $\sigma\Delta$ was added to the $i$th observation, all other observations remaining unchanged. For this situation the expected value of $r_i^*$ was shown to be $\Delta\sqrt{(1 - h_1)}$. This implies that the ability to detect outliers is greatest for observations with values of $h_i$ near zero and decreases with increasing $h_i$. Comparison of Examples 1 and 2 makes this point. For Huber's 'data' observation 6, which may well be a gross outlier due to an incorrect value of $y_6$, gives a value of $r_6^* = 5.32$. For Forbes' data the suspect observation is number 12 which lies near the central value of $x$ with $h_{12} = 0.0639$: the deletion residual $r_{12}^*$ has the appreciable value 12.4. To allow for this effect, Cook and Weisberg (1980) suggest performing the test on the $i$th deletion residual at the level $\alpha_i = \alpha h_i/p$. Since $\sum h_i = p$, the approximate overall level of the test is not affected.

This approximation provides one way in which allowance can be made for selection in the level of the test for one outlier for a given value of $n$. The maximum value of the modified Cook statistic could be used in a similar manner to test for the most influential observation. The distribution of each $C_i$ is that of a multiple of the $r_i^*$, each of which has a Student's $t$ distribution. The allowance for selection will therefore be for finding the maximum of a set of specified multiples of identically distributed random variables.

In addition to the necessity for an allowance for selection of the largest of $n$ observations, a further difficulty arises if the test procedure is to be used sequentially to detect several outlying observations. In the simplest case, even if allowance is made for the selection of the largest absolute value of the $r_i^*$, what is the distribution of the maximum of the remaining $n - 1$ observations after one observation has been deleted?

One approach, due to Dempster and Gasko-Green (1981), is not to interpret the probabilities from the series of tests as hard and fast rules for the acceptance or rejection of observations, but rather to look for patterns in the series of probabilities as observations are successively deleted. Using geometrical arguments, they obtain a highly conditional test which measures the discrepancy of the most extreme observation, as assessed for example by $r_i^*$ or $C_i$, relative to the next most extreme observation. The sequence of probability values obtained by sequentially deleting observations is principally interpreted by noting where in the sequence small values, that is significant results, occur. These indicate that deletion of the next observation will substantially alter the value of the discrepancy criterion. Thus suppose, for example, that $C_i$ were the discrepancy criterion and that the model were regression on one explanatory variable. If there were $m$ observations with appreciable leverage which indicated a slightly different model from the remaining $n - m$ observations for which leverage was low, sequential deletion of the first $m - 1$ observations might give non-significant probability values. Deletion of the last of the group of $m$ would give a small, highly significant value, indicating that the inference to be drawn from the data was about to change appreciably. Thereafter the sequence of non-significant probability values might resume. It is likely that this structure would not be revealed by the calculation of single deletion diagnostic measures for each observation in turn, although it might well be detected by multiple deletion measures. This effect, which has been called 'masking', is discussed further in Section 10.1.

Dempster and Gasko-Green stress that their procedure is an adjunct to the exploration of data and can be used to supplement graphical analysis. In the initial, exploratory, stages of an analysis plots are valuable for suggesting ways in which the data may be contaminated or in which models may be inadequate. Once sharply defined questions have emerged, tests of hypotheses are used to avoid over-interpretation.

A main purpose of the present book is to explore and exhibit graphical techniques for the analysis of data. A graphical alternative to such lists as those of Tables 3.1–3.3 is to provide plots of the diagnostic quantities. One technique, which was introduced in Figs. 2.1 and 2.2, is the index plot in which the quantities are plotted in serial order. Examples are given, amongst others, by Cook and Weisberg (1980, 1982), Pregibon (1981) and Baskerville and Toogood (1982). These plots are much more easily inspected than are lists, for large values in the context of the underlying random variation. Index plots are obviously at their most effective as detectors of pattern if the serial order of the observations corresponds to some physical aspect of the data, such as temporal or spatial ordering. In the absence of such ordering, care may be needed in interpreting trends in the plots. However, as we shall see, index plots are powerful tools for the recognition of single anomalous values.

Before we look at index plots for the two examples considered so far, it must be stressed that plots do not have to be elegant to be useful. The figures in this book were produced by a Kingmatic plotter after the plots had been viewed on a Tektronix graphics terminal. But much the same information would have been obtained from line printer plots produced by a Fortran program without any graphics software.

**Example 1 (continued) Forbes' data 3** To illustrate the comparative advantages of plots as opposed to tables we return to the analysis of Forbes' data, diagnostic quantities for which are given in Table 3.1. Figure 4.1 shows the index plot of $r_i^*$ on which the maximum value for observation 12 is clearly visible: the remaining 16 values are all close to the zero line. The plot of $C_i$, which is not shown here, is similar, as can be confirmed from the entries of Table 3.1. Because observation 12 is near the centre of the experimental

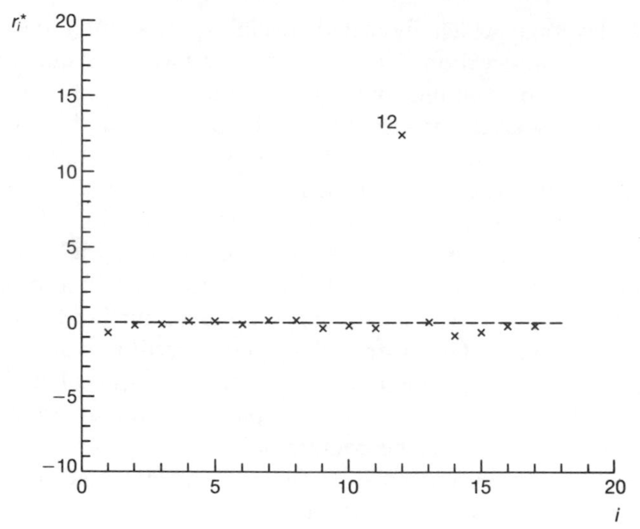


**Fig. 4.1** Example 1: Forbes' data: index plot of deletion residual $r_i^*$

region and so has low leverage, the values of $C_i$ show the outlying value less markedly.

When observation 12 is deleted, the residual sum of squares drops from 2.153 to 0.180. The largest deletion residual in magnitude is $-2.28$ for observation 14, with the largest modified Cook statistic 2.88 for observation

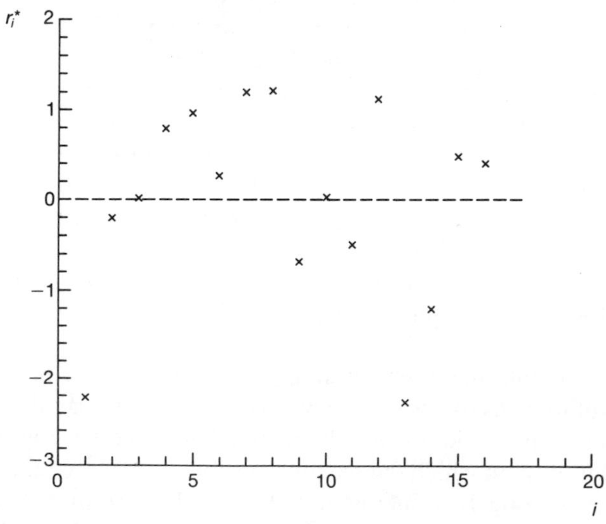


**Fig. 4.2** Example 1: Forbes' data with observation 12 deleted: index plot of deletion residual $r_i^*$

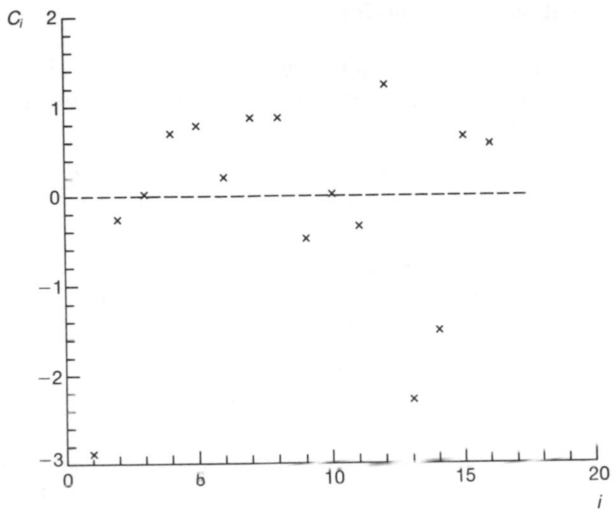


**Fig. 4.3** Example 1: Forbes' data with observation 12 deleted: index plot of signed values of modified Cook statistic $C_i$

1. Figure 4.2 gives the index plot of $r_i^*$ for the remaining 16 observations and Fig. 4.3 the corresponding plot of signed values of $C_i$. From these pictures it seems that all the structure has been removed from the data. This assertion provides an answer to the question posed in Section 3.2 as to whether the value 2.88 was significantly large. □

Although index plots such as these can be useful, they are probably most useful when the order of the observations corresponds to some physical reality, such as the order in which the observations were collected. In that case a structured plot suggests that there is a hitherto unsuspected relationship with time, or with something correlated with time. As was mentioned under point 6 of Section 1.2, a discussion of such unsuspected variables, sometimes called lurking variables, is given by Joiner (1981). In his examples the order of experimentation does often seem to provide unsuspected structure.

In Example 1 the order of the observations corresponds pretty much to increasing $y$ or $x$ values (there are three pairs of $x$ values out of order). It is therefore meaningful to compare Fig. 4.1 with Fig. 1.2, which is a plot of the ordinary residuals $r_i$ against $x$. The two figures are similar in general structure, but very different in the clarity with which observation 12 is exhibited. The values of $r_i^*$ show much more clearly than those of the ordinary residuals $r_i$ the extent to which observation 12 is an outlier. This difference between the plots is caused by the effect of a single outlier on the estimate of $\sigma^2$, which was discussed at the end of Section 3.1.

## 4.2 Half-normal plots with envelopes

Index plots are only one graphical way of presenting the information contained in the values of $r_i^*$ and $C_i$. In Section 1.2 several ways were suggested in which residuals could be plotted to help identify departures from a model. The values of the deletion residuals and the signed values of the modified Cook statistic can also, with advantage, be plotted in many of these ways. In particular it may be helpful to plot:

1. against $x_j, j = 1, \ldots, k$, the explanatory variables in the model;
2. against $x_{\text{out}}$, an explanatory variable the values of which are available, but which is not included in the current fitted model;
3. against the predicted values $\hat{y}$;
4. against the expected values of the order statistics from a standard normal distribution, that is, a normal plot.

The first of these plots will provide information about the need to include further terms, for example a quadratic, in explanatory variables which are already in the model. The second plot provides some information about the need to include an extra explanatory variable. Special graphical techniques for these purposes are described at greater length in Chapter 5. Plots of the first two types of the deletion residuals will indicate non-homogeneity of variance with individual explanatory variables, as will a plot against predicted values for the specific linear combination of the carriers $x^{\mathrm{T}}\hat{\beta}$.

These and similar plots can lead to identification of many ways in which the data and the model do not agree. But they may be too detailed and cumbersome for the routine investigation of a fitted model, particularly if it is necessary to amend the model or the data several times during the process of building a satisfactory model. To reduce the number of possibilities a single summary plot is needed of each of the two diagnostic quantities.

One possibility for the deletion residuals $r_i^*$ is a normal plot similar to those in Figures 1.3 and 1.5 for Forbes' data. These plots make it possible to provide answers to such questions as whether the values of $r_i^*$ in Fig. 4.2 can be taken as arising from a random sample. If they can and the errors are approximately normally distributed, the plot should be a straight line, apart from the fluctuations arising from sampling and the slight distortion which comes from taking the $t$ distribution as normal.

A difficulty with such plots is to know whether they are sufficiently straight or whether the inevitable irregularities are caused by anything more than random fluctuations. A particular difficulty with plots of cumulative distributions is that the ordering of the observations introduces dependence between the values. One or two random departures from expectation may then show up as distortions of the curve, rather than as local irregularities. The ability of the human eye to find patterns in scatters of points is one strong reason for the use of graphical methods. The disadvantage of this responsiveness is that features may be read into the data which are not, in reality, present. One way of overcoming this risk is to use simulation to provide a reference set of fluctuations as did Gentleman and Wilk (1975a, 1975b) in their study of outliers in two-way tables. In this chapter we follow Atkinson (1981) who suggested a form of Monte-Carlo testing in which an envelope is constructed for the plot by simulation. The purpose is not to provide a hard and fast region of acceptance or rejection for observations, but to provide some guidance as to what sort of shape or line can be expected.

The simulation envelopes, of which several examples will be given in this chapter, enhance interpretation of the plots. But the normal plot without envelope remains a useful, and simpler, diagnostic plot. A slight weakness of normal plots is what has been termed supernormality (Gentleman and Wilk 1975a; Gnanadesikan 1977, p. 265). Because the residuals are linear combinations of random variables they will tend to be more normal than the underlying error distribution if this is not normal. Thus a straight plot does not necessarily mean that the error distribution is normal. The main use of the plots in this chapter is to detect unduly influential or outlying observations. But, as we shall see in some specific cases, systematic departures in the error distribution are also revealed by the normal plot.

The simulation to generate the envelope for the plot is not complicated. The matrix $X$ of explanatory variables is taken as fixed. Because $r_i^*$ is a residual its value does not depend on the values of the parameters in the true model, provided this is of the form fitted to the data. Further, since the residual is studentized, the distribution does not depend on the value of $\sigma^2$. To simulate a set of values to act as a benchmark for the observed $r_i^*$ it is therefore only necessary to take a sample of pseudo-random observations at the observed $X$ matrix of explanatory variables and to fit the model by least squares. Since we have assumed, for testing and other purposes, that the observations have normally distributed errors, the sample is generated from the standard normal distribution and the residuals calculated. The purpose of the $X$ matrix is to generate a set of residuals, with covariance given by the matrix $I - H$, which can then be transformed into deletion residuals and Cook statistics.

To find the envelope we need to find the order statistics of the realized values and then order over replications of the simulation. Suppose that on the $m$th simulation of the $n$ observations the values of the deletion residuals are $r_{mi}^*$, $i = 1, \ldots, m$. These are ordered to give the values $r_{m(i)}^*$. The simulation is repeated a fixed number of times, usually 19, and the simulated limits are given by

$$r_{\mathrm{l}(i)}^* = \min_m r_{m(i)}^*$$
$$r_{\mathrm{u}(i)}^* = \max_m r_{m(i)}^* \qquad (4.2.1)$$

where $r_{\mathrm{l}}^*$ and $r_{\mathrm{u}}^*$ form respectively the lower and upper envelope.

The simulation is repeated 19 times to give a chance of 1 in 20 that the observed value of the largest $r_i^*$ lies above the envelope. Because each of the $n$ observations could lie outside the envelope, either above or below, the probability that the envelope contains the whole sample is not 1 in 20, but rather more. Since the ordering of the observed values introduces correlations between the values in addition to those due to the structure of the residuals, the exact probability of the envelope containing the sample is not easily calculated. If it were necessary to calculate the content of the envelope when used as a test, further simulations would be necessary in which the envelope would be kept fixed and the samples varied.

The procedure, used in the analysis of spatial processes by Ripley (1977; 1981, Chapter 8), is quite general. Cook and Weisberg (1982, Fig. 2.3.16) give an example of a normal plot with envelope. Their Fig. 3.5.4 is an example of a plot of the absolute values, that is a half-normal plot, again with envelope. Atkinson (1981, 1982a) reports that half-normal plots seem more effective than full normal plots in detecting outliers and influential observations for moderately sized samples. The maximum number of observations in an example in this book is 60 and only half-normal plots will be given. But, for larger sample sizes, the full normal plots can be expected to be more informative. The signs of the residuals will also often be of importance in the detection of specific departures.

For plotting it is usual to use an approximation to the expected values of the normal order statistics. Given a sample of $n$ observations from a standard normal distribution, let $Y_{(i)}$ be the $i$th ordered observation, that is

$$Y_{(1)} \leqslant Y_{(2)} \leqslant \cdots \leqslant Y_{(i)} \leqslant \cdots \leqslant Y_{(n)}.$$

One approximation to the expected value of the $i$th-order statistic is to take

$$E(Y_{(i)}) = \Phi^{-1}\left(\frac{i - \frac{3}{8}}{n + \frac{1}{4}}\right),$$

where $\Phi$ is the cumulative distribution function of the standard normal distribution. A brief discussion of order statistics is given in Appendix 2 of Cox and Hinkley (1974). Although they are not used to generate the plots in this book, the exact values of the normal order statistics are to be found in Table 28 of the *Biometrika Tables* (Pearson and Hartley, 1966). Algorithms for the calculation of both approximate and exact normal order statistics are given by Royston (1982), with a correction by Königer (1983).

An advantage of the Monte-Carlo approach conditional on the observed $X$ matrix is that it makes possible a test for the values of the modified Cook statistic $C_i$. If the values of the leverage measures $h_i$ have an appreciable spread, then there is no reason why the half normal plot of the $C_i$ should be a straight line. Use of the simulation procedure provides a curved envelope, which reflects the values of the $h_i$, by which the curved plot can be judged. As we shall see in Example 3, this leads to the ability, from only one kind of plot, to distinguish between influential observations where the values of $y$ and $x$ are in reasonable agreement, and those where they are not.

Although addition of the envelopes to the plots provides a useful aid to interpretation, there are disadvantages to Monte-Carlo testing. One is that each regression analysis requires the analysis of twenty sets of data—the original and the nineteen simulated samples. In practice, as computers grow ever faster, more powerful and more readily available, this is not likely to be a serious drawback. A second disadvantage is that the inherent variability of Monte-Carlo testing can lead different analysts to differing conclusions from the same set of data. That is, even when identical techniques are used, except for the random-number generator, different envelopes could, at least in principle, lead to different answers. This disadvantage can be reduced by smoothing the envelope, or by performing a larger number of simulations and taking the envelope not as the extreme order statistics but as the next to smallest and next to largest.

Modifications such as these can make simulation-based techniques less subjective and might perhaps be justified if the purpose were to set up a test procedure for the acceptance or rejection of observations. But such considerations are over elaborate in the provision of an easily used graphical technique to which has been added some guidance as to which departures really are extreme and which are merely due to random fluctuations.

The rest of this chapter gives examples of the use of the half-normal plots with envelopes of $r_i^*$ and $C_i$. In this section we first look at a simulated example which exhibits the structure of the plots and the distinct ways in which outlying values of the response and explanatory variables are exhibited. The section concludes with the two examples from Chapter 1. Both of these sets of data are relatively straightforward and it is at least arguable that the analysis so far has already been excessively laborious. However the purpose, for these two examples, is not to produce model analyses, but to exhibit many of the properties of diagnostic techniques in an unambiguous setting. Two more complicated examples are presented in Section 4.3. These examples serve later to illustrate other aspects of graphical methods. The last section of the chapter contains a few brief remarks on simulation.

**Example 3 A simulated factorial experiment 1** In this example, modified from Atkinson (1981), the data, given in Table 4.1, consist of the results of a $2^4$ factorial experiment with one centre point. The data were simulated from the model

$$Y = x_1 + 2x_2 + 0.5x_3 + 0.5x_4 + Z \tag{4.2.2}$$

with $Z$ drawn from a standard normal generator. In Table 4.1 the data are given in standard order: only those variables which are at the high level are indicated. Thus the factor combination *bc* means that $x_1 = x_4 = 0$ and $x_2 = x_3 = 1$. The values of each $x$ were taken as 0 and 1 with the centre point at 0.5. These values, rather than the more customary $-1$ and 1, were chosen to make plausible the various mistakes to which entry of observation 17, the centre point, is to be subjected.

*3.1 Original data* In this and the other analyses involving these data the fitted model is first order with a constant term. There are therefore five carriers and no systematic departures from (4.2.2).

For the data of Table 4.1 the residual sum of squares for the first-order model is 10.21. The value of the deletion residual $r_{17}^*$ is $-0.55$ and the corresponding modified Cook statistic for the centre point, $C_{17}$, is 0.21. These values are given in Table 4.2. Figure 4.4 shows a half-normal plot of $r_i^*$. The observed values are given by crosses with the envelope from 19 simulations shown by horizontal dashes. The maximum value of $r_i^*$ is 2.82, but the plot shows that this value is not exceptionally large for a sample of 17 observations.

The design is almost, but not quite, D-optimum. Because of the presence of the centre point the value of $h_i$ for the factorial points is 0.3088 rather than $\frac{5}{16} = 0.3125$ for the optimum $2^4$ factorial without centre point. Even so, the values of $C_i$ are practically indistinguishable from those of $r_i^*$. The half normal plot of the $C_i$ is accordingly indistinguishable from Fig. 4.4 and so is not given.

**Table 4.1** Example 3: A simulated factorial experiment

| Observation | Factor combination | Response |
|---|---|---|
| 1 | 1 | 0.005 |
| 2 | a | 0.219 |
| 3 | b | 0.630 |
| 4 | ab | 4.790 |
| 5 | c | −0.218 |
| 6 | ac | 1.290 |
| 7 | bc | 1.351 |
| 8 | abc | 4.123 |
| 9 | d | 0.924 |
| 10 | ad | 1.268 |
| 11 | bd | 4.147 |
| 12 | abd | 3.122 |
| 13 | cd | 1.527 |
| 14 | acd | 1.737 |
| 15 | bcd | 3.541 |
| 16 | abcd | 3.924 |
| 17 | centre point | 1.489 |

**Table 4.2** Example 3: A simulated factorial experiment. Residual sum of squares and diagnostics for observation 17 for original and altered data

| Data | Residual sum of squares | $r_{17}^*$ | $C_{17}$ |
|---|---|---|---|
| $y_{17} = 1.489$ (original) | 10.21 | $-0.55$ | 0.21 |
| $y_{17} = 10.489$ | 77.37 | 8.64 | 3.35 |
| $x_{17} = 5.0$ | 32.02 | $-4.94$ | 34.51 |
| $x_{17} = 5.0$, $y_{17} = 19.489$ | 10.58 | $-0.84$ | 5.87 |

*3.2 Wrong response at the centre point* For the original data the response at the centre point, $y_{17}$, equals 1.489. We now see what happens if this is incorrectly entered as 10.489.

Because the value of $h_{17}$ is very low, 0.0588, an outlying $y$ value at the centre has little effect on estimates of the parameters of the linear model. In fact we know, because of the balance of the design, that the only effect of the centre point is on the estimate of the constant term in the model. The effect of the outlying $y$ value on the residual sum of squares is, however, dramatic. The results of Table 4.2 show that the residual sum of squares is inflated from 10.21 to 77.37.

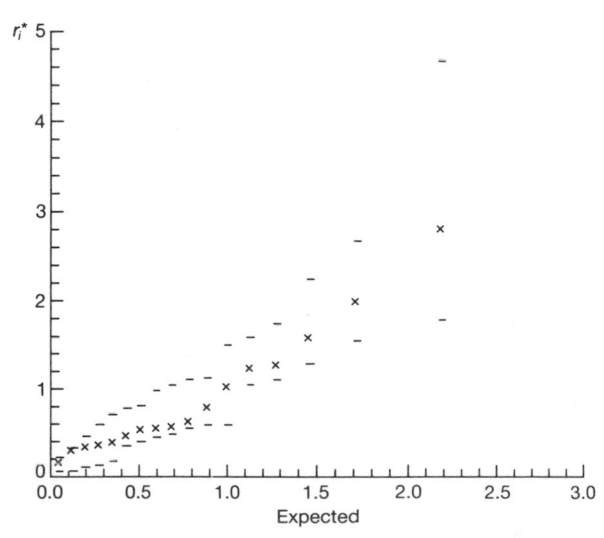


**Fig. 4.4** Example 3.1: A simulated factorial experiment; original data: half-normal plot of deletion residual $r_i^*$

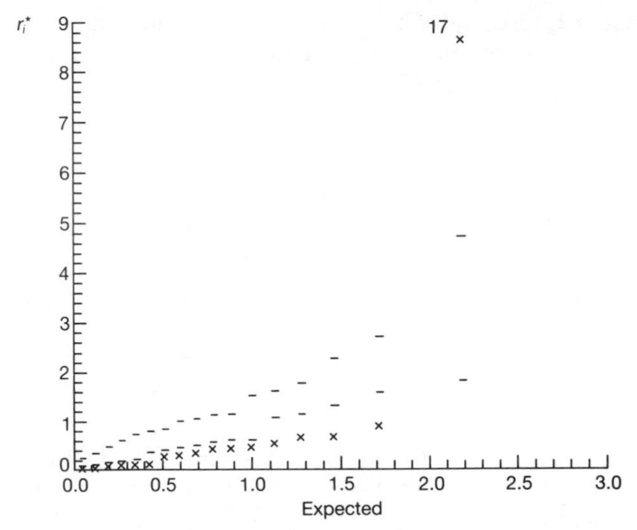


**Fig. 4.5** Example 3.2: A simulated factorial experiment; wrong response at the centre point: half-normal plot of deletion residual $r_i^*$

The observation is not influential as measured by the value of $C_i$. This is clearly shown by comparison of Figs. 4.5 and 4.6. Figure 4.5 shows a half normal plot of $r_i^*$. The maximum value of 8.64 for observation 17 clearly lies outside and above the envelope. All other values lie below the envelope because the estimate of $\sigma^2$ used is inflated by the presence of the outlying

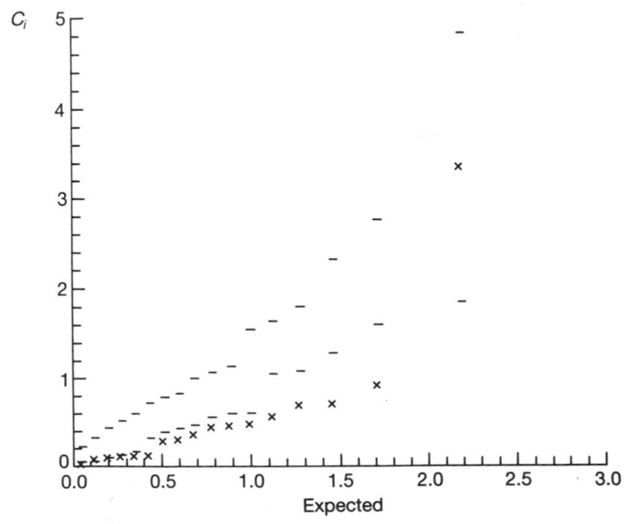


**Fig. 4.6** Example 3.2: A simulated factorial experiment; wrong response at the centre point: half-normal plot of modified Cook statistic $C_i$

observation 17. This shrinkage effect is also present in the plot of $C_i$ (Fig. 4.6). Here the largest value, 3.35, is again for observation 17. Taken on its own this value is not unduly large, lying as it does just about exactly in the middle of the simulated envelope. But the values of $C_i$ for the other observations are all shrunken due to the estimate of $\sigma^2$. The observed values form a dog-legged plot with nearly all points lying below the envelope. This is in line with the discussion of the effect of a single outlier at the end of Section 3.1. The example shows the importance of looking not just at the most discrepant observation, but at quantities calculated for all observations.

*3.3 Wrong centre point* Now suppose that $y_{17}$ is correctly entered as 1.489 but that the $x$ values for the centre point are all incorrectly entered as 5.0 rather than 0.5. As was noted in Section 1.1, such multiplications, or divisions, by factors of ten can easily occur in Fortran when numbers are read in under F format. Mistakes of this kind are one way in which wrongly influential observations can be generated.

The resulting values of $r_{17}^*$ and $C_{17}$ are given in Table 4.2. Because the 'centre' point is now far from the other observations it has high leverage with $h_{17}$ equal to 0.9531. The effect of leverage is reflected in the half-normal plot of $r_i^*$, Fig. 4.7, for which $r_{17}^*$ lies just outside the envelope, with all other values inside. Because the leverage at observation 17 is appreciable the discrepancy between $y$ and $x$ values is reflected in only a moderately large value of the deletion residual. The effect on the modified Cook statistic, Fig. 4.8, is much more dramatic with a huge value of 34.51 for $C_{17}$. This lies far outside the

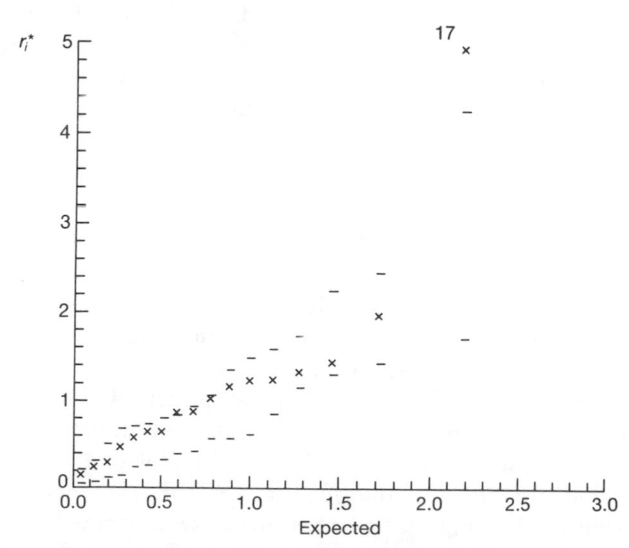


**Fig. 4.7** Example 3.3: A simulated factorial experiment; wrong centre point: half-normal plot of deletion residual $r_i^*$

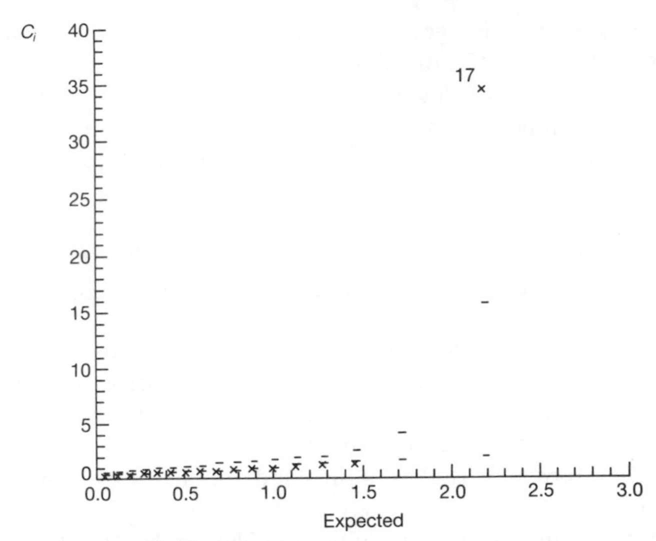


**Fig. 4.8** Example 3.3: A simulated factorial experiment; wrong centre point: half-normal plot of modified Cook statistic $C_i$

envelope, with all other values at or just below the lower boundary of the envelope.

In this example an influential observation was created by the incorrect entry of $x$ values. Such observations may not create a dramatic residual pattern, but the observation will show up clearly as incorrectly influential when the modified Cook statistic is examined. Of course, it is not possible to say whether the plot of $C_i$ is calling attention to incorrect values of the explanatory variables or to an incorrect value of the response variable at a point of high leverage. The next stage of the analysis is to check the data. If the source of errors is not discovered, the experiment yielding the suspect observation may have to be repeated.

*3.4 One influential observation* If the $x$ value of the 'centre' point were indeed 5.0 for all four explanatory variables, the response corresponding to 1.489 would be 19.489. For this set of data the results are, in the main, similar to those for the unaltered set. The residual sum of squares is 10.58, $r_{17}^*$ is $-0.84$ and the maximum deletion residual is 2.69 for observation 4. The half-normal plot of $r_i^*$, Fig. 4.9, shows no significant pattern with all observations except the four smallest lying in the centre of the envelope.

The influential observation is revealed by the plot of $C_i$, Fig. 4.10. The envelope has the same shape as that for Fig. 4.8, with a very wide interval for the largest value, indicating that there is one observation with much higher leverage than the rest. But now the value for $C_{17}$ is firmly in the middle of the interval since the $y$ and $x$ values are in agreement.

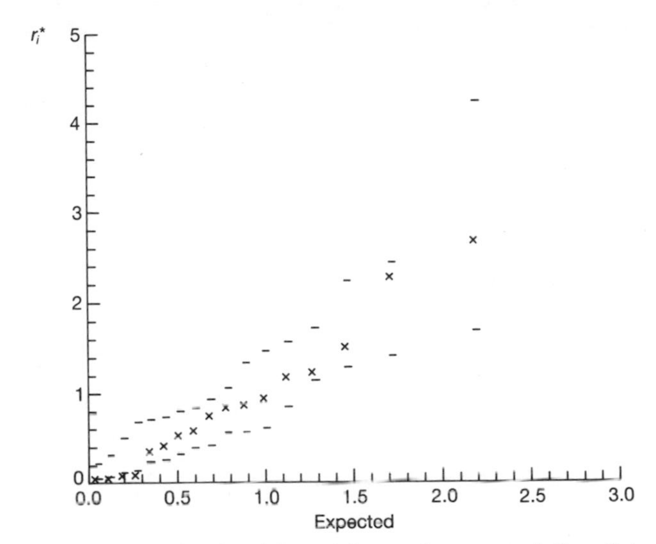


**Fig. 4.9** Example 3.4: A simulated factorial experiment; one influential observation: half-normal plot of deletion residual $r_i^*$

The four parts of this example show how the two plots can be used to detect observations which are outlying or influential. Comparison of Fig. 4.10 and Fig. 4.8 is particularly important as it shows the difference between an influential observation for which the $y$ and $x$ values respectively are and are not in agreement. The values of $r_i^*$ and $h_i$ on their own do not usually allow

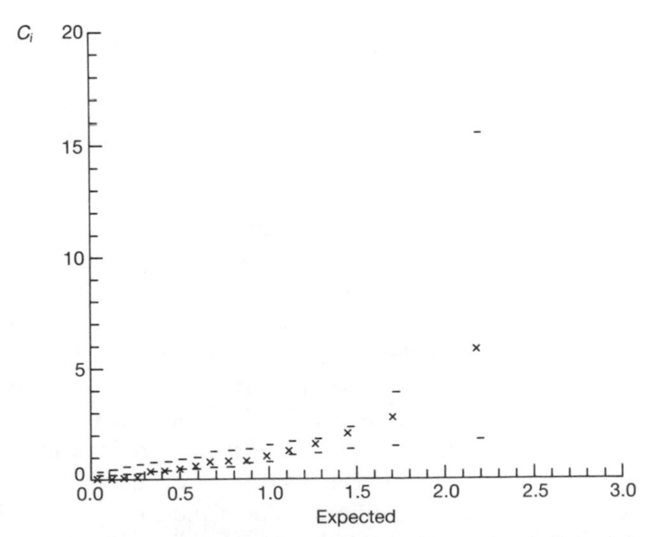


**Fig. 4.10** Example 3.4: A simulated factorial experiment; one influential observation: half-normal plot of modified Cook statistic $C_i$

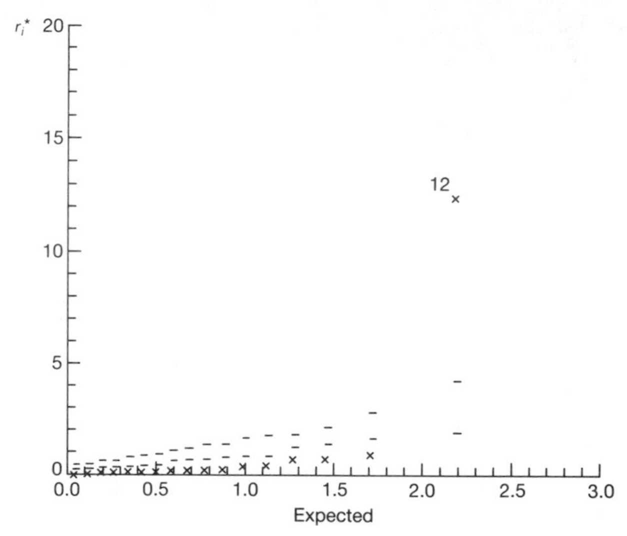


**Fig. 4.11** Example 1: Forbes' data: half-normal plot of deletion residual $r_i^*$

this distinction to be made because high leverage may cause the deletion residual to be not unduly large. □

To conclude this section we consider again the two examples introduced in Chapter 1, but now with particular emphasis on the half normal plots of $r_i^*$ and $C_i$.

**Example 1 (continued) Forbes' data 4** Figure 4.1 gives an index plot for the deletion residuals $r_i^*$ from Forbes' data. The half-normal plot of these quantities, shown in Fig. 4.11, clearly reveals that observation 12 is an outlier. From the results of Table 3.1 it can be inferred that the plot of $C_i$ is similar, but that the outlier is not so prominent. This is indeed the case, and the plot is not shown here.

The index plots for $r_i$ and $C_i$ when observation 12 is deleted (Figs. 4.2 and 4.3) have already been discussed. The half-normal plots of these quantities are given in Figs. 4.12 and 4.13. The plot of $C_i$ shows no peculiarities, whereas there is slight evidence of something strange for the third largest absolute value of $r_i^*$, which is a bit smaller than might be expected. This arises from observation 15. It is a matter of judgement as to what to do about such a point. In the present case the linear relationship between $y$ and $x$ shown in Fig. 1.1 is so overwhelmingly strong that there would seem to be no justification for pursuing this matter further. It is also the case that the data have long since finished their scientific usefulness and so should be allowed to return to the obscurity whence Weisberg rescued them. However it still remains for us to consider whether the decision to work with log pressure as

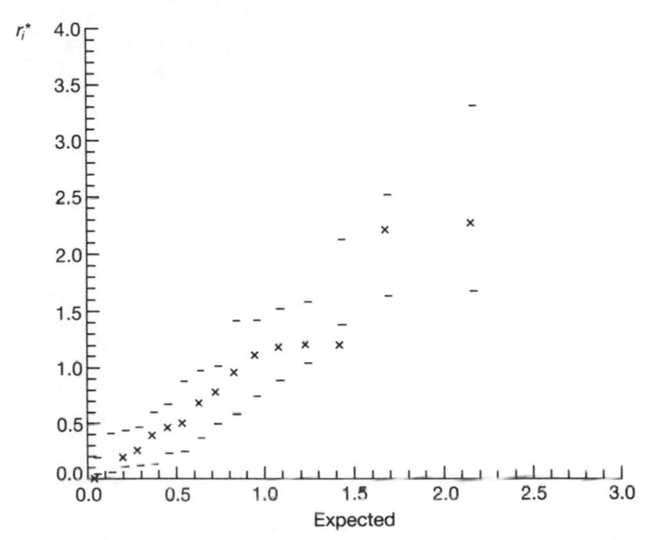


**Fig. 4.12** Example 1: Forbes' data with observation 12 deleted: half-normal plot of deletion residual $r_i^*$

the response, rather than pressure, was correct. This will be undertaken in Chapter 6. □

**Example 2 (concluded) Huber's 'data' 4** The plot of the residuals from linear regression for Huber's 'data', Fig. 1.7, shows clearly what is wrong with

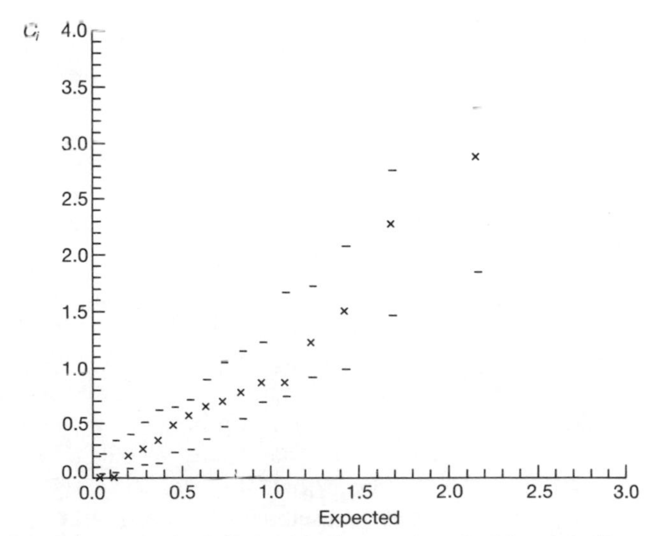


**Fig. 4.13** Example 1: Forbes' data with observation 12 deleted: half-normal plot of modified Cook statistic $C_i$

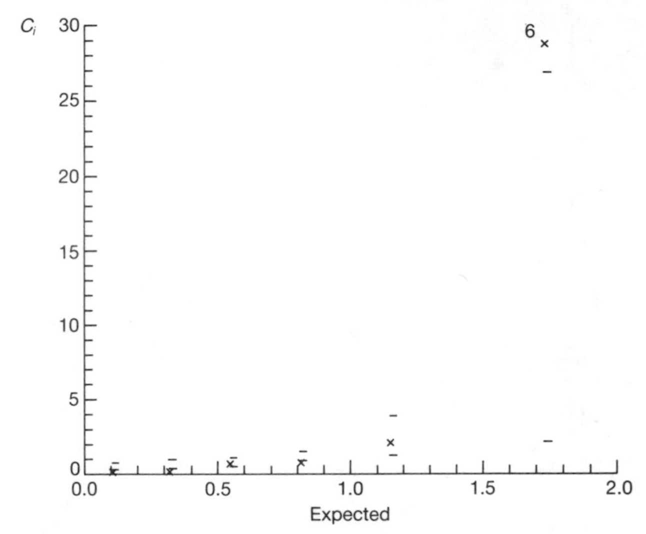


**Fig. 4.14** Example 2: Huber's 'data': half-normal plot of modified Cook statistic $C_i$

the model. The corresponding half-normal plot of $r_i^*$ is similar to the plot of $C_i$, Fig. 4.14, but does not show the departures so clearly. Figure 4.14 indicates that there is one influential observation for which the $y$ and $x$ values do not agree.

If a quadratic model is fitted to these data the $r_i^*$ yield an acceptable half-normal plot, which is shown in Fig. 4.15. The plot of the modified Cook

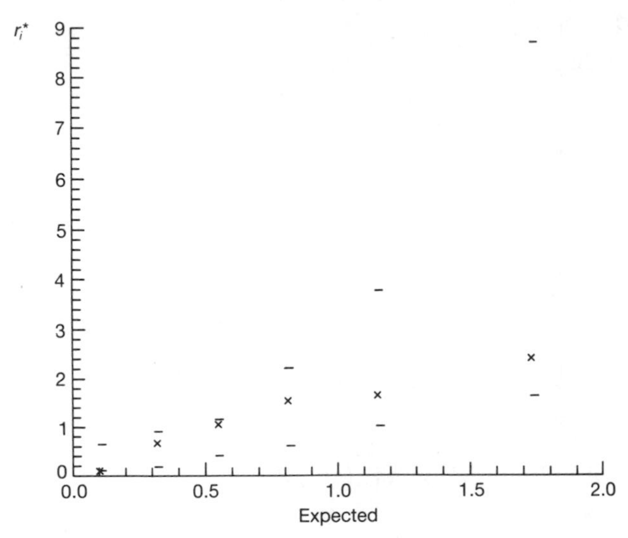


**Fig. 4.15** Example 2: Huber's 'data': half-normal plot of deletion residual $r_i^*$ for quadratic model

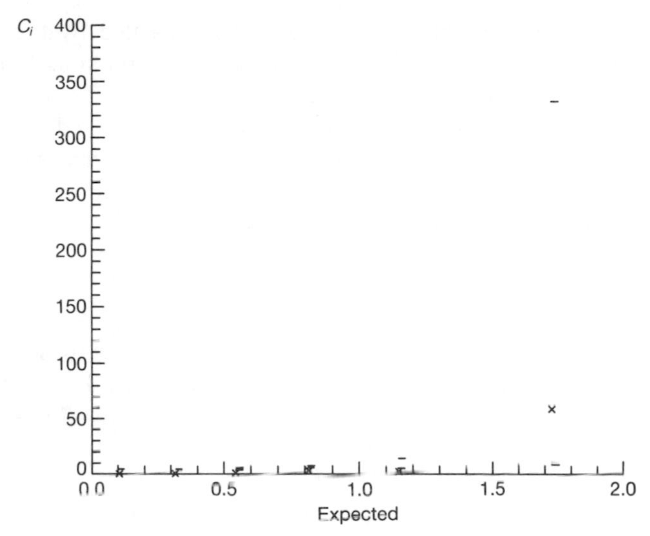


**Fig. 4.16** Example 2: Huber's 'data': half-normal plot of modified Cook statistic $C_i$ for quadratic model

statistics $C_i$, Fig. 4.16, is more difficult to interpret because the very high leverage of observation 6 means that the simulation envelope is extremely wide for this point and correspondingly narrow for the other points. The values of the $C_i$ for observations other than 6 lie just below the envelope, suggesting that there is some evidence that observation 6 is an outlier which is causing variance inflation.

This analysis serves to illustrate one possible abuse of diagnostic plots. Figure 4.14 indicates that there is some discrepancy between linear regression and the data. An obvious way of overcoming this departure is to try to fit a higher-order model. In this case the $t$ value for the second-order term is 7.37 which, even on three degrees of freedom, is significant at the 1 per cent level. Figure 4.15 and the largest value of Fig. 4.16 do not indicate any gross failings of this model. The second-order model might therefore be accepted.

The failure in this strategy is that one term, the quadratic, has been added to the model to accommodate one influential observation. This is suggested by the shape of the envelope in Fig. 4.16. It is confirmed by the value of 0.9993 given in Table 3.3 for the leverage measure $h_6$ in the second-order model. Thus the half-normal plots if discrepancies are indicated, as they are in Fig. 4.14, may with advantage be supplemented by further investigation of the structure of the data. The plots described in Chapter 1 and at the beginning of this section are recommended for this purpose.

These comments conclude our analysis of Huber's 'data'. With one explanatory variable the techniques which have been used are over-elaborate, although they are necessary with three or more explanatory variables. For linear regression the structure of the problem is revealed by the plot of $y$ against $x$ in Fig. 1.6. It is clear that either $y_6$ or $x_6$ (or possibly both) are in error, or are highly informative about the nonlinearity of the model. In this case Huber (1981, p. 155) states that the 'data' were generated from a linear relationship with 12 added to $y_6$ and independent normal errors of standard deviation 0.6 added to the other observations. In the more complicated examples which follow, the true model is unknown. □

## 4.3 Two examples

The experience gained from analysis of the three examples in Section 4.2 is applied in this section to two more complicated examples. Both sets of data have appeared several times in the statistical literature. Some of these analyses are described briefly in Section 12.4.

**Example 4 Salinity data 1** Ruppert and Carroll (1980) give a set of 28 observations on the salinity of water during the spring in Pamlico Sound, North Carolina. Analysis of the data was originally undertaken as part of a project for forecasting the shrimp harvest. The response is the bi-weekly average of salinity. There are three explanatory variables: the salinity in the previous two-week time period, a dummy variable for the time period during March–April and the river discharge. In summary the variables are:

$y$: bi-weekly average salinity
$x_1$: salinity lagged two weeks
$x_2$: trend, a dummy variable for the time period
$x_3$: water flow, that is river discharge.

The data are given in Table 4.3. Use of lagged salinity as one of the explanatory variables in a simple regression model means that we ignore the errors of measurement of this variable. In a full analysis the effect of this approximation should be investigated. Because the readings are taken over only six two-week periods the value of lagged salinity $x_{1i}$ is not necessarily equal to $y_{i-1}$.

We begin by fitting a first-order model in the three explanatory variables; there are therefore four carriers. The half-normal plot of the $r_i^*$ for the first-order model shows no particular features of interest, but the corresponding plot of the $C_i$, Fig. 4.17, shows that there is one observation which is unduly influential. The value of 10.19 for $C_i$ belongs to observation 16. In combination with the unexceptionable value of $r_i^*$, this result suggests that there is something suspect about one of the explanatory variables for this observation. Inspection of the data in Table 4.3 shows that for observation 16 $x_3$ has the value 33.4. The next largest is 29.9 for observation 5, with the remaining $x_3$ values between 20.769 and 26.417. A histogram of these values,

**Table 4.3** Example 4: Salinity data

| Observation | Salinity $y$ | Lagged salinity $x_1$ | Trend $x_2$ | Water flow $x_3$ | Year |
|---|---|---|---|---|---|
| 1 | 7.6 | 8.2 | 4 | 23.005 | 72 |
| 2 | 7.7 | 7.6 | 5 | 23.873 | |
| 3 | 4.3 | 4.6 | 0 | 26.417 | 73 |
| 4 | 5.9 | 4.3 | 1 | 24.868 | |
| 5 | 5.0 | 5.9 | 2 | 29.895 | |
| 6 | 6.5 | 5.0 | 3 | 24.200 | |
| 7 | 8.3 | 6.5 | 4 | 23.215 | |
| 8 | 8.2 | 8.3 | 5 | 21.862 | |
| 9 | 13.2 | 10.1 | 0 | 22.274 | 74 |
| 10 | 12.6 | 13.2 | 1 | 23.830 | |
| 11 | 10.4 | 12.6 | 2 | 25.144 | |
| 12 | 10.8 | 10.4 | 3 | 22.430 | |
| 13 | 13.1 | 10.8 | 4 | 21.785 | |
| 14 | 12.3 | 13.1 | 5 | 22.380 | |
| 15 | 10.4 | 13.3 | 0 | 23.927 | 75 |
| 16 | 10.5 | 10.4 | 1 | 33.443 | |
| 17 | 7.7 | 10.5 | 2 | 24.859 | |
| 18 | 9.5 | 7.7 | 3 | 22.686 | |
| 19 | 12.0 | 10.0 | 0 | 21.789 | 76 |
| 20 | 12.6 | 12.0 | 1 | 22.041 | |
| 21 | 13.6 | 12.1 | 4 | 21.033 | |
| 22 | 14.1 | 13.6 | 5 | 21.005 | |
| 23 | 13.5 | 15.0 | 0 | 25.865 | 77 |
| 24 | 11.5 | 13.5 | 1 | 26.290 | |
| 25 | 12.0 | 11.5 | 2 | 22.932 | |
| 26 | 13.0 | 12.0 | 3 | 21.313 | |
| 27 | 14.1 | 13.0 | 4 | 20.769 | |
| 28 | 15.1 | 14.1 | 5 | 21.393 | |

given as Fig. 4.18, shows how far the value for observation 16 lies from the values for the other observations.

One possibility is to repeat the analysis with observation 16 deleted. As an alternative Atkinson (1982b, 1983) tried to infer the correct value of $x_3$ by assuming 33.443 to be a misprint for 23.443. The result is similar to that of the strategy of Professor Carroll who reports that, in his unpublished analyses of these and similar data, he replaced large values of $x_3$ by 26. The rationale is that the usual effect of increased water flow is to increase mixing in the sound and so reduce salinity. But exceptionally large values of water flow, due to inland thunderstorms, cause some of the fresh water to be taken straight out to sea without mixing. The effect of dilution on salinity is therefore partially

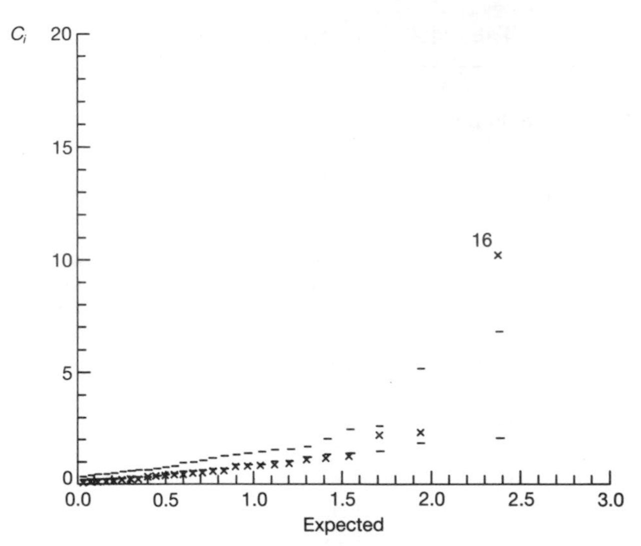


**Fig. 4.17** Example 4: Salinity data: half-normal plot of modified Cook statistic $C_i$

reduced. This truncation of values of water flow at 26 leads to a model in which the effect of $x_3$ consists of two parts, a linear increase and a plateau. An alternative to such a 'bent stick' model would be to include a quadratic term in $x_3$.

If we do replace the value of $x_3$ for observation 16 by 23.443, but leave all other quantities unchanged, the residual sum of squares is reduced from 42.47 to 26.24. The half-normal plot of $C_i$, Fig. 4.19, now shows no unduly influential observations, with the 10 largest values of $C_i$ lying in the envelope. The largest value is 5.15 for observation 5, for which observation $x_3$ equals 29.895. The plot of $r_i^*$ is similar. Both plots show that many of the small values are a little too small. This is difficult to interpret, but may be an indication of an asymmetrical error distribution. We return to this matter when we discuss transformations in Chapter 6.

Three general points emerge from this analysis. The first is that the diagnostic technique has pinpointed one observation as discrepant and indicated that one of the explanatory variables is the culprit. This is to be

| Middle of interval | Number of observations | |
|---|---|---|
| 20. | 1 | * |
| 22. | 13 | ************* |
| 24. | 8 | ******** |
| 26. | 4 | **** |
| 28. | 0 | |
| 30. | 1 | * |
| 32. | 0 | |
| 34. | 1 | * |

**Fig. 4.18** Example 4: Salinity data: histogram of $x_3$

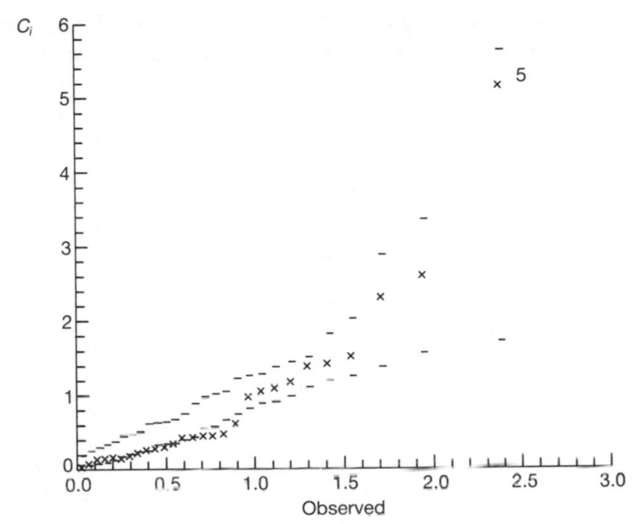


**Fig. 4.19** Example 4: Salinity data: half-normal plot of modified Cook statistic $C_i$ with $x_3 = 23.443$ for observation 16

contrasted with the results of Ruppert and Carroll (1980) who were exemplifying the use of weighted least squares as a method of robust regression. Because observation 16 has high leverage, with $h_{16}$ equal to 0.5467, the ordinary residual is small. The estimation method used by Ruppert and Carroll is iterative with residuals, albeit not necessarily least squares residuals, as the starting point. The method seems unable to recognize observation 16 as the cause of the discrepancy. As a result several other observations are downweighted. This difficulty with leverage points is in agreement with the discussion of robust regression in Section 12.1.

The second general point is that Fig. 4.17 shows there is a discrepancy between the first-order model and the data. It may indeed be that $x_3$ is incorrect for observation 16. But it may be that the model is incorrect. As was mentioned above, an alternative to replacing large values of $x_3$ would be to fit a quadratic term in $x_3$. Table 4.4 gives summary statistics for the four possibilities of first- and second-order models in $x_3$ and original and 'corrected' data. The quantities given are the residual sums of squares and the $t$ values for the individual parameters in the model, which correspond to the $F$ tests for elimination of each variable from the full model. These results suggest that a quadratic term in $x_3$ may provide a better model for the data than a first-order model with the 'corrected' data. In particular the residual sum of squares when the quadratic term is included is further reduced from 26.24 to 20.48 and all terms in the model are significant, although the $t$ value for $x_2$ is exactly on the 5 per cent point of the distribution. The opposite signs of the linear and quadratic coefficients for $x_3$ show that the quadratic is work-

**Table 4.4** Example 4: Salinity data. Residual sums of squares and $t$ values for two models with and without 'correction' to observation 16

| | Original data | Value of $x_3$, observation 16, 'corrected' to 23.443 |
|---|---|---|
| | *First-order model* | |
| Source | $t$ value | $t$ value |
| $x_1$ | 9.01 | 9.85 |
| $x_2$ | −0.16 | −1.17 |
| $x_3$ | −2.76 | −5.22 |
| Residual sum of squares | 42.47 | 26.24 |
| | *Quadratic model* | |
| Source | $t$ value | $t$ value |
| $x_1$ | 10.59 | 10.56 |
| $x_2$ | −2.07 | −2.04 |
| $x_3$ | −5.27 | −2.96 |
| $x_3^2$ | 4.97 | 2.56 |
| Residual sum of squares | 20.48 | 20.43 |

ing as an alternative to the bent stick model induced by truncating $x_3$. Unless there is some good reason to the contrary, smooth models such as polynomials are to be preferred to those with discontinuities in the variables or their derivatives.

Further analysis of this example, including plots of $r_i^*$ and $C_i$ for the quadratic model can be undertaken, but the analysis does not yield any further general points. The third and last general point from this example is that in this case the influential observation was identified as being at a point of high leverage which was detected by a marginal plot of the values of $x_3$. Often such leverage points are not visible from marginal plots. The section concludes with such an example. □

**Example 5 Prater's gasoline data 1** The data in Table 4.5, originally given by Prater, were re-used by Hader and Grandage (1958) to illustrate multiple regression calculations. The response is the percentage of crude oil which, after distillation and fractionation, ends up as gasoline, or petrol. The four explanatory variables are:

$x_1$: crude oil gravity, °API
$x_2$: crude oil vapour pressure, lbf/in$^2$
$x_3$: crude oil ASTM 10 per cent, °F
$x_4$: gasoline end point, °F.

The first three explanatory variables are, obviously, measurements of the properties of the crude oil. Both $x_3$ and $x_4$ measure the temperature at which

**Table 4.5** Example 5: Prater's gasoline data

| Crude oil gravity °API | Crude oil vapour pressure (lbf/in²) | Crude oil ASTM 10% point, °F | Gasoline end point, °F | Gasoline yield, % |
|---|---|---|---|---|
| $x_1$ | $x_2$ | $x_3$ | $x_4$ | $y$ |
| 38.4 | 6.1 | 220 | 235 | 6.9 |
| 40.3 | 4.8 | 231 | 307 | 14.4 |
| 40.0 | 6.1 | 217 | 212 | 7.4 |
| 31.8 | 0.2 | 316 | 365 | 8.5 |
| 40.8 | 3.5 | 210 | 218 | 8.0 |
| 41.3 | 1.8 | 267 | 235 | 2.8 |
| 38.1 | 1.2 | 274 | 285 | 5.0 |
| 50.8 | 8.6 | 190 | 205 | 12.2 |
| 32.2 | 5.2 | 236 | 267 | 10.0 |
| 38.4 | 6.1 | 220 | 300 | 15.2 |
| 40.3 | 4.8 | 231 | 267 | 26.8 |
| 32.2 | 2.4 | 284 | 351 | 14.0 |
| 31.8 | 0.2 | 316 | 379 | 14.7 |
| 41.3 | 1.8 | 267 | 275 | 6.4 |
| 38.1 | 1.2 | 274 | 365 | 17.6 |
| 50.8 | 8.6 | 190 | 275 | 22.3 |
| 32.2 | 5.2 | 236 | 360 | 24.8 |
| 38.4 | 6.1 | 220 | 365 | 26.0 |
| 40.3 | 4.8 | 231 | 395 | 34.9 |
| 40.0 | 6.1 | 217 | 272 | 18.2 |
| 32.2 | 2.4 | 284 | 424 | 23.2 |
| 31.8 | 0.2 | 316 | 428 | 18.0 |
| 40.8 | 3.5 | 210 | 273 | 13.1 |
| 41.3 | 1.8 | 267 | 358 | 16.1 |
| 38.1 | 1.2 | 274 | 444 | 32.1 |
| 50.8 | 8.6 | 190 | 345 | 34.7 |
| 32.2 | 5.2 | 236 | 402 | 31.7 |
| 38.4 | 6.1 | 220 | 410 | 33.6 |
| 40.0 | 6.1 | 217 | 340 | 30.4 |
| 40.8 | 3.5 | 210 | 347 | 26.6 |
| 41.3 | 1.8 | 267 | 416 | 27.8 |
| 50.8 | 8.6 | 190 | 407 | 45.7 |

given amounts of liquid have been vapourized: $x_3$ is the temperature at which 10 per cent of the crude has become vapour and $x_4$ is the temperature at which all of the gasoline has vapourized. It is clear in a general way that light crudes provide more gasoline than heavier crudes and that, as more gasoline is distilled from a given crude, the gasoline end point increases.

Prater's intention was to provide an illustration of the use of multiple regression in building a prediction equation. The paper of Hader and Grandage provides an example of the change of emphasis, mentioned in Chapter 1, in work on regression over the last 25 years. Their concern is, in large part, with the numerical aspects of fitting. Their analysis of the data is unaccompanied by plots.

The numbers for the 32 observations in Table 4.5 are reproduced from a series of data sets distributed to students taking the M.Sc. in statistics at Imperial College, London. We begin by fitting the first-order model

$$E(Y_i) = \beta_0 + \beta_1 x_{1i} + \beta_2 x_{2i} + \beta_3 x_{3i} + \beta_4 x_{4i} \qquad (i = 1, \ldots, 32),$$

for which the residual sum of squares is 348.2. Half-normal plots of $r_i^*$ and $C_i$ are given in Figs. 4.20 and 4.21. Both clearly indicate that there is something strange about the data. The plot of $r_i^*$ shows that there is an outlying observation. This is number 11 with a value of 6.42 for the deletion residual. The value of $C_i$ for this observation is much smaller at 3.68 and, as Fig. 4.21 shows, the value lies within the simulation envelope. But the plot also shows that many other values, including the 15 next smallest, are shrunken below the envelope. Comparison of the data of Table 4.5 with that given by Hader and Grandage shows that there is an error in the value for $x_4$. For observations 9, 10 and 11 the values are 267, 300 and 267 whereas they should be 267, 300 and 367. This error cannot be detected by a plot of the values of $x_4$. The location of the influential observation is thus importantly different from that of observation 16 in the salinity data. This typing mistake is typical of the small errors in data entry which can have large effects on an analysis.

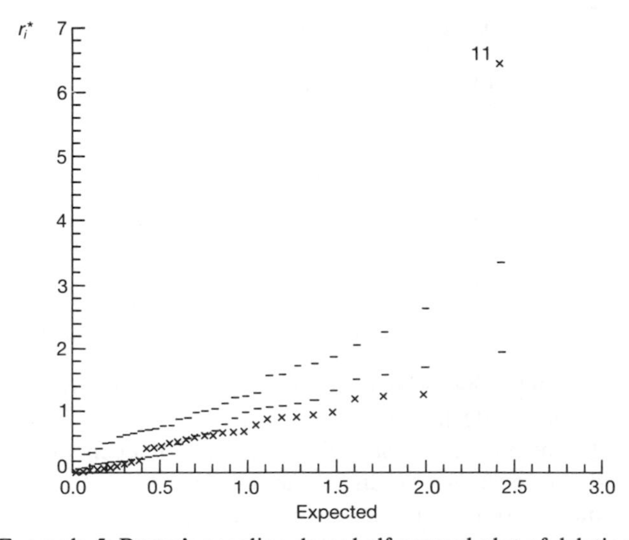


**Fig. 4.20** Example 5: Prater's gasoline data: half-normal plot of deletion residual $r_i^*$

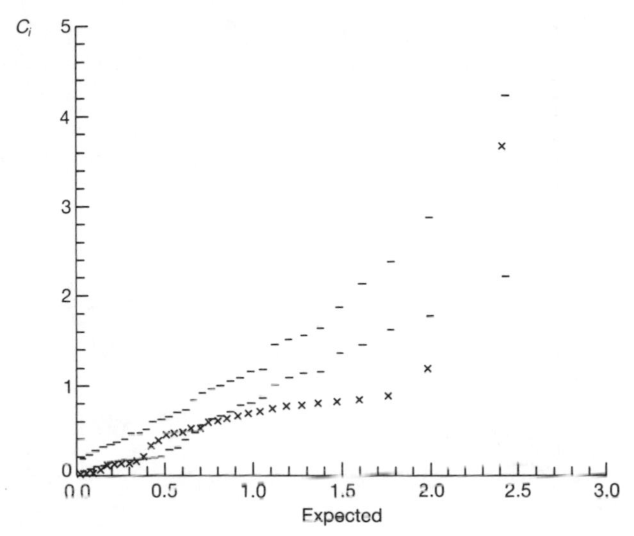


**Fig. 4.21** Example 5: Prater's gasoline data: half-normal plot of modified Cook statistic $C_i$

Once this typing error has been corrected, the residual sum of squares for the first-order model is reduced to 134.8, less than half the previous value of 348.2. The fit of the model is therefore appreciably improved. However there is still some discrepancy between the model and the data. Figure 4.22 shows the half-normal plot of the deletion residuals $r_i^*$. This shows no peculiarities

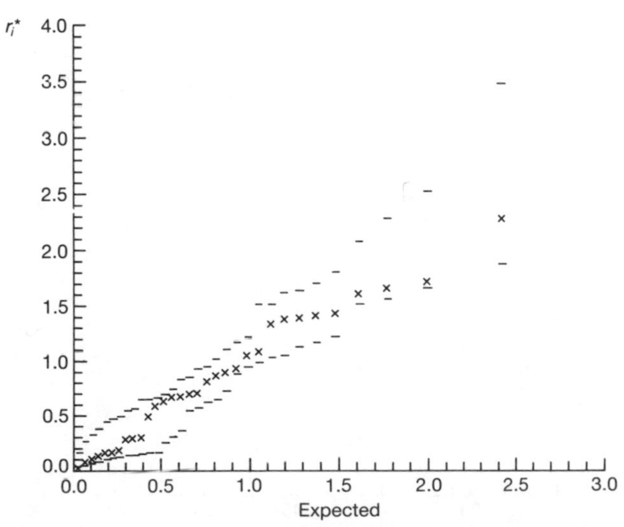


**Fig. 4.22** Example 5: Prater's gasoline data: half-normal plot of deletion residual $r_i^*$ for corrected data

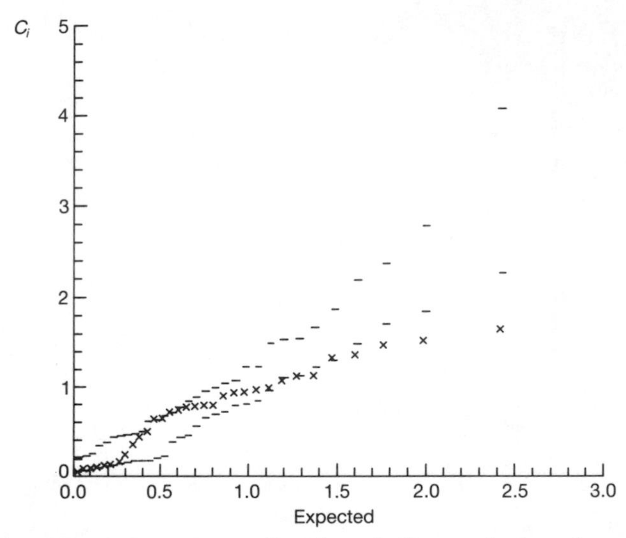


**Fig. 4.23** Example 5: Prater's gasoline data: half-normal plot of modified Cook statistic $C_i$ for corrected data

with, for once, all observed values lying within the simulated envelope. But the plot of $C_i$, Fig. 4.23, is less satisfactory, with the four largest values all lying below the simulated envelope.

To try to explain this hitherto unencountered pattern we return to consideration of the data. Figure 4.24 shows a plot of $y$ against $x_1$ and Fig.

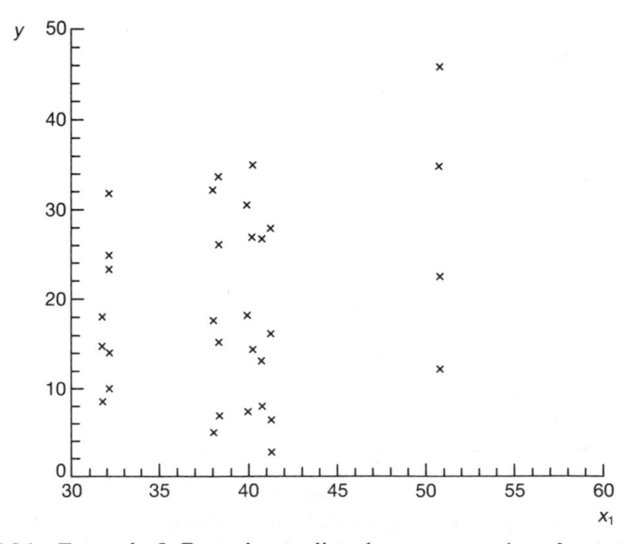


**Fig. 4.24** Example 5: Prater's gasoline data: scatter plot of $y$ against $x_1$

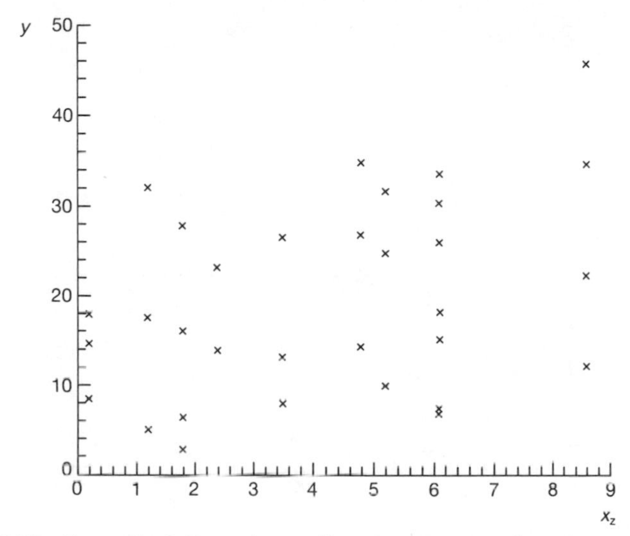


**Fig. 4.25** Example 5: Prater's gasoline data: scatter plot of $y$ against $x_2$

4.25 a plot of $y$ against $x_2$. Both figures, but especially Fig. 4.25, show a pattern of several $y$ values for each value of $x$. The pattern is made explicit in Fig. 4.26 where $x_2$ is plotted against $x_1$. This shows that there are only 10 pairs of values of $x_1$ and $x_2$. Similar plots involving $x_3$, or direct inspection of the data with this clue in mind, show that there are only 10 sets of values of the first three explanatory variables. These three values characterize ten

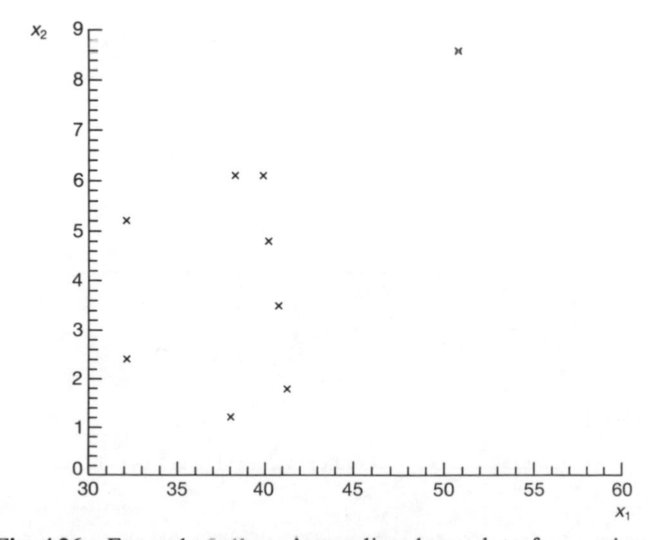


**Fig. 4.26** Example 5: Prater's gasoline data: plot of $x_2$ against $x_1$

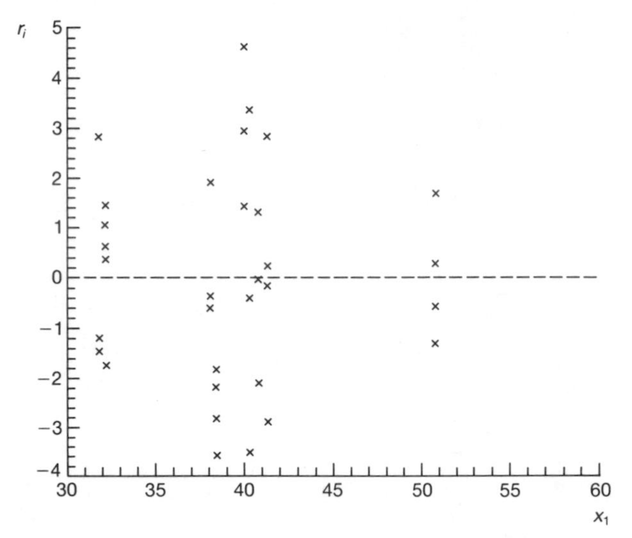


**Fig. 4.27** Example 5: Prater's gasoline data: residuals $r_i$ against $x_1$

different crudes, which were then subjected to experimentally controlled distillation conditions, varying in number from 2 to 4 per crude.

The nested structure of the data may also be seen from Fig. 4.27 and, especially, Fig. 4.28, which show plots of the residuals against $x_1$ and $x_2$. These clearly fall into groups according to the ten crudes. The nested nature of the data was first revealed by Daniel and Wood (1971, Chapter 8). To help

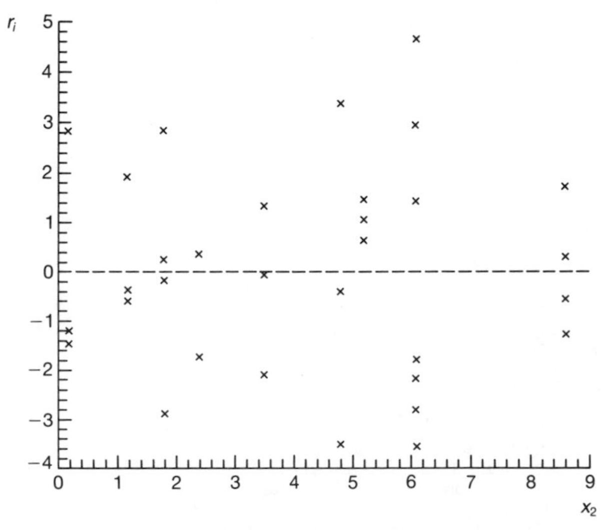


**Fig. 4.28** Example 5: Prater's gasoline data: residuals $r_i$ against $x_2$

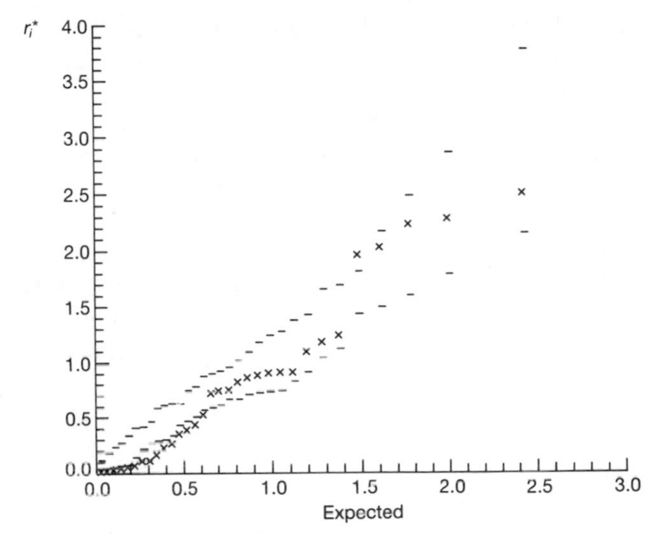


**Fig. 4.29** Example 5: Prater's gasoline data: half-normal plot of deletion residual $r_i^*$

in the detection of such patterns they suggest that the data be sorted on one of the explanatory variables. Provided this is not $x_4$, the repeated values of the explanatory variables are immediately apparent.

The multiple regression model used so far is not appropriate for this structure. We follow Daniel and Wood in considering instead the model

$$E(Y_{ij}) = \alpha_i + \beta x_{4j} \qquad (i = 1, \ldots, 10; j = 1, \ldots, 32). \tag{4.3.1}$$

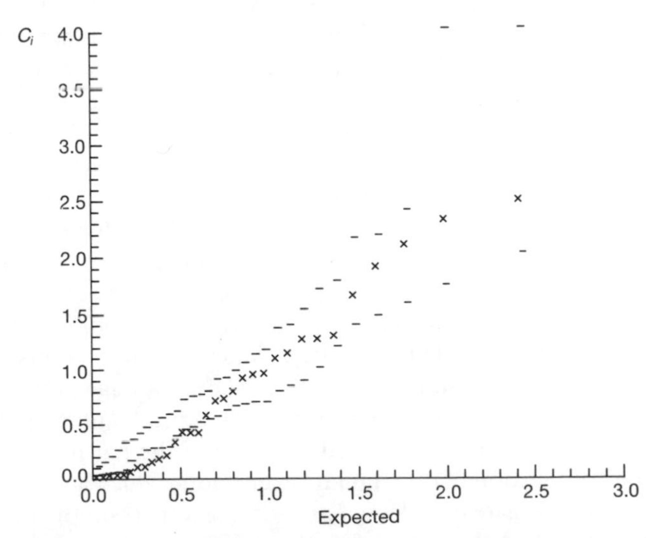


**Fig. 4.30** Example 5: Prater's gasoline data: half-normal plot of modified Cook statistic $C_i$

Here it is assumed that the effect of end point on yield is the same for all 10 crudes, but that the intercept of the regression lines may vary from crude to crude.

The half-normal plots of $r_i^*$ and $C_i$ from fitting this nested model to the data are shown in Figs. 4.29 and 4.30. There seems little structure left, except for one value of $r_i^*$ which is somewhat large. There are, however, an appreciable number of small values on both plots which are smaller than they should be. As with Fig. 4.19, but in a rather more extreme manner, this could be interpreted as an indication that the error distribution is not quite symmetrical, giving rise to some unduly large and small residuals. We return to the discussion of this example when we consider transformations for proportions and percentages in Chapter 7. □

## 4.4 Simulation

The envelopes to the half normal plots used in this chapter require the simulation of $19n$ normal random variables and the calculation and ordering of the $n$ values of $r_i^*$ and $C_i$ for each of the nineteen simulated samples. The required normal random variables were simulated using the convenient algorithm of Marsaglia and Bray (1964). The method makes appreciable use of pseudo-random numbers, requiring on average 3.9 numbers for each normal random variable generated. The comparisons of Atkinson and Pearce (1976) showed that this was the fastest generator on the Cyber system at Imperial College when the in-line CDC RANF random-number generator was used. With a relatively less rapid random-number generator another algorithm might be preferable for sampling the normal distribution.

The residuals needed for the calculation of $r_i^*$ and $C_i$ can be calculated by use of the hat matrix via the relationship $r = (I - H)y$, eqn (2.1.11). If $n$ is not small, storage of the $n \times n$ hat matrix is costly in space, even if advantage is taken of symmetry to reduce the number of elements stored. For calculation of the single deletion diagnostics, as opposed to the residuals using (2.1.11), only the diagonal values $h_i$ are required. In the absence of the off-diagonal elements of $H$, the residuals must be calculated in a less simple way. For example the programs producing the diagrams in this book stored only the matrix $V = (X^TX)^{-1}$, so that the fitted values for the simulations were produced from the relationship $\hat{y} = XVX^Ty$. The advantage of this method is that $V$ is a $p \times p$ matrix, rather than $n \times n$. The disadvantage is that extra multiplications are required compared with use of the hat matrix.

A potential application of the simulation envelope which is not explored in this book is to investigate the effect of non-normal errors on the least squares fit. Sampling from distributions other than the normal can be expected to yield rather differently shaped envelopes. Algorithms for sampling a variety of distributions are described by Atkinson and Pearce (1976), Fishman (1978, Chapter 9), Knuth (1981, Section 3.4.1) and Ripley (1983).

The choice of simulation algorithm and method of calculation of the residuals are unlikely to be important for the examples of this book, where the maximum number of observations is 60. It does however in general seem sensible to avoid inversion of $X^{\mathrm{T}}X$ for each simulated sample. Likewise, for sample sizes less than 100, the method of ordering the values of $r_i^*$ and $C_i$ is also of minor importance. If samples are sufficiently large that the method of ordering does become important, a discussion of suitable methods is given by Knuth (1973).

# 5
# Diagnostic plots for explanatory variables

## 5.1 Residual plots

In the theoretical discussion of leverage and influence in Chapter 3 it is taken for granted that the carriers for a particular set of data are known. With two exceptions, the same assumption was made in the analysis of the examples. The exceptions were in the analysis of Huber's data and of the salinity data, when quadratic terms in one explanatory variable were considered.

In practice it is almost never the case that the carriers are known without doubt. Often there are a large number of potential carriers including some, but not necessarily all, of the given explanatory variables, together with higher-order polynomial terms, such as quadratic and two or more factor interactions. Even without the further complication of non-polynomial terms, the number of potential models resulting from choice of subsets of the carriers is often large. There is an appreciable literature and continuing interest in this problem of variable selection. One introduction is given by Weisberg (1980, Chapter 8).

Methods for choice of a model nearly all depend upon aggregate statistics. One example is the use of $t$ or $F$ tests to determine the inclusion or exclusion of variables. An alternative approach in which carriers are smoothly downweighted, rather than excluded, is provided by shrinkage estimators, reviewed by Vinod and Ullah (1981). Another group of methods leads to the calculation of a measure of model desirability, the model with the most desirable value of the measure being selected. If the measure is the estimated mean squared error of prediction calculated at the observations, the appropriate measure of choice is Mallows' $C_p$ (Mallows 1973), a special case of Akaike's AIC (Akaike 1973). This criterion depends on the individual observations solely through the residual sum of squares for each model and so will hide the effect of individual observations. A graphical means of displaying the contribution of individual observations to the value of $C_p$ is provided by Weisberg (1981).

An advantage of graphical methods is that they can exhibit the effect of each observation. Some standard plots for investigation of fitted models were mentioned in Section 1.2. One way which was suggested of looking at the evidence for inclusion of a particular carrier was a plot of residuals against a carrier not in the model. In the present chapter some extensions of this idea are described and compared. One of these, the added variable plot, will be of appreciable use in the sixth and succeeding chapters when we come to look at transformations of the response $y$. To illustrate the ideas we begin with an example.

**Example 6 Minitab tree data 1** Minitab is a package of computer programs for performing elementary statistical analyses. Because the package is easy to use, it is often employed as a teaching medium. To complement this activity there is an introductory statistics textbook, the *Minitab Student Handbook* (Ryan *et al.* 1976). One of the examples of multiple regression in the handbook is a set of measurements on the volume, girth and height of 31 black cherry trees, where girth is the diameter at 4 ft 6 in above the ground.

The purpose of the measurements on the felled trees was to provide a means of predicting, from easily made measurements, the volume of timber in unfelled trees and hence to predict the value of timber in an area of forest. Because measurements of girth are easier, and more accurate, on unfelled trees than are measurements of height, a formula based solely on girth would be preferred.

The data, taken from Ryan *et al.* (1976, p. 278), are given in Table 5.1. The variables are:

$y$: volume of the tree, in cubic feet.
$x_1$: girth of the tree; diameter in inches at 4 ft 6 in above ground level.
$x_2$: height of the tree in feet.

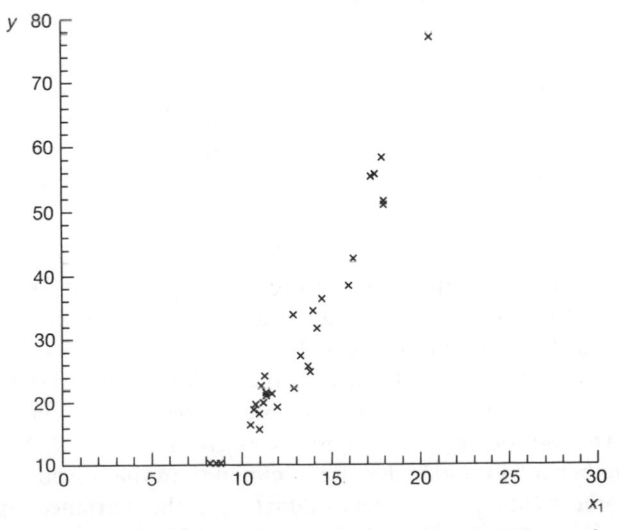


**Fig. 5.1** Example 6: Minitab tree data: scatter plot of $y$ against $x_1$

**Table 5.1** Example 6: Minitab tree data

| Observation $i$ | Diameter $x_1$ | Height $x_2$ | Volume $y$ |
|---|---|---|---|
| 1 | 8.3 | 70 | 10.3 |
| 2 | 8.6 | 65 | 10.3 |
| 3 | 8.8 | 63 | 10.2 |
| 4 | 10.5 | 72 | 16.4 |
| 5 | 10.7 | 81 | 18.8 |
| 6 | 10.8 | 83 | 19.7 |
| 7 | 11.0 | 66 | 15.6 |
| 8 | 11.0 | 75 | 18.2 |
| 9 | 11.1 | 80 | 22.6 |
| 10 | 11.2 | 75 | 19.9 |
| 11 | 11.3 | 79 | 24.2 |
| 12 | 11.4 | 76 | 21.0 |
| 13 | 11.4 | 76 | 21.4 |
| 14 | 11.7 | 69 | 21.3 |
| 15 | 12.0 | 75 | 19.1 |
| 16 | 12.9 | 74 | 22.2 |
| 17 | 12.9 | 85 | 33.8 |
| 18 | 13.3 | 86 | 27.4 |
| 19 | 13.7 | 71 | 25.7 |
| 20 | 13.8 | 64 | 24.9 |
| 21 | 14.0 | 78 | 34.5 |
| 22 | 14.2 | 80 | 31.7 |
| 23 | 14.5 | 74 | 36.3 |
| 24 | 16.0 | 72 | 38.3 |
| 25 | 16.3 | 77 | 42.6 |
| 26 | 17.3 | 81 | 55.4 |
| 27 | 17.5 | 82 | 55.7 |
| 28 | 17.9 | 80 | 58.3 |
| 29 | 18.0 | 80 | 51.5 |
| 30 | 18.0 | 80 | 51.0 |
| 31 | 20.6 | 87 | 77.0 |

The observations given in the table are ordered on diameter and so also pretty much on volume. Any lurking variables to do with time or location in the forest therefore remain undetected.

To begin we look at plots of $y$ against the explanatory variables. Figure 5.1 gives a plot of $y$ against $x_1$, which shows a clear relationship between volume and girth. The plot of $y$ against height, $x_2$, is given in Fig. 5.2. Not only is the relationship between $y$ and $x_2$ less well defined than that given in Fig. 5.1, but there is some evidence of heteroscedasticity, the variance appearing to increase with increasing values of the response. If appreciable heteroscedas-

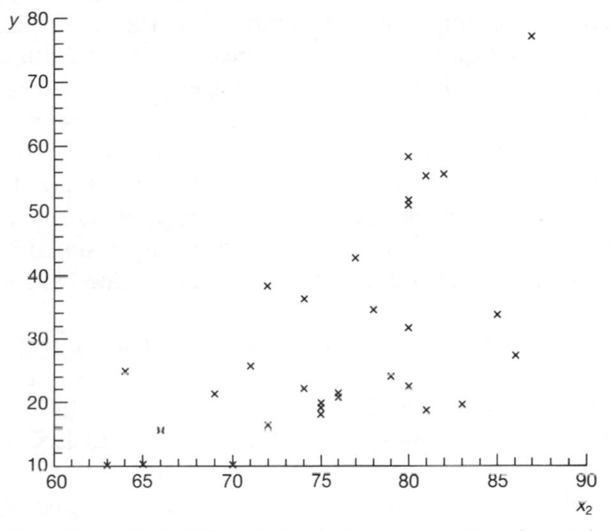


**Fig. 5.2** Example 6: Minitab tree data: scatter plot of $y$ against $x_2$

ticity really is present, it is inefficient to estimate the parameters by ordinary least squares. Instead weighted least squares should be used. We return to some implications of non-constant variance in Chapter 6. For the moment we consider the residuals from linear regression of $y$ on the individual explanatory variables.

In Fig. 5.3 the residuals from regression on $x_1$ alone, which are called $r \mid x_1$, are plotted against $x_2$. The slight slope of this plot suggests that, despite the strong regression on $x_1$ indicated by Fig. 5.1, there is still some residual

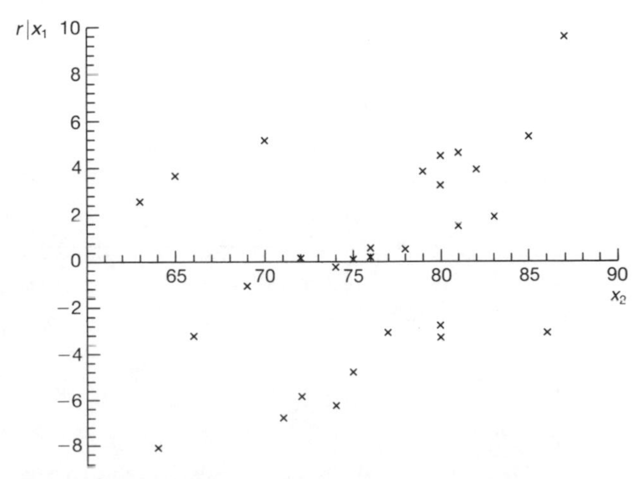


**Fig. 5.3** Example 6: Minitab tree data: residuals from fitting $x_1$ against $x_2$

regression on $x_2$. This impression is confirmed by the value of 2.61 for the $t$ statistic for the coefficient of $x_2$ in the regression on both explanatory variables. Because Fig. 5.1 shows a much stronger relationship between $y$ and $x_1$ than is shown between $y$ and $x_2$ in Fig. 5.2, it is to be expected that the residuals from regression on $x_2$ alone, $r \mid x_2$, would show strong residual regression on $x_1$. This is indeed the case, as is shown in Fig. 5.4. The $t$ value for the coefficient of $x_1$ in the regression on both explanatory variables is a highly significant 17.82, which confirms the visual impression. Figure 5.4 also shows the evidence of heteroscedasticity that was indicated in Fig. 5.2, which was also a plot against $x_2$.

This first analysis of the Minitab tree data shows that, contrary to the hope expressed above, both explanatory variables need to be included in this simple model. In order to check the model it is natural to look at further residual plots and also to consider more complicated models. One possibility is to add higher-order polynomial terms in the two explanatory variables. A second is to follow the hint provided by the increase of variance with volume, and to consider transformations of the response. Additional support for this suggestion comes from dimensional considerations which indicate that the dimensions of the measurements on the two sides of the prediction equation should be the same. A third model, which does satisfy the dimensional requirement, comes from the formula for the volume of a cone, which is a good first approximation to the shape of a tree trunk. These ideas have been explored by Atkinson (1982a) and by Cook and Weisberg (1982, pp. 66, 74). They will also be explored in Chapter 6. But, for the moment, we will continue to use the first-order model in both variables to illustrate graphical methods for choosing explanatory variables.

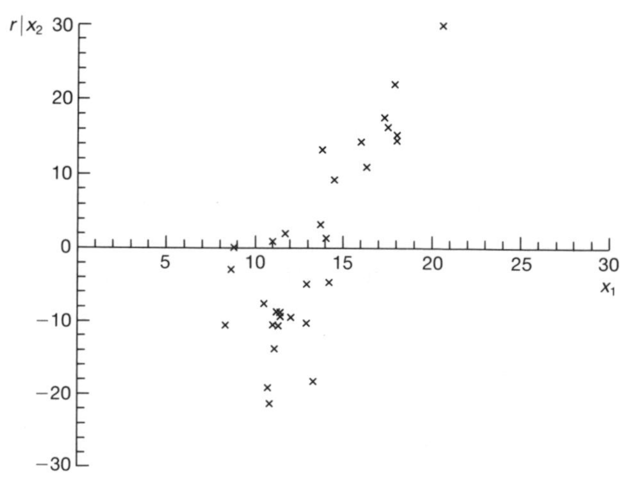


**Fig. 5.4** Example 6: Minitab tree data: residuals from fitting $x_2$ against $x_1$

Before we leave this example, a word on Minitab is in order. Because the regression command is easily combined with line printer graphics, Minitab provides a simple means of implementing some of the ideas described in this book. Version 81.1 of the package calculates the standardized residuals $r_i'$ (2.1.14) as well as the fitted values $\hat{y}_i$. It is therefore straightforward to calculate the leverage measure $1 - h_i$ and hence the diagnostic quantities $r_i^*$ and $C_i$. The details are given by Velleman and Welsch (1981). It is to be expected that later versions of the package will make these, or related, quantities directly available. □

Although the plots of residuals against variables not included in the equation have, in this example, indicated that extra variables should be included in the model, the exact interpretation of the plots is not clear. The difficulty is that, unless the explanatory variable being considered for inclusion is orthogonal to all variables already in the model, the slope of the residual plot is not the estimate of the regression coefficient in the fitted model including the variable in question. In order to detect the effect of individual observations on the evidence for adding extra terms to the model, we devote the rest of this chapter to the description of plots which do have a slope equal in value to the estimated coefficient in the full model.

## 5.2 Added variable plots

To find a graphical display of residuals which has the required slope, we begin by extending the original linear model (2.1.1) to include a set of extra carriers $W$. In place of (2.1.1) the model becomes

$$E(Y) = X^+\beta^+ = X\beta + W\gamma. \tag{5.2.1}$$

Often, but not always, $W$ is a vector corresponding to the addition of one extra carrier. For the moment we assume that the reduced model with matrix of carriers $X$ has been fitted. We want to know whether the additional carriers $W$ should also be included in the model. In the next section we consider the reverse problem of whether, having fitted $X^+$, we can omit $W$ and use a model with terms only in $X$.

The formal test for inclusion of $W$ is, depending upon the dimensionality of $W$, the $t$ or $F$ test for the hypothesis $\gamma = 0$. The $t$ test, that is the test for the addition of one extra carrier, provides the graphical procedure known as the added variable plot. For the partitioned model (5.2.1) the normal equations yielding the least squares estimate $\hat{\gamma}$ are

$$X^TX\hat{\beta} + X^TW\hat{\gamma} = X^Ty \tag{5.2.2}$$

and

$$W^TX\hat{\beta} + W^TW\hat{\gamma} = W^Ty. \tag{5.2.3}$$

If the reduced model with carriers $X$ can be fitted, $(X^T X)^{-1}$ exists and (5.2.2) yields

$$\hat{\beta} = (X^T X)^{-1} X^T y - (X^T X)^{-1} X^T W \hat{\gamma}. \tag{5.2.4}$$

Substitution of this value into (5.2.3) leads, after rearrangement, to

$$W^T\{I - X(X^T X)^{-1} X^T\} W \hat{\gamma} = W^T\{I - X(X^T X)^{-1} X^T\} y,$$

that is to

$$W^T(I - H) W \hat{\gamma} = W^T(I - H) y.$$

In the special case when there is only one extra carrier so that $\gamma$ is a scalar, which is the case of principal interest in the next three chapters,

$$\hat{\gamma} = \frac{w^T(I - H)y}{w^T(I - H)w} = \frac{w^T A y}{w^T A w}. \tag{5.2.5}$$

In (5.2.5) $A = I - H$ is idempotent. Therefore $\hat{\gamma}$ can be expressed in terms of the two sets of residuals

$$r = \overset{*}{y} = (I - H)y = Ay$$

and

$$\overset{*}{w} = (I - H)w = Aw \tag{5.2.6}$$

as

$$\hat{\gamma} = \overset{*}{w}{}^T r / (\overset{*}{w}{}^T \overset{*}{w}). \tag{5.2.7}$$

Thus $\hat{\gamma}$ is the coefficient of linear regression through the origin of the residuals $r$ on the residuals of the new carrier $w$ after regression on the carriers $X$ already in the model. Because the slope of this regression is $\hat{\gamma}$, a plot of $r$ against $\overset{*}{w}$ provides a useful visual assessment of the evidence for regression and will indicate which observations are contributing to the relationship, and which are deviating from it. The plot of the ordinary residual $r$ against $\overset{*}{w}$ is called an added variable plot and provides information about the addition of one further carrier to those already in the model.

**Example 6 (continued) Minitab tree data 2** Figure 5.5 gives, for the Minitab tree data, the added variable plot for the inclusion of $x_2$ in a model which already includes $x_1$. Although there is some slight upward trend in the plot, the evidence for the relationship is not strong, which is in line with the value of 2.61 for the $t$ statistic. This added variable plot of $r$ against the residual carrier $\overset{*}{x}_2$ looks much like the residual plot against $x_2$ of Fig. 5.3. Neither plot reveals any observations which are in any way noteworthy.

The added variable plot for $x_1$ is, on the other hand, appreciably easier to interpret than the residual plot. The plot of $r$ against $\overset{*}{x}_1$ in Fig. 5.6 shows a

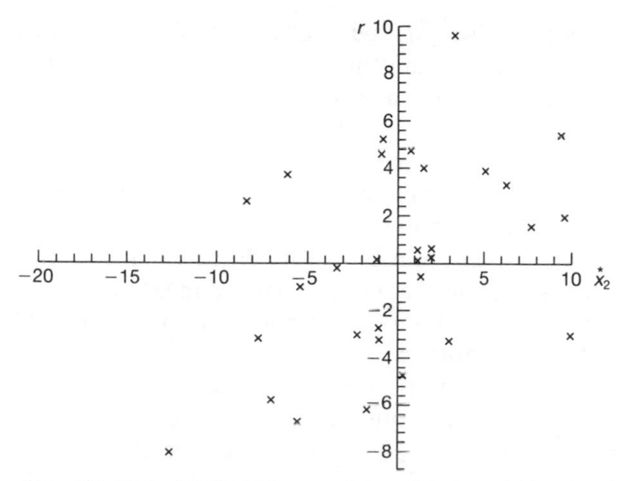


**Fig. 5.5** Example 6: Minitab tree data: added variable plot for $x_2$

strong linear relationship between the two variables, clearly indicating the need to include $x_1$ in the model, a conclusion which reflects the $t$ value of 17.82. The plot of residuals against $x_1$, Fig. 5.4, is, by comparison, less convincing of the need to include $x_1$ in the model. Although there is obviously a relationship between the two variables, the plot has a curved shape which becomes broader at lower values of $x_1$. One conclusion from this example is that added variable plots are easier to interpret than residual plots □

The visual check on the consistency of individual observations provided by the added variable plot is similar to that provided by residual plots for

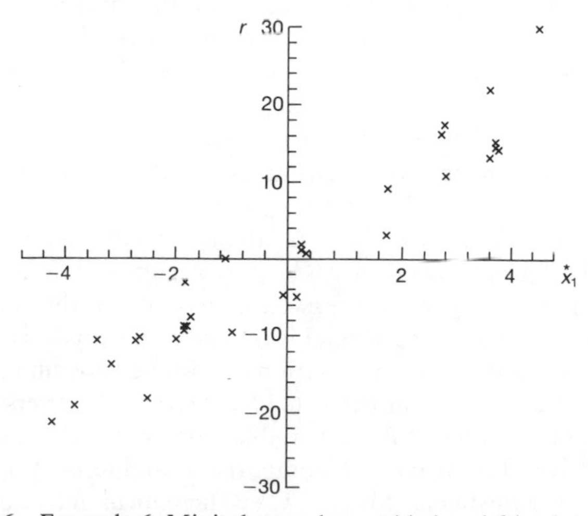


**Fig. 5.6** Example 6: Minitab tree data: added variable plot for $x_1$

univariate regression. It was argued in Chapter 1 that it is relatively easy, in the univariate case, to detect influential observations by inspection. Problems in the visual detection of influence arise only when there are two or more explanatory variables. If more precise information on influence is required, it can be obtained by calculating the modified Cook statistic $C_i$ for the regression in the added variable plot. The information so obtained on the influence of individual observations on estimation of the single parameter $\gamma$ is an approximation to that given by the modified Cook statistic for a single element of the parameter vector, which is derived in Section 10.2. Examples of the use of the modified Cook statistic in conjunction with the added variable plot are described in Chapter 6.

The added variable plot is a graphical expression of the univariate regression relationship underlying the expression for $\hat{\gamma}$ (5.2.7). This relationship is of great statistical importance. In general it leads to the interpretation of multiple regression as a series of linear regressions which is described, for example, by Draper and Smith (1981, Section 4.1). The most important application of such an interpretation is to the analysis of covariance. In the remainder of this section the importance of the interpretation is described and results are derived which will be needed in Chapter 6.

In the canonical example of the analysis of covariance, the matrix of carriers $X$ has a simple structure such as results from a designed experiment. It is required to include in the model a single additional carrier which is not orthogonal to $X$. The technique is discussed by Bliss (1970, Chapter 20) and by Cox and McCullagh (1982). In the language of the analysis of covariance, $w$ is called the concomitant variable. Examples of concomitant variables given by Bliss include the preliminary weight of an animal which is to be used in an experiment on growth hormones and the previous yield, from a uniformity trial, of a plot in an agricultural experiment. In the absence of the concomitant variable the calculations for the analysis of the experiment are usually straightforward, with $X^{\mathrm{T}}X$ a diagonal matrix, so that only totals of groups of observations are required. Formulae such as (5.2.7) and the related formulae for changes in parameter estimates and the residual sum of squares, provide an easy means of adjusting the analysis for the introduction of the concomitant variable. Before the availability of computers for statistical calculation, the formulae were necessary to accommodate a single carrier which gave a non-orthogonal dispersion matrix. When the calculations are performed on a computer, the special formulae of the analysis of covariance are no longer necessary as a computational aid. The concomitant variable $w$ can simply be incorporated in the extended matrix of carriers $X^+$. But the special status of the concomitant variable remains in the analysis. If $\gamma$ is significantly different from zero, all comparisons and tests of significance are performed after adjustment for $w$. A mathematical introduction to the analysis of covariance is given by Seber (1977, §10.1).

For the applications in this book of results on the introduction of an extra carrier we need, in addition to $\hat{\gamma}$, the residual sum of squares from regression on $X$ and $w$ and the new vector of residuals.

Let the residual mean square estimate of $\sigma^2$ from regression on $X$ and $w$ be $s_w^2$. Then the residual sum of squares for the partitioned model (5.2.1) is

$$(n-p-1)s_w^2 = y^T y - \hat{\beta}^T X^T y - \hat{\gamma} w^T y.$$

From the expression for $\hat{\beta}$ (5.2.4) this can be written as

$$\begin{aligned}(n-p-1)s_w^2 &= y^T y - y^T X(X^T X)^{-1} X^T y - \hat{\gamma} w^T y + \hat{\gamma} w^T X(X^T X)^{-1} X^T y \\ &= y^T(I-H)y - \hat{\gamma} w^T (I-H)y.\end{aligned}$$

Substitution of $\hat{\gamma}$ (5.2.7) together with the notation $A = I - H$ leads to the compact expression

$$(n-p-1)s_w^2 = y^T A y - (y^T A w)^2 / w^T A w. \tag{5.2.8}$$

Just as $\hat{\gamma}$ can be interpreted in terms of regression of the residuals $r$ on the residual carrier $\overset{*}{w} = Aw$, so (5.2.8) can be interpreted as the reduction in the residual sum of squares $y^T A y = r^T r$ by regression of $r$ on the residual carrier $\overset{*}{w}$.

The variance of $\hat{\gamma}$ is found from (5.2.7). Since $\hat{\gamma}$ is a linear combination of the observations with var $(Y_i) = \sigma^2$, it follows that

$$\operatorname{var}(\hat{\gamma}) = \sigma^2 \frac{w^T A A^T w}{(w^T A w)^2} = \frac{\sigma^2}{w^T A w} = \sigma^2 / \overset{*}{w}^T \overset{*}{w}. \tag{5.2.9}$$

From the comparison with the expression for univariate regression, (5.2.9) continues the analogy of $\hat{\gamma}$ as the estimated coefficient of the regression of the residuals on the residual variable $w$. The variance of $\hat{\gamma}$ can alternatively be found from the expression for the inverse of a partitioned matrix given in the Appendix. The relevant element of $(X^{+T} X^{+})^{-1}$ is

$$\{w^T w - w^T X(X^T X)^{-1} X^T w\}^{-1}, \quad \text{that is} \quad (w^T A w)^{-1}.$$

The formulae for the least squares estimate $\hat{\gamma}$, its variance and the estimate of $\sigma^2$ given by $s_w^2$ can be combined to test the hypothesis of no regression on $w$. The $t$ test for this hypothesis that $\gamma = 0$ is given by

$$T_w = \frac{\hat{\gamma}}{\text{estimated standard error of } \hat{\gamma}} = \frac{\hat{\gamma}}{s_w \sqrt{(w^T A w)}}, \tag{5.2.10}$$

which will be much used in succeeding chapters for testing hypotheses about the evidence for transformation of the response.

As an aside on the $t$ test, we note that the square of (5.2.10) is the $F$ test for the inclusion of $w$ in the model. If the residual sum of squares for the augmented model (5.2.1) is $R(\hat{\beta}^{+})$ and, just for the moment, the corresponding residual sum of squares for the reduced model without $w$ is called $R(\hat{\beta}^-)$, then, from (5.2.8),

$$R(\hat{\beta}^-) - R(\hat{\beta}^+) = (y^{\mathrm{T}}Aw)^2/w^{\mathrm{T}}Aw.$$

Division by $s_w^2$ as an estimate of $\sigma^2$ leads to the $F$ test which is the square of (5.2.10). This well-known relationship between the two tests is overlooked by Brownlee (1965) who calculates the $t$ test on p. 460 and the $F$ test on p. 461. The advantage of $F$ tests in general is, of course, that they permit inferences about groups of parameters. When only one parameter is of interest, the $t$ test is to be preferred since the sign of the test statistic indicates the direction of departure, if any, from the null hypothesis.

We conclude this section on the algebra of addition of a single carrier by considering the change in the residuals. Let the residuals from the augmented model (5.2.1) be $\overset{+}{r}$. Then

$$\overset{+}{r} = y - X\hat{\beta} - w\hat{\gamma}.$$

The derivation is similar to that leading to (5.2.8) for $s_w^2$. Again (5.2.4) is used to yield

$$\overset{+}{r} = y - X(X^{\mathrm{T}}X)^{-1}X^{\mathrm{T}}y + X(X^{\mathrm{T}}X)^{-1}X^{\mathrm{T}}w\hat{\gamma} - w\hat{\gamma}.$$

From (5.2.6), which defines the residuals $r$ from regression on $X$ alone,

$$\begin{aligned}\overset{+}{r} &= r - (I - H)w\hat{\gamma} \\ &= r - \overset{*}{w}\hat{\gamma}. \qquad (5.2.11)\end{aligned}$$

This expression gives the new residuals from the augmented model in terms of the old residuals and the residual carrier $\overset{*}{w}$.

## 5.3 Partial regression leverage plots

The added variable plots of the previous section are for use if one extra carrier is being considered for addition to a regression model which has already been fitted. The same plot can be used to decide which carrier to omit from a fitted model, but the calculations start from a different point. In the notation of the previous section, we suppose that a model including all the $p + 1$ carriers in $X^+$ has been fitted. The plot is to help determine how the evidence for dropping one of the carriers depends on the individual observations. Added variable plots when deletion is of interest are called partial regression leverage plots by Belsley *et al.* (1980, Chapter 2) who give numerous examples. The computational problem considered in this section is how to form these plots without performing a regression on $p$ out of the $p + 1$ columns of $X^+$ for each plot.

To begin we find a computationally convenient form for the residual carrier $\overset{*}{w}$. The definition of $\overset{*}{w}$ in (5.2.6) can be written in slightly expanded form as

$$\overset{*}{w} = w - X(X^{\mathrm{T}}X)^{-1}X^{\mathrm{T}}w. \qquad (5.3.1)$$

For deletion of a carrier $(X^{\mathrm{T}}X)^{-1}$ is not known, although $(X^{+\mathrm{T}}X^{+})^{-1}$ is. If this inverse is written explicitly in partitioned form as

$$(X^{+\mathrm{T}}X^{+})^{-1} = \begin{bmatrix} X^{\mathrm{T}}X & X^{\mathrm{T}}w \\ w^{\mathrm{T}}X & w^{\mathrm{T}}w \end{bmatrix}^{-1} = \begin{bmatrix} S_{11} & S_{12} \\ S_{12}^{\mathrm{T}} & s_{22} \end{bmatrix}$$

it follows that

$$\begin{bmatrix} X^{\mathrm{T}}X & X^{\mathrm{T}}w \\ w^{\mathrm{T}}X & w^{\mathrm{T}}w \end{bmatrix} \begin{bmatrix} S_{11} & S_{12} \\ S_{12}^{\mathrm{T}} & s_{22} \end{bmatrix} = \begin{bmatrix} I_p & 0 \\ 0 & 1 \end{bmatrix}$$

where $I_p$ is the $p \times p$ identity matrix. In particular, multiplying out leads to

$$X^{\mathrm{T}}XS_{12} + X^{\mathrm{T}}ws_{22} = 0$$

or

$$S_{12} = -(X^{\mathrm{T}}X)^{-1}X^{\mathrm{T}}ws_{22}. \qquad (5.3.2)$$

Taken in combination with (5.3.1), (5.3.2) yields

$$\overset{*}{w} = w + XS_{12}/s_{22}. \qquad (5.3.3)$$

This formula provides the residual carrier $\overset{*}{w}$ in terms of elements of $X^{+}$ and of elements of the inverse $(X^{+\mathrm{T}}X^{+})^{-1}$ which need only be calculated once when the full model is fitted.

The other set of quantities needed for the partial regression leverage plot are the residuals $r$. From (5.2.11)

$$r = \overset{+}{r} + \overset{*}{w}\hat{\gamma}. \qquad (5.3.4)$$

Both $\overset{+}{r}$ and $\hat{\gamma}$ are quantities from the full fit and $\overset{*}{w}$ is given by (5.3.3).

The relationships derived in this section demonstrate how the values for partial regression leverage plots may be calculated without the necessity of refitting the model for each plot. Details of the calculations and of methods of implementation are given by Velleman and Welsch (1981). Because the only difference between added variable plots and partial regression leverage plots is the starting point, which leads to the identical final product, no further examples are given here: Figs. 5.5 and 5.6 suffice.

## 5.4 Partial residual plots

The calculations for the partial regression leverage plot, outlined in Section 5.3, are not particularly heavy. There is however another residual plot which involves even less computing and which also has slope $\hat{\gamma}$. This is the partial residual plot introduced by Ezekiel and Fox (1959) and commended by Larsen and McCleary (1972).

The residuals $\overset{+}{r}$, defined in Section 5.2 as residuals from the augmented model (5.2.1), were given by

$$\overset{+}{r} = y - X\hat{\beta} - w\hat{\gamma}.$$

In this expression $\hat{\beta}$, given by (5.2.4), is the least squares estimate of the first $p$ components of the parameter vector in the augmented model. The partial residual $\tilde{r}$ is defined to be

$$\tilde{r} = y - X\hat{\beta} = \overset{+}{r} + w\hat{\gamma}. \tag{5.4.1}$$

The partial residual is thus the residual from regression on $X$, ignoring $w$, without adjusting the parameter estimate $\hat{\beta}$ for the exclusion of $w$ from the model. To repeat, $\hat{\beta}$ in (5.4.1) is given by (5.2.4), not by (2.1.5). Unless $X$ and $w$ are orthogonal, $\tilde{r}$ is thus not a residual from the least squares fit of a regression model.

Clearly, once the augmented model has been fitted, it is easier to calculate $\tilde{r}$ from $\overset{+}{r}$ than it is to calculate the ordinary residual $r$. In the partial residual plot $\tilde{r}$ is plotted against the carrier $w$. If the model includes a constant, the centred variable $w - \bar{w}$ can be used, where $\bar{w}$ is the mean of $w$.

The slope of the regression line of $\tilde{r}$ against $w$ is given by

$$\tilde{\gamma} = \frac{w^{\mathrm{T}}\tilde{r}}{w^{\mathrm{T}}w} = \frac{w^{\mathrm{T}}(\overset{+}{r} + w\hat{\gamma})}{w^{\mathrm{T}}w}.$$

But $\overset{+}{r}$ is a vector of residuals from a model with carriers $X^{+}$ which include $w$. Therefore

$$w^{\mathrm{T}}\overset{+}{r} = 0$$

whence

$$\tilde{\gamma} = \frac{w^{\mathrm{T}}w\hat{\gamma}}{w^{\mathrm{T}}w} = \hat{\gamma}.$$

The partial residual plot therefore has the desired slope $\hat{\gamma}$.

A weakness of the procedure is that a scatter plot of $\tilde{r}$ against $w$ does not give a correct indication of the variance of $\hat{\gamma}$. The vertical scatter about the line is correct, since, as we have shown, the residuals of the least squares line in the partial residual plot are

$$\tilde{r} - w\tilde{\gamma} = \tilde{r} - w\hat{\gamma} = \overset{+}{r}.$$

The mean squared error estimate of $\sigma^2$ based on the regression line would correctly be $s_w^2$ (5.2.8), provided that, when calculating the degrees of freedom, allowance was made for fitting the $p$ carriers in $X$. However, the horizontal scatter can be far too great, leading to an over-estimate of the precision of the estimate $\hat{\gamma}$. From the regression line in the partial residual plot the variance of $\tilde{\gamma}$, or equivalently $\hat{\gamma}$, would be calculated as $\sigma^2/(w^T w)$, which ignores the effect of fitting the other carriers. The correct formula for the variance of $\hat{\gamma}$ is that given by (5.2.9), namely $\sigma^2/(\mathrm{w}^T A w) = \sigma^2/(\overset{*}{w}{}^T \overset{*}{w})$. If $w$ is highly correlated with some of the carriers $X$ already in the model, the effect can be appreciable.

**Example 6 (continued) Minitab tree data 3** In the data of Table 5.1 there is appreciable correlation between $x_1$, girth, and $x_2$, height. This is because trees of one species growing in one place tend to be similar in shape, so that large trees are large in both girth and height. The effect of this correlation on the apparent precision of estimation of $\hat{\gamma}$ is shown by comparison of the partial residual plots, Figs. 5.7 and 5.8, with the added variable plots, Figs. 5.5 and 5.6. In particular, the partial residual plot for $x_1$ in Fig. 5.8 shows a very much stronger linear trend than does the added variable plot of Fig. 5.6. It has already been argued that this effect is due to the misleadingly great horizontal scatter in the partial residual plot. The ratio of horizontal scatter in the partial residual plot to that in the added variable plot is about 2.5: 1, which quantifies the extent to which Fig. 5.8 over-emphasizes the importance of the relationship between $y$ and $x_1$. □

We shall not be concerned any further with partial residual plots. Some comments on the relationship with partial regression leverage plots are given by Mallows (1982). Perhaps the most important of these in favour of the partial residual plot is that the original value of the carrier $w$ is used in the

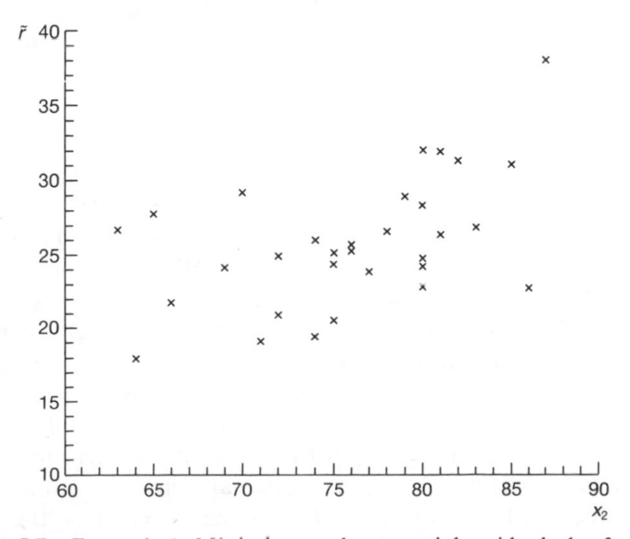


**Fig. 5.7** Example 6: Minitab tree data: partial residual plot for $x_2$

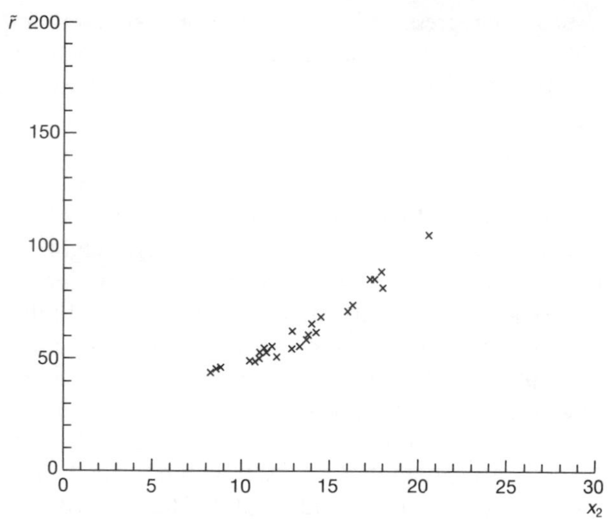


**Fig. 5.8** Example 6: Minitab tree data: partial residual plot for $x_1$

plot, rather than the value of the residual carrier $\overset{*}{w}$. Thus if there is any systematic pattern of departure with $w$, for example inhomogeneity of variance, this may be apparent in the partial residual plot, but not necessarily in the partial regression leverage plot.

In the rest of this book we shall be concerned with tests and plots for a variety of transformations. Many of these reduce to procedures for the addition of explanatory variables and appreciable use will be made of the added variable plot. The tree data will be used to exemplify some of these procedures. But first we consider further aspects of the analysis of the untransformed data.

**Example 6 (continued) Minitab tree data 4** The indication from all plots so far, supported by $t$ values of 17.8 and 2.61, is that both variables should be included in the prediction equation for volume. The plots further indicate that this conclusion is not unduly influenced by one or two outlying observations. But there remains the problem, mentioned in Section 5.1, as to whether this model is adequate.

As a check on the adequacy of the model we look at index plots of the deletion residuals $r_i^*$ and of the modified Cook statistics $C_i$. The interpretation of these plots is aided by the fact that the observations have been arranged in order of increasing $x_1$, that is of diameter. The plot of $r_i^*$ is shown in Fig. 5.9 and that of $C_i$ in Fig. 5.10 for the three-parameter model including both explanatory variables. The plots, which are similar, both show a slight quadratic tendency. This is more marked at the extremes on the plot of $C_i$, because the departure from the first-order model is for the smallest and

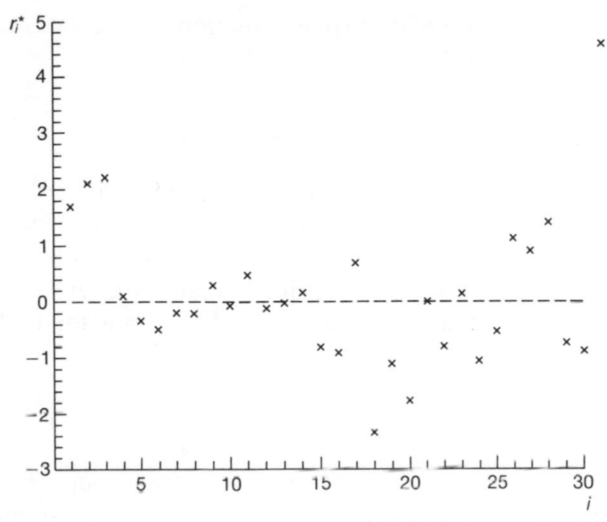


**Fig. 5.9** Example 6: Minitab tree data: index plot of deletion residual $r_i^*$ for model in $x_1$ and $x_2$

largest trees which, being furthest from the centre of the data, have highest leverage. It is not clear from these index plots exactly how the model should be improved. The plots suggest that quadratic terms should be considered, but these could be in $x_1$, or in $x_2$, or in both explanatory variables. These

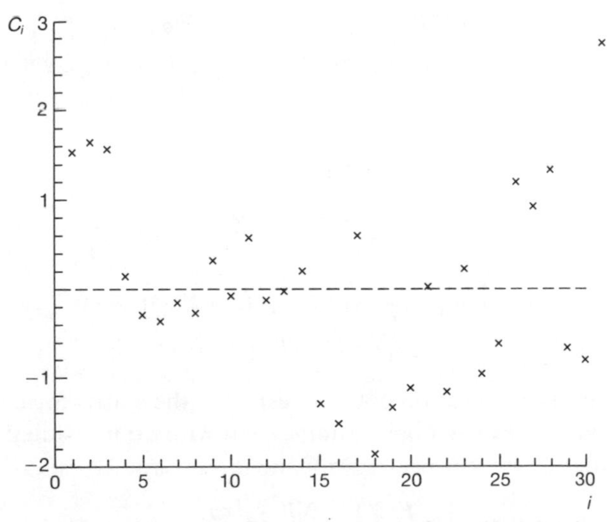


**Fig. 5.10** Example 6: Minitab tree data: index plot of modified Cook statistic $C_i$ for model in $x_1$ and $x_2$

possibilities, together with that of transformations of the data, are explored in Chapter 6. □

## 5.5 Deletion of observations and addition of carriers

As a pendant to the main business of this chapter we explore the relationship between the deletion formulae of Section 2.2 and the results of Section 5.2 on the addition of extra carriers, for the special case in which each of $m$ observations to be deleted has an individual parameter. This model, sometimes called the mean shift outlier model, is that for each of the $m$ specified observations

$$E(Y_i) = x_i^{\mathrm{T}}\beta + \phi_i. \tag{5.5.1}$$

For the remaining $n - m$ observations $E(Y_i) = x_i^{\mathrm{T}}\beta$.

The test of the hypothesis $\phi_i = 0$ is equivalent to testing whether the $i$th observation is an outlier. Least squares estimation of the parameter $\phi_i$ will give a value of zero for the $i$th residual in the model (5.5.1). The particular observation will then make no contribution to the residual sum of squares, which will take the value given in Section 2.2, namely

$$(n - p - m)s_{(I)}^2 = (n - p)s^2 - r_I^{\mathrm{T}}(I - H_I)^{-1}r_I. \tag{2.2.7}$$

In this section we obtain an expression for the test for $m$ outliers and relate it to the residuals of earlier sections. We also show, as a matter of confirmation, that deletion and the addition of extra parameters yield the same residual sum of squares.

We begin by developing the results of Section 5.2 for the partitioned model (5.2.1) in which the matrix of additional carriers $W$ is quite general. If, as in (5.2.5) and succeeding equations, we let $A = I - H$, the least squares estimate of the vector parameter $\gamma$ is

$$\hat{\gamma} = (W^{\mathrm{T}}AW)^{-1}W^{\mathrm{T}}Ay, \tag{5.5.2}$$

which is the generalization of (5.2.5). The residual sum of squares is, analogously to (5.2.8), given by

$$\begin{aligned}(n - p - m)s_W^2 &= y^{\mathrm{T}}Ay - y^{\mathrm{T}}AW(W^{\mathrm{T}}AW)^{-1}W^{\mathrm{T}}Ay \\ &= y^{\mathrm{T}}Ay - \hat{\gamma}^{\mathrm{T}}(W^{\mathrm{T}}AW)\hat{\gamma}.\end{aligned} \tag{5.5.3}$$

We consider the application of these results to the matrix form of the mean shift outlier model. For $m$ observations, each with an individual parameter, (5.5.1) becomes

$$E(Y) = X\beta + D\phi, \tag{5.5.4}$$

where $D$ is an $n \times m$ matrix with a one in the $i$th row for each of the $m$

observations belonging to the set $I$ and with only one non-zero element in each column.

We now substitute $D$ for $W$ in (5.5.3) to obtain

$$(n - p - m)s_D^2 = y^T Ay - y^T AD(D^T AD)^{-1} D^T Ay, \qquad (5.5.5)$$

$$= y^T Ay - \hat{\phi}^T (D^T AD)\hat{\phi}, \qquad (5.5.6)$$

where

$$\hat{\phi} = (D^T AD)^{-1} D^T Ay.$$

The effect of premultiplication of a vector by $D^T$ is to pick out the $m$ elements in the set $I$. Since, from (5.2.6), the vector of residuals $r = Ay$, $D^T Ay = D^T r = r_I$. Similarly since $A = I - H$, $D^T AD$ is equal to $I - H_I$. Further, since $y^T Ay$ is the residual sum of squares from regression on $X$ alone, which can be written as $(n - p)s^2$, (5.5.5) and (2.2.7) are indeed equal.

For testing hypotheses about the values of the $\phi_i$ (5.5.6) is a more revealing form of the adjusted residual sum of squares. The hypothesis that all elements of $\phi$ are zero, that is that there are no outliers, is tested, when $\sigma^2$ is known, by comparing $\hat{\phi}^T (D^T AD)\hat{\phi}/\sigma^2$ to the $\chi^2$ distribution on $m$ degrees of freedom. Since we can re-write $\hat{\phi}$ as

$$\hat{\phi} = (I - H_I)^{-1} r_I$$

an alternative form for the test statistic is

$$r_I^T (I - H_I)^{-1} r_I / \sigma^2, \qquad (5.5.7)$$

a quadratic form in the residuals. In practice $\sigma^2$ will not be known and will have to be estimated. Either $s^2$ or $s_{(I)}^2$ can be used to yield a statistic which, in the latter case will, after scaling by the degrees of freedom, follow the $F$ distribution. For one suspected outlier these two possibilities lead respectively to the squared standardized residual and the squared deletion residual. As has already been argued in Section 3.1, the latter statistic is to be preferred. But the main point of this section is that the mean shift outlier model, in which an extra parameter is included for each suspected outlier, leads to recovery of the results for deletion which were obtained in Section 2.2.

# 6
# Transformations and constructed variables

## 6.1 The need for transformations

The assumptions behind the linear model which has been used in the preceding chapters of this book were stated in Section 2.1. These included:

(a) homogeneity of variance;
(b) simplicity of structure for the expected value of the response; and
(c) at least approximate normality of the additive errors.

It was also assumed that the errors were independent.

If it is not possible to satisfy these three requirements in the original scale of measurement of the response $y$, it may be that there is a nonlinear transformation of the response which will yield homogeneity of variance and, at least approximate, normality. The implications of this hope form the subject not only of this chapter but also of the greater part of the remainder of the book.

Given a set of data, there are several empirical indications as to whether a transformation might be helpful. One indication is if the response is non-negative. As examples, times until an event occurs and the measured diameters of particles are both non-negative and so cannot strictly follow a normal distribution. For such random variables it is quite likely that the log of the response will be more nearly normally distributed than the response itself. The original response will then have a lognormal distribution. Of course, if all values of the response are far from zero and the scatter in the observations is relatively small, the transformation will have little effect: the heights of adult men, for example, could be modelled either by the normal or by the lognormal distribution. On the other hand, if the ratio of the largest observation to the smallest is one or more powers of ten, so that the data cover several cycles, it is most unlikely that a model of additive variance will be appropriate. Under such conditions a transformation is often desirable.

In the context of regression models, simplicity of structure implies that a first-order model is adequate with few, if any, second-order terms, either quadratic or interaction. The combination of normality of errors and constancy of variance implies that, after transformation, the data should not contain any outliers. Particularly if replication is absent from the data, it may be difficult to obtain precise information on distributional form. But the presence of either outliers or, more particularly, of systematic departures from assumption in the residuals is, as we shall see, sometimes an indication of the need for a transformation.

There is here the makings of a paradox. If the presence of one or two outliers is suggestive of the need for a transformation, it could be that the evidence of a need for a transformation is being unduly influenced by one or two outliers. We therefore develop diagnostic plots for the influence of individual observations on the evidence for a transformation.

The result of a successful transformation is that the three requirements of homogeneity of variance, additivity of structure and normality of errors are all satisfied. As we shall see, it is a matter of empirical observation that all three criteria are often satisfied for the same transformation. However, in regression examples, there is often little evidence about the distribution of the errors: due to supernormality the residuals will tend to appear normally distributed for a variety of transformations. A normal plot of residuals is therefore often not an informative tool for detecting the need for a transformation.

To begin this chapter we look at an example and consider plots and analysis of variance tables for the original data and for the logarithmic and reciprocal transformations. In line with the desire for a simple additive structure for the linear model, we compare these three scales using the $F$ test for second-order terms. Because the transformation of the original observations leads to a change in scale of the response fitted to the linear model, it is not meaningful to compare residual sums of squares directly for the various transformations. We accordingly look at the parametric family of transformations analysed by Box and Cox (1964). This brings the choice of a transformation within the framework of standard statistical theory. The greater part of this chapter is concerned with exploring the implications of this idea. One main development leads to regression on an extra carrier, a special case of the variable $w$ of Chapter 5.

Before we start the exploration of transformations, an exploration which will occupy much of the rest of this book, a word of caution is in order. Because the assumptions for a simple linear model outlined at the beginning of this section are violated, it does not necessarily follow that a transformation of the response will make all well. It may be that a generalized linear model, as described in Section 10.3.2, would be more appropriate. Or it might be that the response is more complicated. For example if the response consists of a mixture of two normal distributions, transformation will not be helpful. This and related examples are discussed by Bradley (1982).

**Example 7 Wool data 1** In their analysis of a parametric family of transformations, Box and Cox (1964) use two examples, one of which records the number of cycles to failure of a worsted yarn under cycles of repeated loading. The results from a single replicate of a full $3^3$ experiment are given in Table 6.1. The three factors and their levels are:

$x_1$: length of test specimen (250, 300, 350 mm)
$x_2$: amplitude of loading cycle (8, 9, 10 mm)
$x_3$: load (40, 45, 50 g).

The number of cycles to breakage ranges from 90, for the shortest specimen subjected to the most severe test conditions, to 3636 for observation 19 which comes from the longest specimen subjected to the mildest conditions. Given this great range in values of the non-negative response it is natural to consider analysis in terms of log $y$. Box and Cox outline a rather informal analysis in which consideration is given to a model with logarithms of both response and explanatory variables. The properties of models in which the

**Table 6.1** Example 7: Wool data

| | Factor levels | | | |
|---|---|---|---|---|
| Observation | $x_1$ | $x_2$ | $x_3$ | Cycles to failure |
| 1 | −1 | −1 | −1 | 674 |
| 2 | −1 | −1 | 0 | 370 |
| 3 | −1 | −1 | 1 | 292 |
| 4 | −1 | 0 | −1 | 338 |
| 5 | −1 | 0 | 0 | 266 |
| 6 | −1 | 0 | 1 | 210 |
| 7 | −1 | 1 | −1 | 170 |
| 8 | −1 | 1 | 0 | 118 |
| 9 | −1 | 1 | 1 | 90 |
| 10 | 0 | −1 | −1 | 1414 |
| 11 | 0 | −1 | 0 | 1198 |
| 12 | 0 | −1 | 1 | 634 |
| 13 | 0 | 0 | −1 | 1022 |
| 14 | 0 | 0 | 0 | 620 |
| 15 | 0 | 0 | 1 | 438 |
| 16 | 0 | 1 | −1 | 442 |
| 17 | 0 | 1 | 0 | 332 |
| 18 | 0 | 1 | 1 | 220 |
| 19 | 1 | −1 | −1 | 3636 |
| 20 | 1 | −1 | 0 | 3184 |
| 21 | 1 | −1 | 1 | 2000 |
| 22 | 1 | 0 | −1 | 1568 |
| 23 | 1 | 0 | 0 | 1070 |
| 24 | 1 | 0 | 1 | 566 |
| 25 | 1 | 1 | −1 | 1140 |
| 26 | 1 | 1 | 0 | 884 |
| 27 | 1 | 1 | 1 | 360 |

carriers are transformations of the original explanatory variables are investigated in Chapter 8. For this example were merely note that the values of the uncoded factors do not cover an appreciable relative range, so it is to be expected that there will not be much difference between a model in $x_j$ and one in $\log x_j$.

We begin by fitting a first-order model in the three explanatory variables. Figure 6.1 shows a half-normal plot of the resulting deletion residuals $r_i^*$. There is some indication of disagreement between the model and the data. The largest value of $r_i^*$ is 3.58 for observation 19, which lies just outside and above the simulation envelope. Five of the medium-sized values lie below the envelope and the general impression is of a plot like a broken stick with two different slopes. One cause for the shape of this plot might be non-homogeneity of variance, which could be investigated by a plot of residuals against predicted values. Given the range of the data such an effect is likely. We look instead at the effect of fitting more complicated models.

One possibility is to add quadratic and interaction terms in the three explanatory variables to give the ten-carrier second-order model

$$y_i = \beta_0 + \sum_{j=1}^{3} \beta_j x_{ij} + \sum_{j=1}^{3} \beta_{jj} x_{ij}^2 + \sum_{j=1}^{2} \sum_{k=j+1}^{3} \beta_{jk} x_{ij} x_{ik} + \varepsilon_i. \quad (6.1.1)$$

The $F$ value for the six additional terms is 9.52 which, on 6 and 17 degrees of freedom, is significant at the 0.1 per cent level. The half-normal plot of $r_i^*$, as well as the plot of the modified Cook statistic $C_i$, show no departures from the model. Addition of the quadratic terms has thus significantly improved the model.

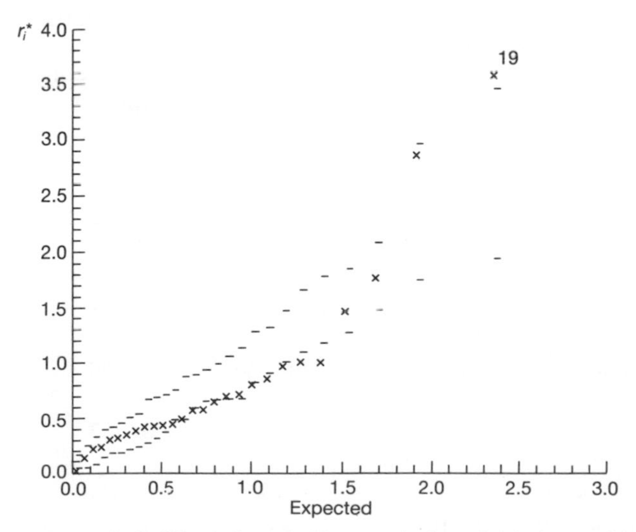


**Fig. 6.1** Example 7: Wool data: half-normal plot of deletion residuals $r_i^*$

There are two drawbacks to this approach. One is that the improvement has been achieved by the addition of six extra parameters. In fact only four of the six extra terms are significant at the 5 per cent level, so some slight simplification of the model is possible. But the two most significant terms correspond to two interactions. The interpretation of such interactions is not as straightforward as that of models in which interactions are absent, for the effect of one factor depends upon the levels of one or more other factors. The presence of interactions is sometimes an indication either of defective data, being caused by undetected outliers, or of a defective model. In this case the indication is of analysis in the wrong scale of the response.

The analysis of the data was repeated with log $y$ and with the reciprocal $1/y$ as the response. The resulting analysis of variance tables are summarized in Table 6.2. The important statistic for our present purposes is the $F$ value for the second-order terms. For the original analysis and the log and reciprocal analyses this is respectively 9.52, 0.68 and 12.04. Thus on the logarithmic scale, but not on the other two scales, a first-order model without interactions is adequate. On this scale, due to the lack of interactions, physical interpretation of the model is appreciably simplified. □

In the example in this section analysis of the original data and of two transformed sets has led to selection of the logarithmic transformation via inspection of the value of the $F$ statistic for second-order terms. This reflects the interest in additivity, which is only one of the reasons for choosing a particular transformation. Further, in this analysis, only a few transforma-

**Table 6.2** Example 7: Wool data. Analysis of variance tables for original data and two transformations

| | Response | | |
|---|---|---|---|
| | Original data $y$ | Logarithm log $y$ | Reciprocal $1/y$ $(\times 10^{-4})$ |
| Sum of squares for first-order terms (3 degrees of freedom) | 14 748 475 | 22.4112 | 1.2922 |
| Sum of squares for second-order terms (6 degrees of freedom) | 4 224 300 | 0.1538 | 0.3201 |
| Mean square for second-order terms | 704 050 | 0.0256 | 0.0534 |
| Residual sum of squares (17 degrees of freedom) | 1 256 681 | 0.6387 | 0.0754 |
| Residual mean square | 73 922 | 0.0376 | 0.0044 |
| $F$ statistic for second-order terms | 9.52 | 0.68 | 12.04 |

tions were considered. Other possibilities would have been, for example, the square root and its reciprocal. In the next section we describe a method for bringing the choice of a transformation within the framework of parametric statistical inference.

## 6.2 A parametric family of power transformations

At the beginning of this chapter the hope was expressed that, perhaps as a result of a transformation of the response, a model would be found which would satisfy the three desiderata of homogeneity of variance, additivity and normality. The analysis of the wool data by $F$ tests for second-order terms is one way of seeking to achieve this. A more general method, which leads to standard statistical inferences about the choice of transformation, was analysed by Box and Cox who considered the parametric family of power transformations

$$y(\lambda) = \begin{cases} \dfrac{y^\lambda - 1}{\lambda} & (\lambda \neq 0) \\ \log y & (\lambda = 0)\,. \end{cases} \tag{6.2.1}$$

In the absence of a transformation $\lambda = 1$. The value of the transformation for $\lambda = 0$ is found as the limit of (6.2.1) as $\lambda$ tends to zero.

The intention is that, for some $\lambda$,

$$E\{Y(\lambda)\} = X\beta \tag{6.2.2}$$

with $Y(\lambda)$ satisfying the second-order assumptions of constancy of variance, independence and additivity. It is also hoped that, to an adequate degree of approximation, $Y(\lambda)$ will be normally distributed. If the requirement of additivity is met, then the structure of $X\beta$ will be relatively simple. For the transformation (6.2.1) to be generally applicable it is necessary, as Schlesselman (1971) stressed, that the model include a constant. It will be assumed throughout that this condition is met.

The power transformation (6.2.1) is only one of many parametric families of transformations which have been suggested for the analysis of data. A more general development of the theory for parametric families of transformations is given in Chapter 7 and applied to data in which the response is either a percentage or a proportion. In this chapter the theory is developed for power transformations of responses which are non-negative and without an upper bound.

For a single given value of $\lambda$ comparisons of linear models are made using the residual sum of squares of $y(\lambda)$, just as the residual sum of squares of the untransformed observations, $R(\hat{\beta})$, is used for $\lambda = 1$. But between different $\lambda$ values allowance has to be made for the effect of $\lambda$ on the scale of the observations. As an example, the residual sums of squares for the second-order model in Table 6.2 are, for $\lambda = 1$, 0 and $-1$ respectively 1 256 681, 0.6387 and $0.0754 \times 10^{-4}$. As we have seen from examination of the values of the $F$ statistic, this does not mean that the inverse transformation provides the best model.

To compare different values of $\lambda$ it is necessary to compare the likelihood in relation to that of the original observations $y$ which is

$$(2\pi\sigma^2)^{-n/2} \exp\{-(Y(\lambda) - X\beta)^{\mathrm{T}}(y(\lambda) - X\beta)/2\sigma^2\} J,$$

where the Jacobian is

$$J = \prod_{i=1}^{n} \left| \frac{\partial y_i(\lambda)}{\partial y_i} \right|. \tag{6.2.3}$$

The Jacobian allows for the change of scale of the response due to the operation of the power transformation $y(\lambda)$.

The maximum likelihood estimates of the parameters are found in two stages. For fixed $\lambda$ (6.2.3) is, apart from the Jacobian which does not involve $\beta$ or $\sigma$, the likelihood for a least squares problem with response $y(\lambda)$. The maximum likelihood estimates of $\beta$ for given $\lambda$, which we shall call $\hat{\beta}(\lambda)$, are therefore the least squares estimates given by

$$\hat{\beta}(\lambda) = (X^{\mathrm{T}}X)^{-1}X^{\mathrm{T}}y(\lambda).$$

The residual sum of squares of the $y(\lambda)$ is

$$\begin{aligned} S(\lambda) &= y(\lambda)^{\mathrm{T}}\{I - X(X^{\mathrm{T}}X)^{-1}X^{\mathrm{T}}\}y(\lambda) \\ &= y(\lambda)^{\mathrm{T}}(I - H)y(\lambda) = y(\lambda)^{\mathrm{T}}Ay(\lambda). \end{aligned} \tag{6.2.4}$$

Division of (6.2.4) by $n$, rather than by $(n - p)$, yields the maximum likelihood estimate of $\sigma^2$ as

$$\hat{\sigma}^2(\lambda) = S(\lambda)/n. \tag{6.2.5}$$

Replacement of this estimate by the least squares estimate in which the divisor is $(n - p)$ does not affect the development which follows.

For fixed $\lambda$ we can now find the log-likelihood maximized over both $\beta$ and $\sigma^2$. Substitution of the expressions for $S(\lambda)$ (6.2.4) and $\hat{\sigma}^2(\lambda)$ (6.2.5) into the logarithm of the likelihood given by (6.2.3) yields, apart from a constant,

$$L_{\max}(\lambda) = -(n/2)\log\hat{\sigma}^2(\lambda) + \log J.$$

This partially maximized or profile log-likelihood is therefore a function of $\lambda$ which depends both on the residual sum of squares $S(\lambda)$ and on the Jacobian $J$.

A simpler, but equivalent, form for $L_{\max}(\lambda)$ is found by working with the normalized transformation

$$z(\lambda) = y(\lambda)/J^{1/n}.$$

For the power transformation (6.2.1)

$$\partial y_i(\lambda)/\partial y_i = y_i^{\lambda - 1}$$

so that

$$\log J = (\lambda - 1) \sum \log y_i.$$

If we let $\dot{y}$ be the geometric mean of the observations, the normalized power transformation is

$$z(\lambda) = \begin{cases} \dfrac{y^\lambda - 1}{\lambda \dot{y}^{\lambda - 1}} & (\lambda \neq 0) \\ \dot{y} \log y & (\lambda = 0). \end{cases} \tag{6.2.6}$$

Apart from a constant, the partially maximized log-likelihood of the observations can then be written as

$$L_{\max}(\lambda) = -(n/2) \log \{R(\lambda)/n\}, \tag{6.2.7}$$

where

$$R(\lambda) = z(\lambda)^{\mathrm{T}} A z(\lambda) \tag{6.2.8}$$

is the residual sum of squares of the $z(\lambda)$.

The maximum likelihood estimate $\hat{\lambda}$ is the value of the transformation parameter for which the partially maximized log-likelihood is a maximum. Equivalently, $\hat{\lambda}$ is the value for which the residual sum of squares $R(\lambda)$ is minimized.

To compare the log-likelihood for various values of $\lambda$, $L_{\max}(\lambda)$ can be plotted over a range of plausible values. Usually values of $\lambda$ between $-1$ and 1 suffice. An approximate $100(1 - \alpha)$ per cent confidence region for $\lambda$ is found from those values for which

$$2\{L_{\max}(\hat{\lambda}) - L_{\max}(\lambda)\} \leqslant \chi^2_{1,\alpha} \tag{6.2.9}$$

Thus an approximate 95 per cent confidence interval for $\lambda$ would include all those values for which the difference in twice the log-likelihood on the left-hand side of (6.2.9) was less than 3.84.

**Example 7 (continued) Wool data 2** Figure 6.2 shows plots of $2L_{\max}(\lambda)$ (6.2.7) against $\lambda$ for both first- and second-order models fitted to the wool data. For the first-order, that is additive, model the log-likelihood is a maximum at $\hat{\lambda} = -0.0593$. The approximate 95 per cent confidence interval for $\lambda$ defined by (6.2.9) covers $-0.183$ to 0.064 and so firmly excludes 1. For the additive model there is therefore strong evidence that the data should be analysed after a logarithmic transformation.

For any fixed value of $\lambda$ the second-order model, because it contains extra parameters, will have a higher likelihood than the first-order model, as can be seen in Fig. 6.2. The difference in the log-likelihoods may however not be statistically significant. The results of Table 6.2 show a significant difference between the two models for $\lambda = 1$ and $-1$. But for the log transformation, $\lambda = 0$, there is little difference between them. For the second-order model the maximum likelihood estimate is $\hat{\lambda} = -0.219$ with 95 per cent confidence interval $-0.486$ to 0.115. Again the confidence interval excludes the value one.

The conclusion of this analysis, based on aggregate statistics, is that the first-order model with response log $y$ is appropriate. A more general point is that the 95 per cent confidence interval for $\lambda$ from the quadratic model has width 0.601 as opposed to width 0.247 for the first-order model. The extra polynomial terms are therefore, to some extent, acting as an alternative to a transformation. This is reflected in Fig. 6.2 in the flatness near $\lambda = 0$ of $2L_{\max}(\lambda)$ for the quadratic model relative to the more curved plot for the first-order model. □

The model (6.2.2) is that, after transformation, the observations follow a normal theory linear model. We have already seen that, because of the change of scale on transformation, the residual sum of squares $S(\lambda)$ (6.2.4) cannot be used to compare the adequacy of models for various values of $\lambda$. However $R(\lambda)$, the residual sum of squares of the $z(\lambda)$ defined by (6.2.8), which is a monotone function of $L_{\max}(\lambda)$, can be used as the basis of an approximate method for making such comparisons. The advantage is that inference about $\lambda$ is brought within the framework of regression models. The disadvantage is

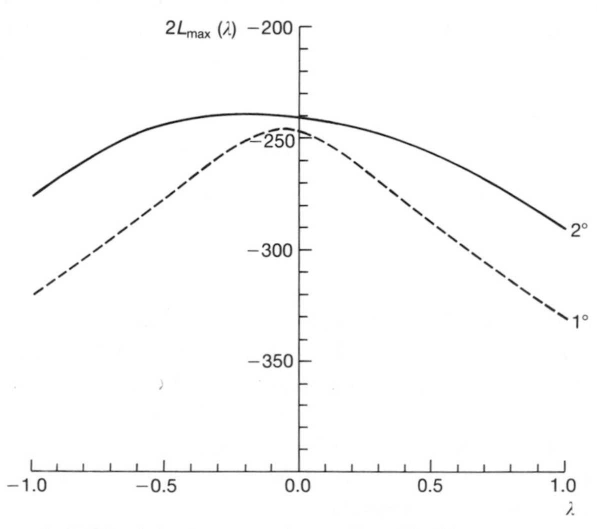


**Fig. 6.2** Example 7: Wool data: comparison of log-likelihoods; $2L_{\max}(\lambda)$ for first- and second-order models

the approximation of ignoring the sampling variability of $\dot{y}$ by assuming that $Z(\lambda)$ (6.2.6), rather than $Y(\lambda)$, follows a normal theory linear model. Asymptotically, sampling fluctuations in $\dot{y}$ will become less important and the distribution of $z(\lambda)$ will converge to that of $y(\lambda)$. It would then be possible to base inferences on $F$ tests applied to changes in the residual sum of squares. As an example we compare confidence intervals for $\lambda$ based on $2L_{\max}(\lambda)$ and on $R(\lambda)$.

In terms of $L_{\max}(\lambda)$ the $100(1-\alpha)$ per cent confidence interval for $\lambda$ is given by (6.2.9). If we use (6.2.7) to rewrite this expression in terms of $R(\lambda)$ we obtain the interval as those values satisfying

$$n \log \{R(\lambda)/R(\hat{\lambda})\} \leqslant \chi^2_{1,\alpha}. \tag{6.2.10}$$

To find an interval based on the residual sum of squares of $z(\lambda)$, let $s^2$ be the residual mean square estimate of the variance of $z(\lambda)$ given by

$$s^2 = R(\hat{\lambda})/(n-p). \tag{6.2.11}$$

From normal linear model theory the $100(1-\alpha)$ per cent confidence interval for $\lambda$ is given by those values for which

$$R(\lambda) - R(\hat{\lambda}) \leqslant s^2 F_{1,n-p,\alpha},$$

that is,

$$R(\lambda) - R(\hat{\lambda}) \leqslant \frac{F_{1,n-p,\alpha} R(\hat{\lambda})}{n-p}. \tag{6.2.12}$$

The likelihood confidence region for $\lambda$ can be put in a form analogous to (6.2.12) by rewriting (6.2.10) as

$$n \log [1 + \{R(\lambda) - R(\hat{\lambda})\}/R(\hat{\lambda})] \leqslant \chi^2_{1,\alpha}.$$

In the neighbourhood of $\hat{\lambda}$ the difference $R(\lambda) - R(\hat{\lambda})$ should be small relative to $R(\hat{\lambda})$. If we call this small quantity $\delta$ and replace $\log(1+\delta)$ by the series expansion $\delta$, we obtain the approximate region

$$R(\lambda) - R(\hat{\lambda}) \leqslant (\chi^2_{1,\alpha}/n) R(\hat{\lambda}). \tag{6.2.13}$$

The two intervals are then asymptotically equivalent. In small samples the major difference is due to the factors $F_{1,n-p,\alpha}/(n-p)$ and $\chi^2_{1,\alpha}/n$, the numerators and denominators of which both work in the direction of making the interval derived from the residual sum of squares broader than that from the likelihood.

**Example 7 (continued) Wool data 3** Figure 6.3 shows plots of the residual sums of squares for the first- and second-order models. For the first-order model the residual sum of squares is approximately parabolic, whereas for the

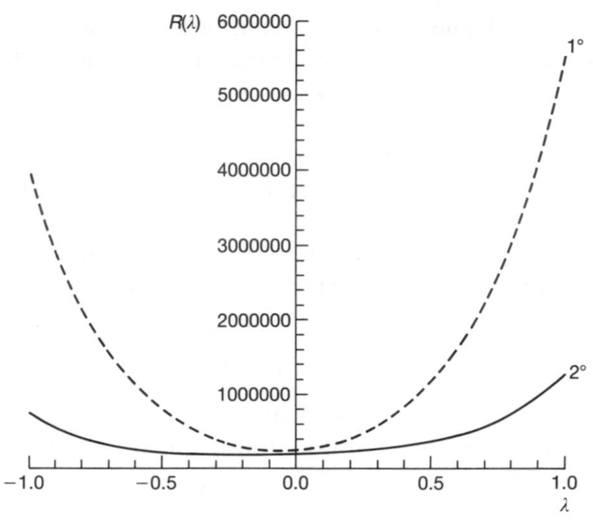


**Fig. 6.3** Example 7: Wool data: residual sum of squares $R(\lambda)$ for first- and second-order models

second-order model the shape, which is relatively flat at the centre, curves up more sharply at the ends of the plotted range of $\lambda$ values.

The results of Table 6.3 show that, for both models, the confidence intervals for $\lambda$ based on the residual sums of squares are similar to, but slightly broader than, the likelihood intervals. This would be expected from the preceding theoretical development. To discover the small sample coverage probabilities of these two intervals would require further investigation, either by simulation or by consideration of asymptotic expansions. □

To find the estimate of $\lambda$ and the boundaries of either confidence interval it is necessary to perform three iterative calculations. The first is to find the value of $\lambda$ which maximizes the log-likelihood or, equivalently, minimizes the

**Table 6.3** Example 7: Wool data. Maximum likelihood estimates of and 95 per cent confidence intervals for the transformation parameter $\lambda$

| | Model | |
|---|---|---|
| | First order | Second order |
| Maximum likelihood estimate $\hat{\lambda}$ | −0.059 | −0.219 |
| Likelihood interval for $\lambda$ (6.2.9) | −0.183, 0.064 | −0.486, 0.115 |
| Interval from sum of squares $R(\lambda)$ (6.2.12) | −0.195, 0.077 | −0.553, 0.215 |
| Interval from approximation at $R(\hat{\lambda})$ (6.4.9) | −0.202, 0.083 | −0.908, 0.470 |

residual sum of squares. The other two iterations are to find the two end values of the region where the log-likelihood is a specified amount below its maximum value, or the residual sum of squares a specified amount above its minimum. There is nothing, as far as this calculation is concerned, to choose between the two procedures. However, as we shall see in Section 6.4, the residual sum of squares $R(\lambda)$ has advantages in providing relatively simple procedures, particularly for testing whether a transformation is necessary.

Although precise calculation of the estimate $\hat{\lambda}$ requires iteration, it is unlikely that the exact value of $\hat{\lambda}$ will be used in a statistical model. Rather the value of $\hat{\lambda}$ is used to suggest a suitable simple model. For example, in the analysis of the wool data, the value of $-0.0593$ for $\hat{\lambda}$ was taken to suggest the appropriateness of the log transformation. Plausible values for $\lambda$ are usually $-1$, $-\frac{1}{2}$, 0, $\frac{1}{2}$ and 1, with the log transformation $\lambda = 0$ perhaps the most frequently encountered when a transformation is required. Occasionally values outside this range are encountered, as are also other simple values such as the $\frac{1}{3}$ which occurs in the analysis of the Minitab tree data in Section 6.5.

Once a value of $\lambda$ has been determined, Box and Cox (1964) argue that inference about the parameters in the linear model should be made in the transformed scale, ignoring the fact that the transformation has been estimated. Thus, for example in Table 6.2, the $F$ tests for the second-order terms were calculated without allowance for the effect of estimating $\lambda$.

Bickel and Doksum (1981) point out that the estimates of $\lambda$ and $\beta$ can be highly correlated, so that the marginal variance of the estimated $\beta$'s can be hugely inflated by not knowing $\lambda$. Box and Cox (1982) comment that this fact is both obvious and irrelevant. The obviousness is illustrated by the residual sums of squares in Table 6.2. The original observations are of order about 1000. The logged values are of magnitude about 7 and the reciprocal values are of order about 0.001. The parameters of the linear models, since all $x_j$ are scaled between $-1$ and 1, are of the same orders. The numerical values thus depend strongly on the unknown value of $\lambda$. To give, under such conditions, a numerical value for a parameter estimate or treatment effect without specifying the value of $\lambda$ to which it corresponds, is scientifically meaningless.

In accord with this argument we shall, where appropriate, make inferential statements about the parameters of linear models without numerical allowance for the effect of estimating $\lambda$. Two subsidiary arguments in favour of this course are that if, for example, prediction is of importance, transformation of the predicted response back to the original scale renders irrelevant the effect of the transformation on the parameter estimates, a point investigated by Carroll and Ruppert (1981). The second argument is that working with $z(\lambda)$ rather than $y(\lambda)$ removes the effect of the transformation on the scale of the response. The residual sums of squares $R(\lambda)$ for different $\lambda$ values are directly comparable and so will be the values of the parameter estimates. One way to see this is to consider the dimension of $z(\lambda)$ which is that of $y$ for all $\lambda$ including zero when $z(0) = \dot{y} \log y$.

## 6.3 Asymptotically equivalent tests of hypotheses

To determine whether a transformation was needed for the example of Section 6.2 we used the fact that the confidence region for $\lambda$ excluded the value $\lambda = 1$. The usual starting point in an analysis is to fit a model to the untransformed data and to determine whether a transformation would be beneficial, formulated as a test of the null hypothesis $\lambda_0 = 1$. The likelihood ratio test related to the confidence region (6.2.9) rejects the null hypothesis at the $100\alpha$ per cent level if

$$2\{L_{\max}(\hat{\lambda}) - L_{\max}(\lambda_0)\} > \chi^2_{1,\alpha}. \tag{6.3.1}$$

Usually, but not always, we shall be interested in the hypothesis of no transformation, that is we take $\lambda_0 = 1$.

An appreciable disadvantage of the test given by (6.3.1) is that it requires calculation of the maximum likelihood estimate $\hat{\lambda}$. In Section 6.4 an approximate score test is developed which is asymptotically equivalent to (6.3.1) but which does not require maximization of the likelihood. The procedure also leads to a useful diagnostic plot for the effect of individual observations on the evidence for a transformation. The development of the statistic is within the framework of least squares regression and does not rely explicitly on asymptotic statistical theory. In this section a sketch is given of the theory of asymptotic tests of hypotheses, which raises some interesting open questions related to the theory of tests of transformations. The reader whose concern is with the methods which form the major part of this book might prefer, on a first reading, to go straight to Section 6.4.

To begin we consider likelihood inference about all $p$ elements of a vector of parameters $\theta$. The general theory is discussed by Cox and Hinkley (1974, Section 9.3). Shorter summaries can be found in several recent books on applied statistics such as Kalbfleisch and Prentice (1980, Section 3.4). The log-likelihood of the $n$ observations is $L(\theta)$, so that the likelihood ratio test (6.3.1) rejects the null hypothesis $\theta = \theta_0$ at the $100\alpha$ per cent level if

$$2\{L(\hat{\theta}) - L(\theta_0)\} > \chi^2_{p,\alpha}. \tag{6.3.2}$$

The $p$ partial derivatives of the log-likelihood form the vector of efficient scores

$$U(\theta) = \frac{\partial L(\theta)}{\partial \theta} = \left\{\frac{\partial L(\theta)}{\partial \theta_j}\right\}_{p \times 1}. \tag{6.3.3}$$

Under regularity conditions which are discussed later in this section, the maximum likelihood estimate of $\theta$ is defined by the $p$ equations

$$U(\hat{\theta}) = 0. \tag{6.3.4}$$

The distribution of $\hat{\theta}$ is then asymptotically normal with mean $\theta$ and variance $\mathscr{I}^{-1}(\theta)$, where

$$\mathscr{I}(\theta) = E\{U(\theta)U(\theta)^{\mathrm{T}}\} = -\left\{E\,\frac{\partial^2 L(\theta)}{\partial\theta_j\,\partial\theta_k}\right\}_{p\times p} \tag{6.3.5}$$

is the expected information. In some cases it may be easier to use the observed information

$$I(\theta) = -\left\{\frac{\partial^2 L(\theta)}{\partial\theta_j\,\partial\theta_k}\right\}_{p\times p}. \tag{6.3.6}$$

In practice the value of $\theta$ will not be known and so the estimate $\hat{\theta}$ will be substituted to give the values $\mathscr{I}(\hat{\theta})$ and $I(\hat{\theta})$. Asymptotically these four forms are equivalent.

A test of the null hypothesis $\theta = \theta_0$ which is asymptotically equivalent to the likelihood ratio test (6.3.2) is given by the Wald test which rejects the null hypothesis if

$$(\hat{\theta} - \theta_0)^{\mathrm{T}}\mathscr{I}(\theta_0)(\hat{\theta} - \theta_0) > \chi^2_{p,\alpha}. \tag{6.3.7}$$

Although the kernel of the quadratic form in the Wald test is calculated under the null hypothesis, the test statistic, like the likelihood ratio statistic, does require the estimate $\hat{\theta}$.

Knowledge of the maximum likelihood estimate is not required for the score test, which is based on the observed values of the efficient scores at $\theta_0$. If $\theta_0$ is near $\hat{\theta}$, it follows from (6.3.3) that these observed values will be small. Asymptotically the efficient scores are normally distributed with variance $\mathscr{I}(\theta)$. The score test accordingly rejects the null hypothesis if

$$U(\theta_0)^{\mathrm{T}}\mathscr{I}^{-1}(\theta_0)U(\theta_0) > \chi^2_{p,\alpha}. \tag{6.3.8}$$

The asymptotic properties of the score test are not changed if the observed information $I(\theta_0)$ is substituted for $\mathscr{I}(\theta_0)$. Unlike the other two tests, (6.3.2) and (6.3.7), the score test requires only quantities calculated under the null hypothesis.

The asymptotic equivalence of the three tests is shown by Cox and Hinkley. The geometric relationship between the three tests depends on the log-likelihood being, at least approximately, a paraboloid in $p$ dimensions. Asymptotic normality of the maximum likelihood estimate means that the log-likelihood becomes more nearly parabolic in the neighbourhood of $\hat{\theta}$ as the number of observations increases. The likelihood ratio test (6.3.1) measures the vertical distance between $L(\hat{\theta})$ and $L(\theta_0)$. The Wald test looks at the horizontal distance between $\hat{\theta}$ and $\theta_0$, whereas the score test is based on the gradient of $L(\theta)$ evaluated at $\theta_0$. For an exactly parabolic log-likelihood the three statistics are identical.

A discussion of the conditions for the limiting normality of the maximum likelihood estimator to hold is given by Cox and Hinkley. In broad terms there are two main requirements. The first is that the conditions for the central limit theorem hold, so that the efficient score vector, which for independent observations is a sum of individual contributions from $n$ observations, is not dominated by a few observations. In non-linear regression such violations can occur if the distribution of the explanatory variables becomes degenerate in some way as $n$ increases, for example by converging on one point in the $p$-dimensional space of the carriers. The second requirement is that the regularity conditions which permit the interchange of the order of differentiation and integration should be satisfied. This seemingly simple requirement is often violated in quite elementary cases, for example if the range of the observations depends upon the parameters of the distribution. One simple case is the uniform distribution on $(0, \theta)$, where the maximum likelihood estimator of $\theta$ is the maximum observed value, which has the beta distribution with density $(n/\theta)(y/\theta)^{n-1}$. As $n$ goes to infinity, the asymptotic distribution of $\hat{\theta}$ is not normal but such that $\theta - \hat{\theta}$ has an exponential distribution with mean $\theta/n$. A more general case of greater importance is that the formulation excludes threshold parameters. As we shall see in Chapter 9, this complicates inference in cases where a power transformation is applied after addition or subtraction of a quantity which has to be estimated.

The asymptotic theory considered so far has been for all the parameters in a model. But in the model (6.2.2) for transformations, inferences are to be made about a single parameter $\lambda$ in the presence of a vector of nuisance parameters $\beta$. More generally we assume that the vector of parameters can be broken into two parts, $\theta$ and $\phi$, where $\theta$ is, as before, the vector of $p$ parameters of interest but $\phi$ is a vector of $s$ nuisance parameters. The likelihood ratio, Wald and score statistics are analogous to those given above in the absence of nuisance parameters. As an example we derive the score test.

Both the likelihood and the score test are functions of $\theta$ and $\phi$. Under the null hypothesis $\theta = \theta_0$, let the constrained maximum likelihood estimator of $\phi$ be $\hat{\phi}_0$. Let the vector of score statistics for $\theta$ evaluated at the null hypothesis be $U(\theta_0, \hat{\phi}_0)$. The information matrix partitioned according to the partition of the parameter vector is

$$\mathscr{I}(\theta, \phi) = \begin{bmatrix} \mathscr{I}_{11}(\theta, \phi) & \mathscr{I}_{12}(\theta, \phi) \\ \mathscr{I}_{12}^{\mathrm{T}}(\theta, \phi) & \mathscr{I}_{22}(\theta, \phi) \end{bmatrix}.$$

From the expression for the inverse of a partitioned matrix given in the

Appendix, the score test for the hypothesis $\theta = \theta_0$ is given by calculating

$$U(\theta_0, \hat{\phi}_0)^{\mathrm{T}}\{\mathscr{I}_{11} - \mathscr{I}_{12}\mathscr{I}_{22}^{-1}\mathscr{I}_{12}^{\mathrm{T}}\}^{-1}U(\theta_0, \hat{\phi}_0), \tag{6.3.9}$$

where all $\mathscr{I}_{ij}$ are calculated at $(\theta_0, \hat{\phi}_0)$. If, as in the case of the power transformation parameter $\lambda$, the parameter of interest $\theta$ is scalar, the signed square root of (6.3.9) provides an asymptotically normal test statistic.

In any of the statistics described in the section, either the observed or expected information can be used to give a pair of statistics with the same asymptotic properties. For finite samples the choice between the two measures of information is often dictated by numerical convenience. A discussion of the inferential properties of observed as opposed to expected information is given by Efron and Hinkley (1978).

Some guidance on the choice between the three asymptotically equivalent tests for finite samples is given by Cox and Hinkley (1974, Section 9.3.vi). The Wald test is not recommended, because it depends upon the parameterization of the problem. In the discussion in the present chapter of this book, the choice between the score test and the likelihood ratio test has been presented in terms of numerical convenience. But there are also the questions of the null distribution of the test statistics and of the statistical power of the two procedures once any necessary adjustments have been made for differences in the size of nominally identical tests. The comparisons of Harris and Peers (1980) are typical in that they demonstrate that which test is to be preferred for higher power depends upon the region of the parameter space which is studied. For the specific problem of testing hypotheses about transformations, simulation studies reported in Chapter 9 fail to show any difference in power between the score and likelihood ratio statistics. The results do however suggest that the null distribution of the likelihood ratio statistic is more nearly $\chi^2$ than is that of the score statistic. Cox and Hinkley discuss a method due to Bartlett for obtaining a modification of the likelihood ratio test which has improved distributional properties. A general discussion of the 'Bartlett correction' is given by Barndorff-Nielsen and Cox (1984).

We conclude this section with an example to illustrate the derivation of the two tests, for which the likelihood ratio test is to be preferred for its null distribution. The problem is that of testing for a single outlier. We start from the mean shift outlier model (5.5.4). For only one anticipated outlier this can be re-written, in accordance with the parameterization of (6.3.9), as

$$E(Y) = d\theta + X\phi, \tag{6.3.10}$$

where $d$ is a vector of zeros except for a one in the $i$th position. Under the null hypothesis of no outlier, that is $\theta = 0$, the model is that for standard linear regression. Estimation of $\theta$ by least squares is equivalent to deletion of the $i$th observation.

If we assume that $\sigma^2$ has the known value $\sigma_0^2$, the log-likelihood, apart from a constant, is given by

$$L(\theta, \phi) = -(y - d\theta - X\phi)^{\mathrm{T}}(y - d\theta - X\phi)/2\sigma_0^2. \tag{6.3.11}$$

The likelihood ratio test is then proportional to the difference of two residual sums of squares, which can be found from the results of Section 5.5. Under the null hypothesis, that is at $(\theta_0, \hat{\phi}_0)$, the residual sum of squares is $(n-p)s^2$ or, in the standard notation for the linear model, $R(\hat{\beta})$. At $(\hat{\theta}, \hat{\phi})$ the residual sum of squares is $(n-p-1)s_{(i)}^2$. From the expression for the effect of deletion on the residual sum of squares (2.2.7) it follows that

$$\begin{aligned} 2\{L(\hat{\theta}, \hat{\phi}) - L(\theta_0, \hat{\phi}_0)\} &= \{(n-p)s^2 - (n-p-1)s_{(i)}^2\}/\sigma_0^2 \\ &= r_i^2/\{\sigma_0^2(1-h_i)\}. \end{aligned} \tag{6.3.12}$$

This statistic is the square of the $i$th residual divided by its known variance. Under the null hypothesis it has a chi-squared distribution on one degree of freedom.

To derive the score statistic we continue with the assumption that the variance has the known value $\sigma_0^2$. Then the efficient score is given by

$$U(\theta, \phi) = \frac{\partial L(\theta, \phi)}{\partial \theta} = (y - d\phi - X\theta)^{\mathrm{T}} d/\sigma_0^2$$

and, under the null hypothesis of no outlier,

$$U(\theta_0, \hat{\phi}_0) = r_i/\sigma_0^2. \tag{6.3.13}$$

The information matrix $I(\theta, \phi)$ is found by differentiation to be given by

$$\sigma_0^2 I(\theta, \phi) = \begin{bmatrix} d^{\mathrm{T}}d & d^{\mathrm{T}}X \\ X^{\mathrm{T}}d & X^{\mathrm{T}}X \end{bmatrix}. \tag{6.3.14}$$

Because of the linearity of this model the observed information $I(\theta, \phi)$ does not depend upon the observations and so (6.3.14) is also the expression for the expected information. From (6.3.9) the score statistic is therefore given by

$$(r_i^2/\sigma_0^2)\{d^{\mathrm{T}}d - d^{\mathrm{T}}X(X^{\mathrm{T}}X)^{-1}X^{\mathrm{T}}d\}^{-1}. \tag{6.3.15}$$

From the results of Section 5.5 the score statistic (6.3.15) reduces straightforwardly to the expression given in (6.3.12) for the likelihood ratio test.

Thus when the variance $\sigma^2$ is known, the score test and the likelihood ratio test are identical, as they must be for this normal likelihood with its consequently quadratic log-likelihood. The difference between the two procedures emerges when an estimate is substituted for $\sigma^2$. For the score statistic all quantities are evaluated at the null hypothesis, which in this case is that there are no outliers. The appropriate estimate is $s^2$, so that the score test gives rise to the squared standardized residual which, as was shown in

Section 3.1, has a scaled beta distribution. For the likelihood ratio test, $\sigma^2$ is estimated at the maximum likelihood estimate $(\hat{\theta}, \hat{\phi})$ which corresponds to deletion of the $i$th observation. The likelihood ratio statistic is therefore the scaled deletion residual $r_i^{*2}$. Since $r_i^*$ follows a Student's $t$ distribution, the likelihood ratio test produces, for this particular example, a statistic which is closer to its asymptotic distribution than does the score test.

## 6.4 Score tests and constructed variables

For the likelihood for transformations given by (6.2.6), calculation of the score statistic (6.3.9) is complicated. In this section a simple approximate score test is derived, which is the $t$ test for the significance of a regression coefficient. Various applications of this approximate test form one of the main diagnostic tools of the remainder of this book.

The starting point is to replace the model for the transformed observations $y(\lambda)$ (6.2.2) by a model for the normalized transformation $z(\lambda)$. This approximation is the same as that made in the derivation of confidence intervals for $\lambda$ based on the residual sum of squares $R(\lambda)$. We then hope that, for some $\lambda$,

$$z(\lambda) = X\beta + \varepsilon.$$

The value of $\lambda$ for which this applies is to be estimated. Expansion of $z(\lambda)$ in a Taylor series about the known value $\lambda_0$ yields

$$z(\lambda) \simeq z(\lambda_0) + (\lambda - \lambda_0)w(\lambda_0)$$

and the approximate linear model

$$\begin{aligned} z(\lambda_0) &= X\beta - (\lambda - \lambda_0)w(\lambda_0) + \varepsilon \\ &= X\beta + \gamma w + \varepsilon, \end{aligned} \tag{6.4.1}$$

where $w(\lambda_0) = \partial z(\lambda)/\partial\lambda$, evaluated at $\lambda = \lambda_0$.

The variable $w(\lambda_0)$ in (6.4.1) was called a constructed variable by Box (1980). The term 'derived carrier' is also used. Comparison of (6.4.1) with the partitioned linear model (5.2.1) shows that Section 5.2 provides all the results required for inference about the importance of this new carrier. For example, from (5.2.5) the least squares estimate of the regression coefficient of $w(\lambda_0)$ is

$$\hat{\gamma} = w^{\mathrm{T}}(\lambda_0)Az(\lambda_0)/w^{\mathrm{T}}(\lambda_0)Aw(\lambda_0). \tag{6.4.2}$$

In (6.4.2) and similar expressions, if dependence on $\lambda$ is important we write $z(\lambda)$ and $w(\lambda)$. Otherwise we shall write $z$ and $w$.

The variance of $\hat{\gamma}$ is, from (5.2.9), $\sigma^2/(w^{\mathrm{T}}Aw)$. To form a $t$ test for the significance of $\hat{\gamma}$ we require an estimate of $\sigma^2$. At the end of Section 6.3, when a score test was derived for the presence of a single outlier, the estimate that was used was that under the null hypothesis, namely $s^2$. The purpose of the score test is to avoid calculation of $\hat{\lambda}$, so that the maximum likelihood estimate of $\sigma^2$ is not available. But, from (5.2.8), an approximation to this estimate is given by

$$(n - p - 1)s_z^2 = z^{\mathrm{T}}Az - (z^{\mathrm{T}}Aw)^2/(w^{\mathrm{T}}Aw).$$

The $t$ test for the hypothesis $\gamma = 0$ is the approximate score statistic (Atkinson 1973)

$$T_{\mathrm{p}}(\lambda_0) = -\frac{z(\lambda_0)^{\mathrm{T}}Aw(\lambda_0)}{s_z\{w(\lambda_0)^{\mathrm{T}}Aw(\lambda_0)\}^{1/2}}. \qquad (6.4.3)$$

The negative sign arises because, in (6.4.1), $\gamma = -(\lambda - \lambda_0)$. Use of $s_z^2$ in (6.4.3), rather than $s^2$, to estimate $\sigma^2$ yields a test with higher power. To the extent that the linear approximation which leads to (6.4.1) is exact, $\hat{\gamma} = -(\hat{\lambda} - \lambda_0)$. $T_{\mathrm{p}}(\lambda)$ (6.4.3) therefore provides an approximate test of the hypothesis $\lambda = \lambda_0$. This score test is an approximation to the likelihood ratio test. In some of the examples in this chapter and in Chapter 8 we investigate how good this approximation is.

The score test can usefully be interpreted in terms of the univariate regression through the origin of the residual values of the $z(\lambda)$ on the residual constructed variables. If, as in (5.2.6), we define

$$\overset{*}{w} = (I - H)w = Aw \qquad \text{and} \qquad \overset{*}{z} = (I - H)z = Az, \qquad (6.4.4)$$

the added variable plot of $\overset{*}{z}$ against $\overset{*}{w}$, as described in Section 5.2, can be used to see whether information for the transformation is spread throughout the data, or whether it depends on one or a few observations.

Because the slope of the added variable plot is the estimated regression coefficient $\hat{\gamma}$, the residual constructed variables can be used to assess the influence of individual observations by calculation of a modified Cook statistic. The modified Cook statistic resulting from deletion of a single observation for inference about a single parameter $\beta_j$ will be derived in Chapter 10 and is given in (10.2.12). An approximation to this can be found by treating $\overset{*}{z}$ and $\overset{*}{w}$ as variables, rather than as residuals, in the univariate regression which leads to $\hat{\gamma}$ and forming the modified Cook statistic for this regression. Because of the dependence of the statistic on the constructed variable, this version of the modified Cook statistic will be denoted $C_i(w)$. Experience has shown that this quantity provides a useful diagnostic which confirms and strengthens the visual impression of influential observations from the added variable plot. The approximation to (10.2.12) in this derivation arises because the hat matrix which defines the residual variables $\overset{*}{z}$ and $\overset{*}{w}$ depends on the $i$th observation. The effect that this can have is discussed in Section 12.3. In that section we also develop approximations to

$\hat{\lambda} - \hat{\lambda}_{(i)}$, the change in the maximum likelihood estimate of the transformation parameter when observation $i$ is deleted.

The difference between the two procedures is likely to be appreciable only for observations which have high leverage for the linear model. In this chapter we shall use added variable plots and half-normal plots of the modified Cook statistic to identify seemingly influential observations. Identification of influential observations will be followed by deletion and recalculation of the score test and other diagnostic quantities and plots. Because both constituents of the modified Cook statistic, $\overset{*}{z}$ and $\overset{*}{w}$, involve observational error, the half-normal plots are presented without simulation envelopes.

For the power transformation (6.2.6) the constructed variable

$$w_{\mathrm{p}}(\lambda) = \frac{\partial z(\lambda)}{\partial \lambda} = \frac{y^{\lambda} \log y}{\lambda \dot{y}^{\lambda-1}} - \frac{y^{\lambda} - 1}{\lambda \dot{y}^{\lambda-1}} (1/\lambda + \log \dot{y}). \qquad (6.4.5)$$

We shall usually be interested in testing hypotheses about a few special values of $\lambda$. For $\lambda_0 = 1$, that is the null hypothesis of no transformation,

$$w_{\mathrm{p}}(1) = y\{\log (y/\dot{y}) - 1\} + \log \dot{y} + 1. \qquad (6.4.6)$$

Presence of regression on this constructed variable would therefore be evidence that a transformation was required. Similarly for the hypothesis of a log transformation, the limit of (6.4.5) as $\lambda$ approaches zero is

$$w_{\mathrm{p}}(0) = \dot{y} \log y(\log y/2 - \log \dot{y}). \qquad (6.4.7)$$

Calculation of the score statistic does not require $w_{\mathrm{p}}(\lambda)$ directly, but rather the residual constructed variable $\overset{*}{w}_{\mathrm{p}}(\lambda)$ which is formed in (6.4.4) by the operation of the matrix $A = I - H$. In some cases a simpler form can be found for $w$ which yields the same value of $\overset{*}{w}$. For example when $\lambda = 1$ and the linear model contains a constant, which it must for the power transformation (6.2.1) to be applicable, the residuals from (6.4.6) are identical to the residuals from the constructed variable

$$w_{\mathrm{p}}(1) = y\{\log (y/\dot{y}) - 1\}. \qquad (6.4.8)$$

The remainder of the chapter is concerned with application of these diagnostic techniques to several examples. The main purpose is to assess the evidence for a transformation and the contribution of individual observations to that evidence. If a transformation is indicated, then iterative calculation of the maximum likelihood estimate $\hat{\lambda}$ will have to be undertaken and an interval statement for $\lambda$ may be required. The confidence intervals given by (6.2.9) and (6.2.12) both, as has already been remarked, require iterative calculations to find their endpoints. We can however use the results of this section to obtain an approximate interval which depends only on quantities calculated at $\hat{\lambda}$. We have already used the fact that the variance of $\hat{\lambda}$ is $\sigma^2/(w^T Aw)$. At $\hat{\lambda}$ the value of $\hat{\gamma}$ is identically zero. From the linearized model (6.4.1) an approximation to the variance of $\hat{\lambda}$ is therefore $s^2/(w^T Aw)$ and the corresponding approximate $100\alpha$ per cent confidence interval for $\lambda$ accordingly has limits

$$\hat{\lambda} \pm t_{\alpha, n-p-1} s / \surd\{w(\hat{\lambda}) A w(\hat{\lambda})\}. \tag{6.4.9}$$

Here all quantities are calculated at $\hat{\lambda}$ and $s^2$ is the residual mean square estimate of $\sigma^2$.

**Example 7 (continued) Wool data 4** We now examine the evidence for a transformation of the wool data afforded by the techniques developed in this section. The results are collected in Table 6.4.

The likelihood ratio test for the hypothesis of no transformation (6.3.1) has values of 84.09 for the first-order model and of 50.97 for the second-order model. Both values are highly significant when compared with the 0.1 per cent point of the $\chi_1^2$ distribution which is 10.83. There is thus no doubt that the data should be transformed, although the evidence is slightly reduced by the presence of second-order terms in the linear model.

To compare the likelihood ratio tests with the score statistics it is convenient to use the signed square roots of the likelihood ratio values which asymptotically have a standard normal distribution. For the first- and second-order models these respectively have the values $-9.17$ and $-7.14$, the

**Table 6.4** Example 7: Wool data. Tests for transformations and measures of influence

(a) Tests for transformations

| | Model | |
|---|---|---|
| | First order | Second order |
| Likelihood ratio test (6.3.1) | 84.09 | 50.97 |
| Signed square root of likelihood ratio test | $-9.17$ | $-7.14$ |
| Score test for no transformation $T_p(1)$ | $-18.55$ | $-8.22$ |
| Score test for log transformation $T_p(0)$ | $-0.91$ | $-2.19$ |

(b) Measures of influence for first-order model

| | $T_p(1)$ | $i_{max}$† | $C_{i\,max}(w_p)$ |
|---|---|---|---|
| All observations | $-18.55$ | 19 | 7.66 |
| Observation 19 deleted | $-17.00$ | 20 | 12.73 |
| Observations 19 and 20 deleted | $-12.11$ | 24 | 2.44 |

† The observation for which the modified Cook statistic $C_i(w_p)$ is a maximum is denoted by $i_{max}$.

negative values indicating that $\hat{\lambda}$ is less than $\lambda_0$. The score statistics $T_p(1)$ given by (6.4.3) have values $-18.55$ and $-8.22$. There is thus no disagreement between the conclusions on the evidence for a transformation drawn from the two sets of tests. There is however appreciable disagreement in the numbers themselves.

The disagreement is in line with the plots of Figs. 6.2 and 6.3. In Fig. 6.2 the likelihood ratio statistic is the vertical distance between $2L_{max}(\hat{\lambda})$ and $2L_{max}(1)$. These distances are quite similar for the two models although the value of $\hat{\lambda}$ is much more precisely determined for the first-order model. The score statistics $T_p(1)$ are, as has been suggested, derived from the slope of the residual sum of squares $R(\lambda)$. Figure 6.3 shows that at $\lambda = 1$ the slopes are very different for the two models.

For these data the value of $\lambda = 1$ is far from the maximum likelihood value, so it is perhaps not surprising that the values of the two statistics are not in close agreement. The two procedures will become closer together for larger values of $n$ and for values of $\lambda_0$ nearer $\hat{\lambda}$. For the null hypothesis $\lambda_0 = \hat{\lambda}$ the two statistics are both identically zero.

Maximum likelihood estimates of the transformation parameters for both models are given in Table 6.3 along with confidence intervals, including those based on the approximation (6.4.9). For the first-order model this approximation gives the limits of the 95 per cent confidence interval as $-0.202$ to $0.083$, in adequate agreement with the other two sets of limits. But for the second-order model the limits are $-0.908$ to $0.470$, about twice as long as the intervals yielded by the other two procedures. As was suggested in explanation of the much smaller disagreement between the intervals from (6.2.9) and (6.2.12), this is probably due to the non-quadratic shape of $R(\lambda)$ for the second-order model revealed by Fig. 6.3. Because the surface is too flat in the centre, an estimate of curvature based on local evidence at $\hat{\lambda}$ will underestimate the rate at which $R(\lambda)$ increases and so yield an erroneously long interval.

The evidence of this example suggests that the score test may not be an exact guide to the evidence for a transformation if $\lambda_0$ is far from $\hat{\lambda}$. Since the purpose of a diagnostic test is to call attention to possible departures from assumptions, a test which is too liberal is to be preferred to one which is conservative and causes evidence of departures to be overlooked. The further analysis prompted by the diagnostic test serves as a check on the presence of the indicated departure.

We now consider the second purpose of the developments of this section, which is to provide graphical representations of the contribution of individual observations to the evidence for a transformation. Figure 6.4 shows the added variable plot of $\overset{*}{z}(1)$, that is the ordinary residual $r$, against the residual constructed variable $\overset{*}{w}_p(1)$ for the first-order model. In its straight-line relationship which all observations follow, this plot shows that

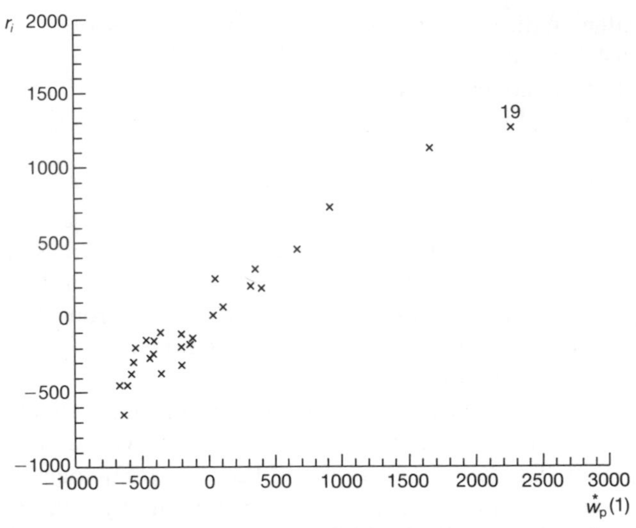


**Fig. 6.4** Example 7: Wool data: added variable plot for constructed variable $w_p(1)$

evidence for the transformation is spread throughout the data. The half-normal plot of the modified Cook statistic $C_i(w_p)$ (Fig. 6.5) shows that observation 19 is by far the most influential, with $C_{19}(w_p) = 7.66$. The value of $y_{19}$, 3636, is the largest observed response. With the non-uniformity of variance which suggests the log transformation, it is to be expected that information for the need for a transformation would come from large observed values.

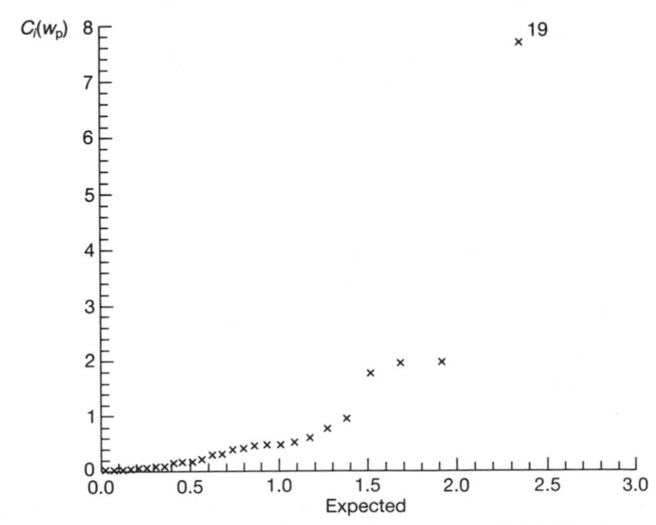


**Fig. 6.5** Example 7: Wool data: half-normal plot of modified Cook statistic $C_i(w_p)$ for Fig. 6.4

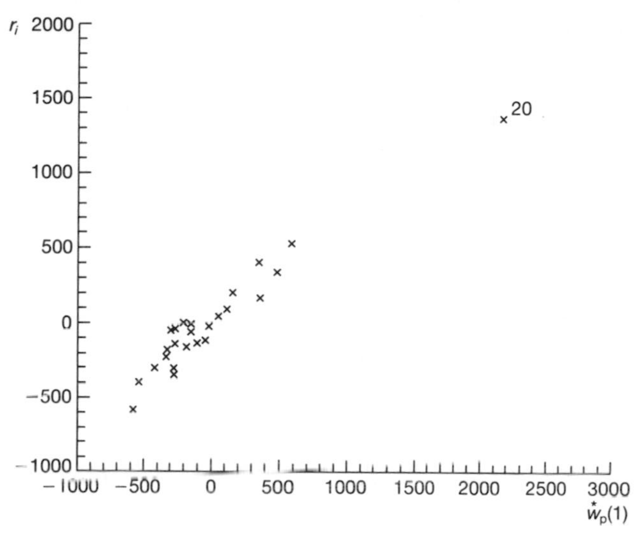


**Fig. 6.6** Example 7: Wool data, observation 19 deleted: added variable plot for constructed variable $w_p(1)$

But observation 19 is not determining the transformation on its own. The value of $-17.00$ given in Table 6.4 for the score statistic $T_p(1)$ when observation 19 is deleted shows that the evidence for the transformation remains overwhelming. The added variable plot for $w_p(1)$ and the related half-normal plot of $C_i(w_p)$ for the 26 observations after observation 19 is

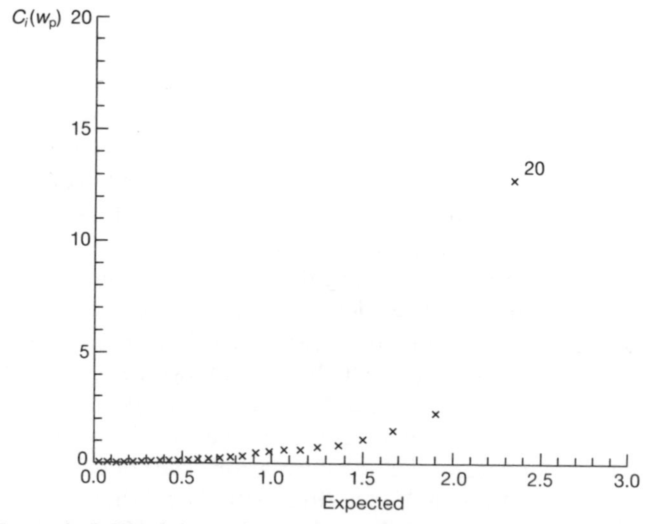


**Fig. 6.7** Example 7: Wool data, observation 19 deleted: half-normal plot of modified Cook statistic $C_i(w_p)$ for Fig. 6.6

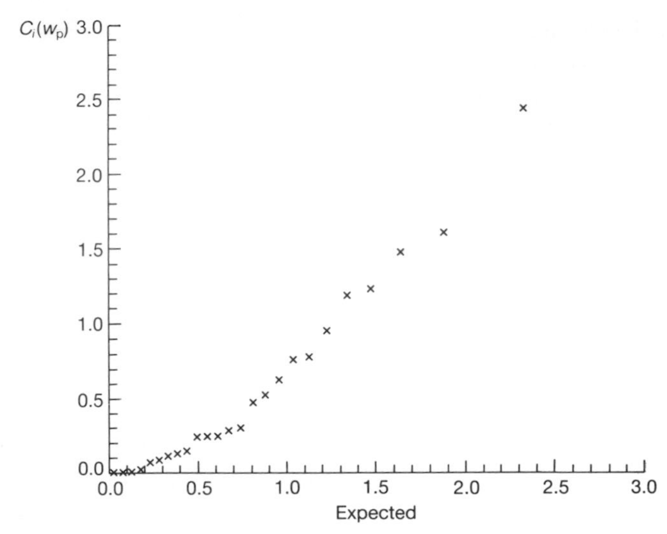


**Fig. 6.8** Example 7: Wool data, observations 19 and 20 deleted: half-normal plot of modified Cook statistic $C_i(w_p)$

deleted (Figs. 6.6 and 6.7) show that evidence for the transformation is still spread throughout the data. Now observation 20, for which $y = 3184$, is the most influential. If these two observations are deleted the results of Table 6.4 show that the evidence for the transformation, as measured by the score statistic, is reduced to $-12.11$ and the largest value of $C_i(w_p)$ is 2.44. The half-normal plot of $C_i(w_p)$, Fig. 6.8, is unexceptionable, showing no features deserving comment.

The purpose of this analysis is not to suggest that observations 19 and 20 are in any way suspect, but to demonstrate the properties of the plots. These have led to detection of the influence of the two largest observations on the transformation. The next largest response, after these two values of over 3000, is exactly 2000. These large values are obviously important in determining the transformation, but are not crucial. In two examples to be studied later, the opposite situation is encountered in which the plots reveal that all the evidence for a transformation is being determined by a single observation.

Once a transformation has been determined, the added variable plot can be used to check the structure of the transformed data. For $\lambda = 0$ the added variable plot of $\overset{*}{z}(0)$ against $\overset{*}{w}(0)$, which is not shown here, has no structure. The implication is that the log transformation is satisfactory.

A final point on this example, until Chapter 9, is on the relationship between the linear model and the transformation. The added variable plot for $w_p(1)$ from the second-order model (Fig. 6.9) shows that evidence for the transformation is again spread throughout the data. However, comparison with Fig. 6.4 for the first-order model shows that the evidence for the second-order model depends less extremely on a few observations. This is because fitting the second-order model reduces the residuals at the larger observaations, so reducing their impact on the transformation. The expression of the evidence for a transformation as regression on a constructed variable makes explicit the way in which a transformation may be an alternative to the addition of extra carriers to the model. In the present case the evidence is however clear that a transformation coupled with a first-order model provides a parsimonious representation of the data. □

## 6.5 Examples

In this section further analyses are given of three examples from earlier sections and three new examples are introduced. One theme that emerges from these analyses is that seeming outliers may indicate the need for a transformation, as in the preceding analysis of the wool data, or may just indeed be outliers, indicative of nothing. Conversely, the apparent need for a transformation may be determined by the presence of a single outlier. A second theme is that elaboration of the linear model is sometimes an alternative to the rejection of outliers or to the selection of a transformation. In a rather general way, the relationship between the three possibilities of outlier rejection, transformation of the response and of elaboration of the linear model is not surprising. As we saw in Section 5.5, rejection of an outlier is equivalent to elaboration of the model by an extra carrier. And we have also seen that

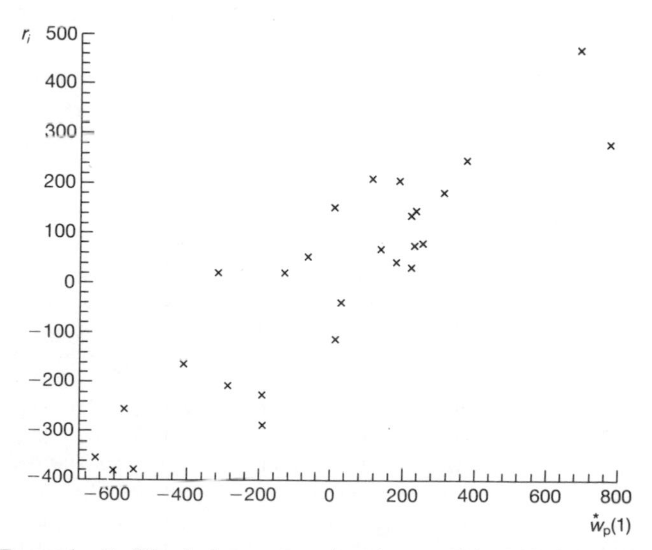


**Fig. 6.9** Example 7: Wool data, second-order model: added variable plot for constructed variable $w_p(1)$

transformation of the response is, at least locally, equivalent to regression on an extra carrier. The choice between the three possibilities is then sometimes the choice between regression models with highly correlated carriers. The important non-mathematical difference between the three possibilities is that the three sets of variables have very different physical interpretations.

**Example 8 John's $3^{4-1}$ experiment 1** This example, and the one that follows, are taken from John (1978) where interest is in outliers in factorial experiments. In the first example the purpose of the original experiment was to study the properties of an industrial yarn made from nylon. This was produced over a heated draw roll with subsequent relaxation of the yarn. The response is the 2 per cent modulus, that is the slope of the line joining the

**Table 6.5** Example 8: John's $3^{4-1}$ experiment

| Observation | Factor levels $A$ | $B$ | $C$ | $D$ | 2 per cent modulus $y$ |
|---|---|---|---|---|---|
| 1 | −1 | −1 | −1 | −1 | 29.3 |
| 2 | 0 | −1 | −1 | 0 | 30.6 |
| 3 | 1 | −1 | −1 | 1 | 31.8 |
| 4 | −1 | 0 | −1 | 1 | 26.8 |
| 5 | 0 | 0 | −1 | −1 | 30.0 |
| 6 | 1 | 0 | −1 | 0 | 29.3 |
| 7 | −1 | 1 | −1 | 0 | 27.6 |
| 8 | 0 | 1 | −1 | 1 | 29.2 |
| 9 | 1 | 1 | −1 | −1 | 29.5 |
| 10 | −1 | −1 | 0 | 1 | 30.4 |
| 11 | 0 | −1 | 0 | −1 | 26.4 |
| 12 | 1 | −1 | 0 | 0 | 30.7 |
| 13 | −1 | 0 | 0 | 0 | 28.7 |
| 14 | 0 | 0 | 0 | 1 | 30.0 |
| 15 | 1 | 0 | 0 | −1 | 28.6 |
| 16 | −1 | 1 | 0 | −1 | 29.0 |
| 17 | 0 | 1 | 0 | 0 | 29.6 |
| 18 | 1 | 1 | 0 | 1 | 29.8 |
| 19 | −1 | −1 | 1 | 0 | 27.6 |
| 20 | 0 | −1 | 1 | 1 | 28.8 |
| 21 | 1 | −1 | 1 | −1 | 28.8 |
| 22 | −1 | 0 | 1 | −1 | 28.1 |
| 23 | 0 | 0 | 1 | 0 | 29.6 |
| 24 | 1 | 0 | 1 | 1 | 28.9 |
| 25 | −1 | 1 | 1 | 1 | 27.0 |
| 26 | 0 | 1 | 1 | −1 | 30.6 |
| 27 | 1 | 1 | 1 | 0 | 29.4 |

origin to the load/extension curve at an extension of 2 per cent. The data are given in standard order in Table 6.5 where the variables are:

$y$: 2 per cent modulus
$x_1$ (Factor A): draw ratio—ratio of speeds of draw and feed rolls
$x_2$ (Factor B): relax ratio—ratio of speeds of relax and draw rolls
$x_3$ (Factor C): draw roll temperature
$x_4$ (Factor D): temperature of hot plate between feed and draw rolls.

The response given in Table 6.5 is the 2 per cent modulus reconstructed from the response

$$y' = 10(y - 25)$$

used, for numerical convenience, by John. If power transformations of the response are to be considered, it is necessary to work in the original units.

The data form a one third replicate of a $3^4$ experiment, so that there are 27 observations. The two degrees of freedom for each factor can be split into linear and quadratic contrasts. The effects of factor $D$ were, according to John, known to be relatively small and no interaction with the other three factors was considered likely. Following John we consider a model in which in addition to the constant and two degrees of freedom for each of the main factors, we also include the linear × linear interactions $AB$, $AC$ and $BC$. This model, which we shall call model 1, therefore has twelve carriers.

A half-normal plot of $r_i^*$ for this twelve-carrier model is shown in Fig. 6.10. There is clearly one outlying observation, number 11. Summary statistics for

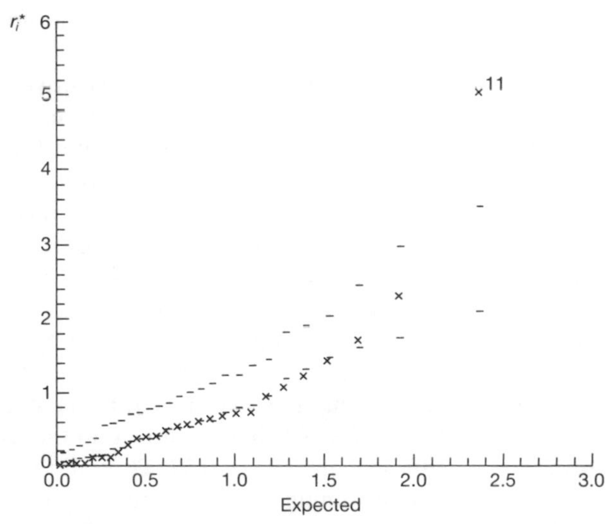


**Fig. 6.10** Example 8: John's $3^{4-1}$ experiment: half-normal plot of deletion residual $r_i^*$

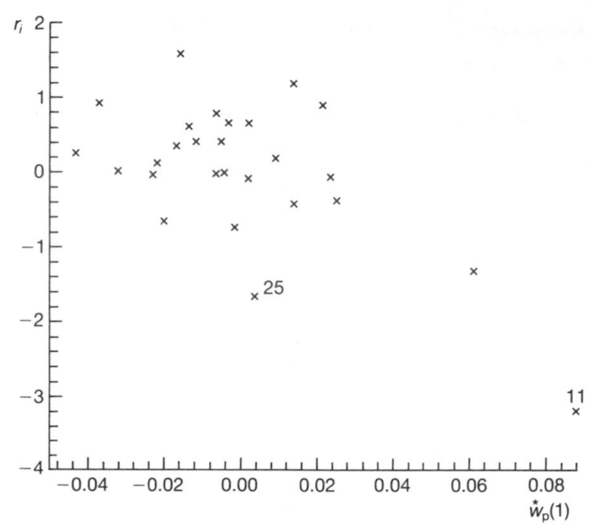


**Fig. 6.11** Example 8: John's $3^{4-1}$ experiment: added variable plot for constructed variable $w_p(1)$

this and other models in Table 6.6 show that $r_{11}^* = -5.06$ and $C_{11} = 4.0$. One possibility would be to delete this outlying observation and to refit the model. We leave this for the moment and notice instead that the score statistic for the power transformation $T_p(1) = 2.98$. Since model 1 has 12 carriers, the estimate of $\sigma^2$, $s_z^2$ given in Section 6.4 is on $n - p - 1 = 14$ degrees of freedom. The value of the score statistic is therefore just significant at the 1 per cent level, which suggests that a power transformation be considered.

A surprising feature of the statistic is that the value is positive, indicating a value of $\lambda > 1$. To see how this indication is being influenced by individual observations we show, in Fig. 6.11, the added variable plot for $w_p(1)$. This reveals a scatter of points following a seemingly downward trend with observation 11 most influential. The downward trend with a positive value of the score statistic arises because $T_p(\lambda)$ is the test for regression on $-w_p(\lambda)$.

To elucidate the influence of observation 11 we look at the values of the approximate modified Cook statistic $C_i(w_p)$. The maximum value is 6.85 for observation 11. The half-normal plot of these values in Fig. 6.12 shows how relatively large is this value. The inference from the figures is that the evidence for the transformation is being highly influenced by this one observation. If the observation is deleted, the results of Table 6.6 show that the score statistic for the transformation is reduced from 2.98 to 0.57. Thus all the evidence for the need for a transformation is coming from the one outlying observation. The added variable plot for $w_p(1)$ in Fig. 6.13 shows, by comparison with Fig. 6.11, that after deletion of observation 11, there is no evidence of a trend.

**Table 6.6** Example 8: John's $3^{4-1}$ experiment. Tests for transformations and measures of influence

| | Model | | | |
|---|---|---|---|---|
| | Model 1† (John 1978) | Model 1 − observation 11 | Aitkin and Wilson (1980) | Discussion (Aitkin 1982) |
| Residual sum of squares $R(\lambda)$ | 24.10 | 8.53 | 25.30 | 19.42 |
| $C_{i\,max}$ | 4.00 | 3.75 | 2.57 | 2.33 |
| $i_{max}$‡ | 11 | 25 | 1 | 11 |
| $r^*_{j\,max}$ | −5.06 | −2.82 | −2.23 | −2.68 |
| $j_{max}$ | 11 | 25 | 11 | 4 |
| $T_p(1)$ | 2.98 | 0.57 | 1.26 | 2.81 |
| $C_{k\,max}(w_p)$ | 6.85 | 2.56 | 4.67 | 2.87 |
| $k_{max}$ | 11 | 4 | 3 | 11 |

† Model 1 includes linear and quadratic terms in all four factors with the addition of the linear × linear interactions $AB$, $AC$ and $BC$. Aitkin and Wilson's model contains only the linear terms in $A$, $B$ and $D$ and the linear × linear interaction $BD$. The 'discussion' model augments this with terms in $C$ and the $BC$ interaction.

‡ The observation for which the modified Cook statistic $C_i$ is a maximum is denoted by $i_{max}$. Similarly, $j_{max}$ is the observation for which the deletion residual $r_i^*$ is a maximum and $k_{max}$ that for which $C_i(w_p)$ is a maximum.

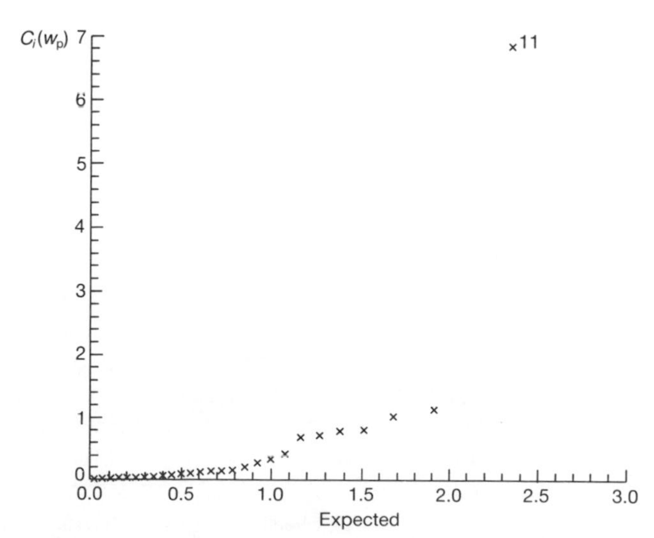


**Fig. 6.12** Example 8: John's $3^{4-1}$ experiment: half-normal plot of modified Cook statistic $C_i(w_p)$ for Fig. 6.11.

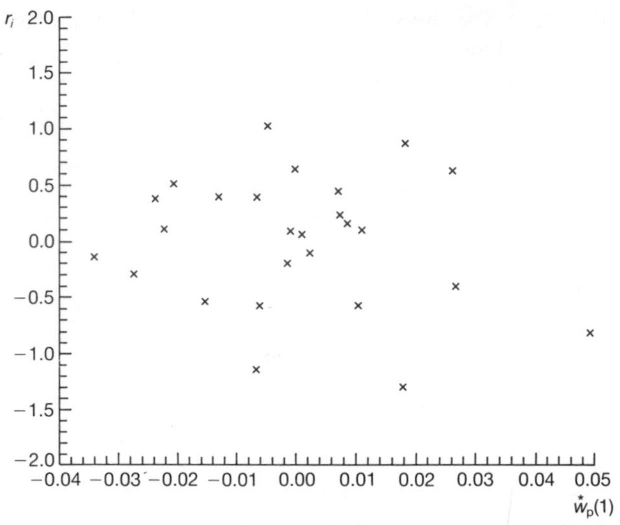


**Fig. 6.13** Example 8: John's $3^{4-1}$ experiment, observation 11 deleted: added variable plot for constructed variable $w_p(1)$

Because of the effect of the deleted observation on the projection (6.4.4), Fig. 6.13 is not quite Fig. 6.11 with observation 11 deleted. But the difference is not such as to suggest any different conclusions.

Once observation 11 has been deleted, the residual sum of squares for model 1 is dramatically reduced from 24.10 to 8.53. This deletion does not

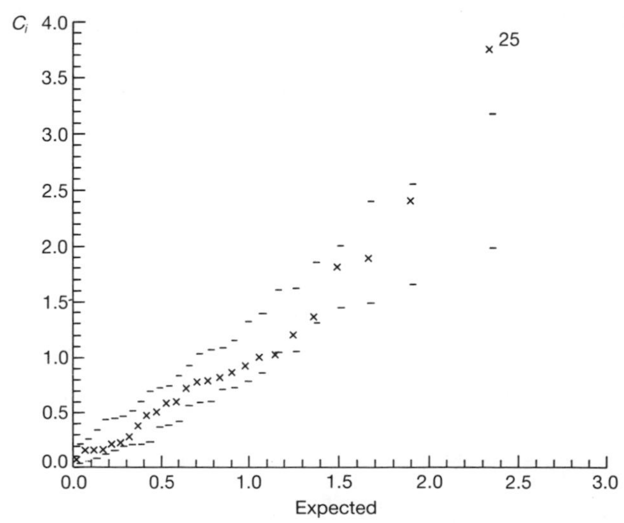


**Fig. 6.14** Example 8: John's $3^{4-1}$ experiment, observation 11 deleted: half-normal plot of modified Cook statistic $C_i$

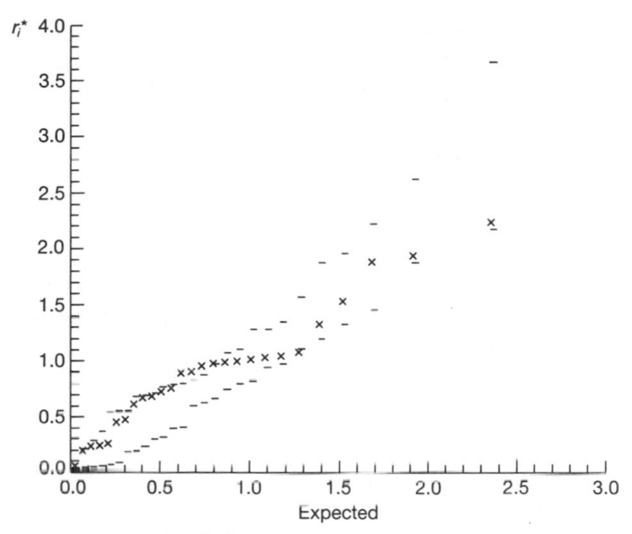


**Fig. 6.15** Example 8: John's $3^{4-1}$ experiment: half-normal plot of deletion residuals $r_i^*$ for five-carrier model of Aitkin and Wilson

necessarily remove all the outliers in the data. Figure 6.14 shows a half-normal plot of the modified Cook statistic $C_i$ which suggests that there may still be an outlying observation. The results of Table 6.6 show this to be observation 25. The procedures of Chapter 4 can be used for further investigation of this matter, but do not yield any new insights on transformations or on any of the other procedures of this book.

As an alternative to the twelve-carrier model 1, Aitkin and Wilson (1980) suggest a five-carrier model with terms in $A$, $B$, $D$ and $BD$. This is in opposition to the model suggested by John, who stated that interactions with $D$ could be ignored. The residual sum of squares for this model is 25.30, much the same as that for model 1. As the half-normal plot of $r_i^*$ in Fig. 6.15 shows, there is no evidence of an outlier. Further, the results of Table 6.6 give a value of 1.26 for $T_p(1)$, so that a transformation is not needed. Apart from the meaning of the $BD$ interaction, a strange feature of this model is that only the terms in $A$ and $BD$ are significant.

The example is discussed further by Aitkin (1982) who suggests a model with terms in $A$, $B$, $C$, $D$, $BC$ and $BD$. As the results of Table 6.6 show, this model has a residual sum of squares of 19.42. The half-normal plots of $r_i^*$ and $C_i$ look much like Figure 6.15. But now $T_p(1)$ is 2.81, with observation 11 most influential in determining this value.

The conclusion is that the example illustrates two ways in which an undetected outlier may affect the choice of model. One is that due to the inflation of the residual sum of squares caused by the outlier, a correct model may seem over-elaborate when the terms are assessed by significance tests.

Hence, perhaps, the too simple model of Aitkin and Wilson. The second is that, conversely, the inadequacies of an over-simple model may be masked by residuals which are too large in a seemingly unstructured way. □

**Example 9 John's confounded $2^5$ experiment** In the two previous examples deletion of observations led to a reduction in the evidence for a transformation. In this example, on the contrary, deletion of an observation leads to an increase in the evidence for a transformation. Otherwise the example has many features in common with Examples 7 and 8. The treatment of diagnostic plots will therefore be correspondingly brief. However the analysis does provide a good opportunity to exemplify the use of half-normal plots of effects in the analysis of $2^k$ factorial experiments.

The data, given in Table 6.7, arise from a $2^5$ experiment with the five-factor interaction *ABCDE* confounded with operators, two of whom were needed for the thirty-two trials. The response is the abrasion loss of a coating for metal when the coating was abraded under standard conditions. The five factors, each at two levels, are:

$x_1$ (Factor A): method of cleaning or preparing the substrate
$x_2$ (Factor B): composition of the substrate
$x_3$ (Factor C): pre-heat temperature of the substrate before application of the coating powder
$x_4$ (Factor D): type of coating powder
$x_5$ (Factor E): size of powder particles.

Abrasion loss is clearly a non-negative quantity, so it seems reasonable to consider power transformation of the response. The range of $y$ is 1.0 to 12.6, sufficientiy great that additive errors of constant variance are an unlikely model. John's analysis of the untransformed data used a model with main effects and all first-order interactions in addition to an effect for operators. However, as with the textile data of Example 7, a complicated model is not needed on the log scale. Following Atkinson (1982a), we use only a first-order model. We also omit the operator effect, which is not required either on the transformed or untransformed scales.

Figure 6.16 shows a plot of residuals against predicted values when a first-order model is fitted to the untransformed data. There is a strangely systematic trend for the lower residuals and some evidence of increasing scatter with $\hat{y}$. There seems no evidence for the two outliers which John suspected from a similar plot for the model including interactions. The half-normal plot of $r_i^*$ shows no particular outliers.

However the score statistic for the power transformation, $T_p(1)$, has a value of $-9.67$. If a log transformation is taken, the score statistic $T_p(0)$ equals $-0.58$, which supports such a transformation. One effect of the power transformation is to reduce the residual sum of squares $R(\lambda)$ from 80.06 to

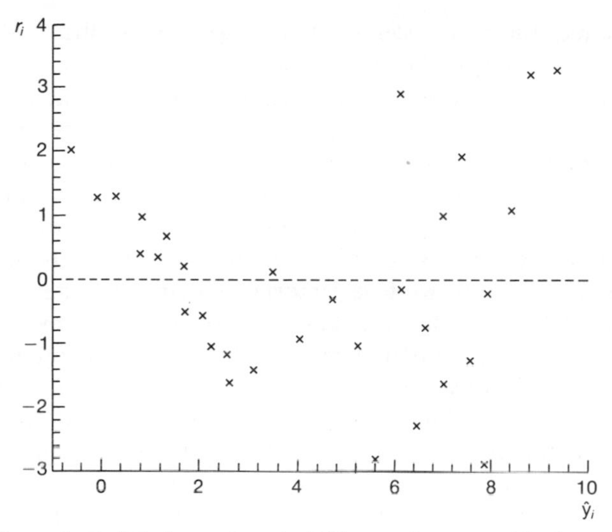


**Fig. 6.16** Example 9: John's confounded $2^5$ experiment: residuals $r_i$ against fitted values $\hat{y}_i$

27.51. The added variable plot for $w_p(1)$ (Fig. 6.17) shows that the evidence for the transformation is spread through the data. The most influential observation is number 3, which stands rather apart from the rest of the data. If this observation is deleted the value of $T_p(1)$ increases in magnitude to

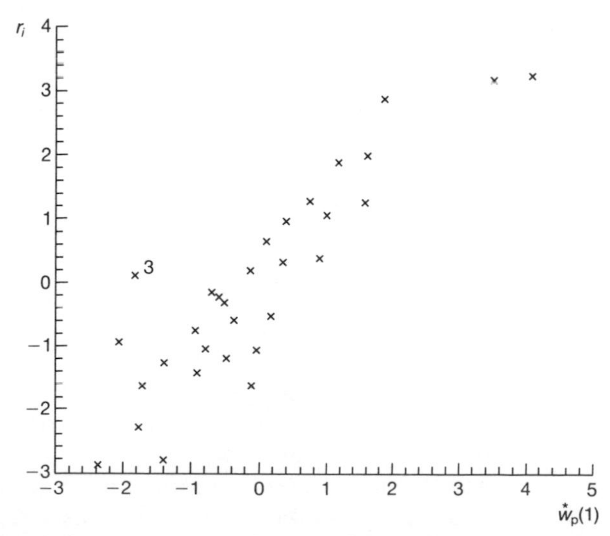


**Fig. 6.17** Example 9: John's confounded $2^5$ experiment: added variable plot for constructed variable $w_p(1)$

$-11.51$. The log transformation is still appropriate and the residual sum of squares $R(0)$ is reduced to 24.20.

The increase in the evidence for a transformation following deletion of an observation is the opposite of the effect observed in the previous two examples. Although deletion of an influential observation usually decreases the evidence for a transformation, there are, as we shall see, other examples in which deletion of an observation does lead to increased evidence.

We now consider the use of half-normal plots of effects as an aid in determining the model for two-level factorial experiments, a development due to Daniel (1959). So far we have assumed a simple structure for the first-order model involving only first-order terms. A difficulty in checking this model is the lack of an estimate of $\sigma^2$ against which to test hypotheses about the values of the parameters. The difficulty arises because the thirty-two degrees of freedom can be uniquely decomposed into a grand mean, main effects and a series of two-, three-, four- and five-factor interactions.

One customary solution is to assume all interactions above a certain order to be zero and to pool the sums of squares associated with these interactions to provide an estimate of error. This procedure is aided by use of the half-normal plot of effects, which indicates which terms to pool. In the absence of any real effects, the $2^k - 1$ estimated effects are each independently normally distributed with zero mean and the same variance. A normal, or half-normal, plot should therefore indicate which effects are real by the presence of points which depart from the otherwise linear plot.

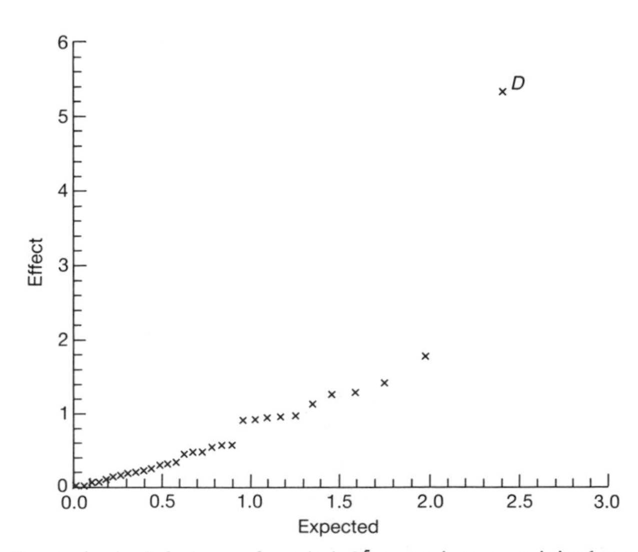


**Fig. 6.18** Example 9: John's confounded $2^5$ experiment, original response: half-normal plot of effects

**Table 6.7** Example 9: John's confounded $2^5$ experiment

| Obs.† | Treat.‡ | $y$ | Obs. | Treat. | $y$ | Obs. | Treat. | $y$ | Obs. | Treat. | $y$ |
|---|---|---|---|---|---|---|---|---|---|---|---|
| 1 | (1) | 1.4 | 9 | *d* | 5.0 | 17 | *e* | 1.7 | 25 | *de* | 9.5 |
| 2 | *a* | 1.2 | 10 | *ad* | 9.0 | 18 | *ae* | 2.0 | 26 | *ade* | 5.9 |
| 3 | *b* | 3.6 | 11 | *bd* | 12.0 | 19 | *be* | 3.1 | 27 | *bde* | 12.6 |
| 4 | *ab* | 1.2 | 12 | *abd* | 5.4 | 20 | *abe* | 1.2 | 28 | *abde* | 6.3 |
| 5 | *c* | 1.5 | 13 | *cd* | 4.2 | 21 | *ce* | 1.9 | 29 | *cde* | 8.0 |
| 6 | *ac* | 1.4 | 14 | *acd* | 4.4 | 22 | *ace* | 1.2 | 30 | *acde* | 4.2 |
| 7 | *bc* | 1.5 | 15 | *bcd* | 9.3 | 23 | *bce* | 1.0 | 31 | *bcde* | 7.7 |
| 8 | *abc* | 1.6 | 16 | *abcd* | 2.8 | 24 | *abce* | 1.8 | 32 | *abcde* | 6.0 |

† Obs. = Observation number.
‡ Treat. = Treatment combination.

Figure 6.18 shows a half-normal plot of effects for the original thirty-two observations. In line with other plots in this book, such as plots of residuals against predicted values, the observed values of the effects are plotted as $y$ with the expected values as $x$. This is the reverse of Daniel's practice. The overwhelming impression from Fig. 6.18 is of the importance of factor $D$. This has been stressed by Williams (1982) who comments that when the data are displayed as in Table 6.7, the observations at the high level of factor $D$, which are in columns 2 and 4, have both a higher mean and a higher variance than the readings at the low level of $D$ which are in the other two columns. The half-normal plot of effects after a log transformation is shown in Fig. 6.19. This again shows the overwhelming effect of $D$.

Since $D$ is so clearly a real effect, it can be removed from the set of effects under test, and a half-normal plot made of the remaining 30 effects. If only $D$ is significantly different from zero, the slope of this line can be used to provide an estimate of the standard error of the estimated effects. The revised plot, omitting $D$, is shown in Fig. 6.20. This shows a curious break, the plot appearing to be in two parts. The absolute values of the ten largest effects range from 0.913 to 1.775. The value of the next largest is much smaller at 0.575. In the figure, the five largest effects have been labelled: several are interactions of factors which are not individually significant.

This plot shows two properties which often characterize the presence of outliers. One is an appreciable number of interaction terms. A second is a jump in the plot, which falls approximately into two parallel but displaced parts. Such a plot can be caused by the presence of a single outlier, which causes an amount $\Delta$ to be added to half the estimated effects and the same amount $\Delta$ to be subtracted from the other half. Depending on the sign of the effect, its absolute value is either increased or decreased by an amount which will be $\Delta$, unless the sign is changed. The net effect is to yield plots not unlike Fig. 6.20.

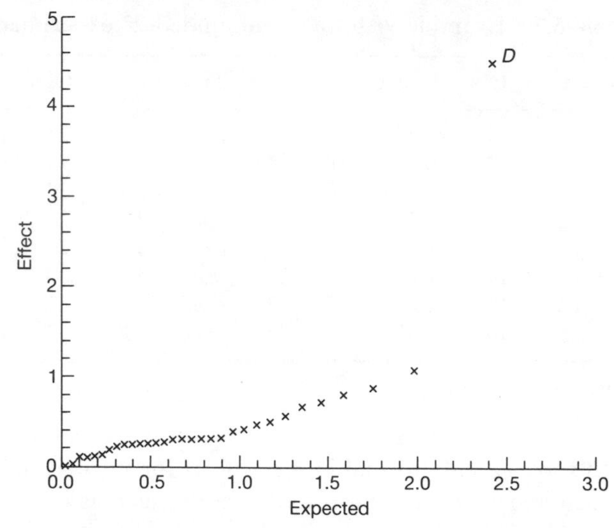


**Fig. 6.19** Example 9: John's confounded $2^5$ experiment, logged response: half-normal plot of effects

The effect, if there is indeed one, is much less marked on the plot for the log scale, Fig. 6.21. The strong suggestion here is that $D$ is the only effect which need be considered, although the plot does seem to be of two halves with slightly different slopes. The conclusion of this analysis is rather simpler than

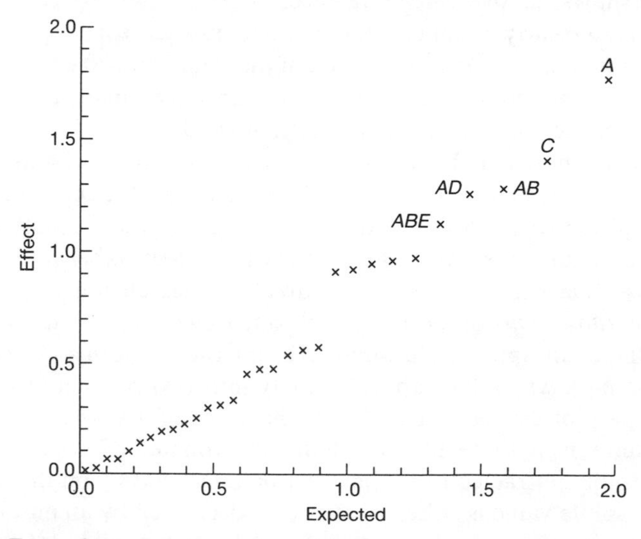


**Fig. 6.20** Example 9: John's confounded $2^5$ experiment, original response: half-normal plot of all effects save $D$

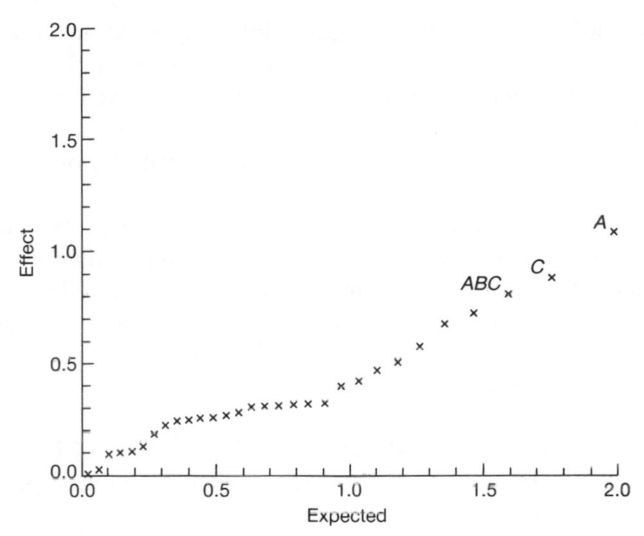


**Fig. 6.21** Example 9: John's confounded $2^5$ experiment, logged response: half-normal plot of all effects save *D*

that of Williams (1982) who mentions the large *ABC* interaction on the log scale.

A virtue of Williams's analysis is that he stresses the importance of looking at the data before charging (his word) into an analysis. If we do look again at Table 6.7 we see that most of the strange features of the data are concentrated in the third line. In column 2 the largest observation is 12.0, which is in the third row, with 9.3 the next largest in the column. In column 4 the row 3 value is 12.6, with 9.5 next largest. In the analysis following from Fig. 6.18 it was found that observation 3, in column 1, was the most influential at 3.6, with 1.6 the next largest observation in this column. The third observation in column 3 is also the largest with the value 3.1 against 2.0 for the next largest. This feature, which is not easily interpretable in a factorial structure may, as Williams suggests, be best accommodated by dividing the experiment into two parts on factor *D* and analysing the parts separately. It is clear that much of the evidence for a transformation comes from the increased yield and variance at the higher level of *D*. It is however discouraging for any more serious analysis that John reports that the values of 12.0 and 12.6, which figure prominently in the argument of this paragraph, were provided by the experimenter to replace misrecorded values.

Although it is accordingly not worthwhile to pursue any further the analysis of these data, the analysis given here does exhibit the usefulness of the half-normal plots of effects. Further examples are given in Chapter 12 of Box *et al.* (1978). The technique, which is not confined to the analysis of $2^k$ experiments, is quite general. It can be used whenever the total sum of squares of the observations can be broken down into physically meaningful single degrees of freedom. For example in the three-level factorial experiments of Examples 7 and 8, linear and quadratic terms and their interactions yield such a decomposition. As an aid in model building, the square roots of these sums of squares could then be presented as half normal plots. □

Use of normal, or half normal, plots requires single degrees of freedom. If this breakdown cannot be achieved in a meaningful manner, a similar graphical technique using the gamma distribution is available, provided the sums of squares or squared contrasts are all on the same number of degrees of freedom. Examples of gamma plots of squared contrasts for a $2^k$ experiment are given by Gnanadesikan (1977, pp. 244–6).

**Example 1 (concluded) Forbes' data 5** In the analysis of Forbes' boiling point and pressure data we have so far worked with the log response, as was suggested by Forbes' theoretical results. To conclude the analysis of these data we now consider briefly whether there is evidence from the data to support this power transformation.

For illustrative purposes the question of a transformation will be approached in two ways. Both start from the first-order model fitted to the untransformed response, which is pressure. This model can be elaborated either by a transformation of the response or by the addition of a new carrier, a quadratic in the explanatory variable. From our earlier analysis we know that observation 12 is an outlier. For the moment we do not delete this

**Table 6.8** Example 1: Forbes' data. Tests for transformations and measures of influence

| | (a) All observations First-order model | | Second-order model | |
|---|---|---|---|---|
| | $\lambda = 1$ | $\lambda = 0$ | $\lambda = 1$ | $\lambda = 0$ |
| Residual sum of squares $R(\lambda)$ | 0.813 | 0.708 | 0.694 | 0.629 |
| $T_p(\lambda)$ | −1.83 | 1.06 | −3.55 | −5.62 |
| $C_{i\max}(w_p)$ | 5.90 | 5.40 | 11.66 | 15.62 |
| $i_{\max}$ | 12 | 12 | 12 | 12 |
| | (b) Observation 12 deleted | | | |
| Residual sum of squares $R(\lambda)$ | 0.362 | 0.059 | 0.063 | 0.054 |
| $T_p(\lambda)$ | −8.96 | 1.01 | −3.12 | −0.61 |
| $C_{i\max}(w_p)$ | 3.18 | 3.21 | 3.00 | 3.60 |
| $i_{\max}$ | 1 | 1 | 14 | 14 |

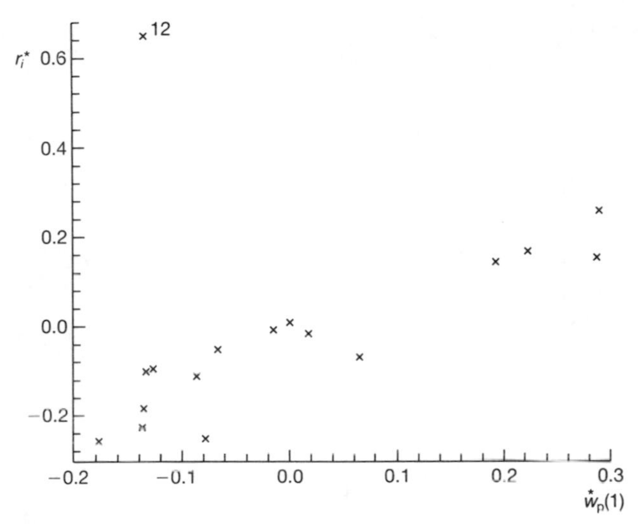


**Fig. 6.22** Example 1: Forbes' 'data': added variable plot for constructed variable $w_p(1)$

observation, but behave as if the analysis of the untransformed data were starting afresh.

We begin with the evidence for a transformation. The value for the score statistic for the first-order model is given in Table 6.8 as $-1.83$, evidence for a transformation which is not quite significant at the 5 per cent level. The added variable plot for the constructed variable $w_p(1)$ for the power transformation is shown in Fig. 6.22. The plot consists of a steady upward trend for all observations except observation 12, which is again revealed as an outlier. The maximum value of the modified Cook statistic $C_i(w_p)$ for this regression is 5.90, which is for observation 12. If observation 12 is deleted, the score statistic changes to a highly significant $-8.96$. The added variable plot for $w_p(1)$, Fig. 6.23, now shows the clear trend of Fig. 6.22 without any distortion. The score statistic $T_p(0)$ for the null hypothesis of a log transformation equals 1.01; the log transformation is therefore acceptable. The analysis of transformations thus leads to the model analysed in earlier chapters, once the outlier has been detected and omitted.

The effect of the outlying observation 12 is extreme and would not be detected in the framework of transformations without use of the influence measure $C_i(w_p)$ or of the added variable plot of Fig. 6.22. In the absence of such aids the transformation would be rejected. However, as we shall see, the outlier is immediately apparent from the analysis of residuals on the untransformed scale.

A plot of the residuals of the original data against the predicted values $\hat{y}$ is shown in Fig. 6.24. This clearly shows that observation 12 is an outlier on the

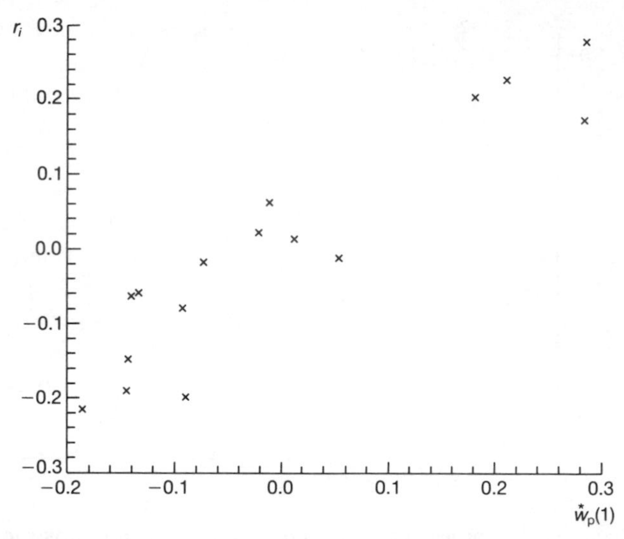


**Fig. 6.23** Example 1: Forbes' 'data', observation 12 deleted: added variable plot of constructed variable $w_p(1)$

untransformed scale, just as Fig. 1.2 did for the logged observations. For the first-order model $\hat{y}$ is a linear function of $x$, so that the plot of the residuals against the explanatory variable is, apart from the scaling of one axis, the same as Fig. 6.24. The curvature visible in Fig. 6.24 therefore suggests the addition to the model of a quadratic term in $x$.

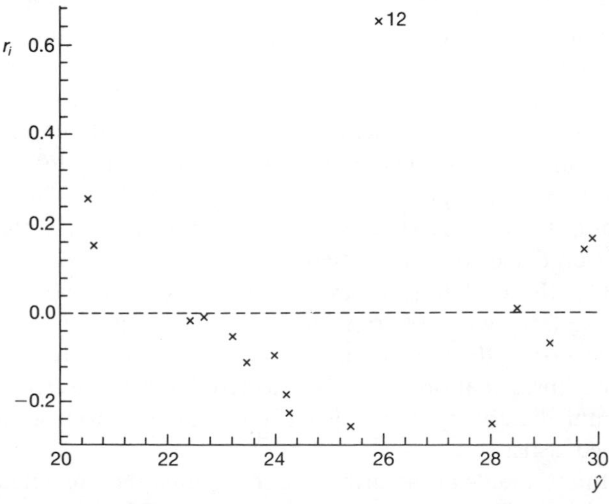


**Fig. 6.24** Example 1: Forbes' 'data': residuals $r_i$ against fitted values

Deletion of observation 12 reduces the residual sum of squares for the first-order model from 0.813 to 0.362. Addition of the quadratic term causes a further, and more dramatic, drop to 0.063. Plots of residuals show no particular discrepancy between this model and the data, except that the normal plot of residuals, Fig. 6.25, seems to consist of two straight halves of different slopes.

This indication of something strange about the error distribution might suggest investigation of a power transformation. For the quadratic model $T_p(1) = -3.12$, significant at the 1 per cent level, but not at 0.1 per cent. This value is much less significant than the $-8.96$ calculated for the first-order model, so this example again shows the way in which elaboration of a linear model may serve as an alternative to a power transformation. If a power transformation is used with the second-order model, the log transformation is acceptable with $T_p(0) = -0.61$. The residual sum of squares is 0.054, compared with 0.059 for the first-order model with logged response. On the log scale the quadratic term is therefore not required.

These results vindicate the previous analysis which led to a first-order model with logged response. Of course, if theoretical considerations suggest a particular form of model, this information should always be respected. However the theory should equally be tested against the data. The most important aspect of this final analysis of Forbes' data, from the standpoint of this book, is the impact of the one outlying observation on the evidence for transformation, an impact which is revealed by Fig. 6.22. □

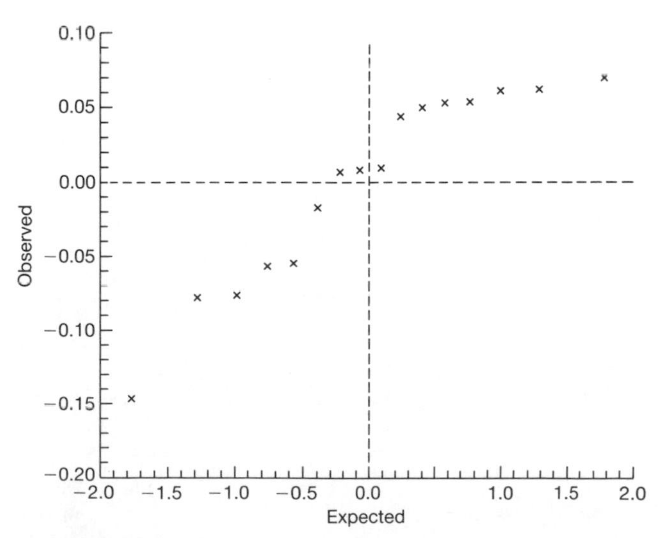


**Fig. 6.25** Example 1: Forbes' 'data', observation 12 deleted: normal plot of residuals from quadratic model

**Example 4 (concluded) Salinity data 2** In Chapter 4 the analysis of the salinity data led to the detection of one highly influential observation, number 16, for which the value of $x_3$ was 33.443. The analysis proceeded by taking this outlying value to be a misprint for 23.443. The half-normal plots of $r_i^*$ and $C_i$ (Fig. 4.19) then gave little indication of any misfit between model and data.

We now consider the evidence for a power transformation of the response. For the original set of data there is no evidence of the need for a transformation, with the score statistic $T_p(1)$ at $-0.08$, close to zero. Once the data have been 'corrected' the value increases in magnitude to $-1.61$. An added variable plot for $w_p(1)$ is given in Fig. 6.26. This shows a cloud of points which plausibly have some regression, except for observation 3 which is somewhat separate. This point has a value of 6.71 for the modified Cook statistic $C_i(w)$. The half-normal plot of $C_i(w)$ (Fig. 6.27) reveals how influential this one observation is in denying the need for a transformation. If observation 3 is deleted the score statistic for the power transformation becomes $-2.50$. The maximum likelihood estimate of the transformation parameter is $-0.15$ and a log transformation is in agreement with the data.

These results are summarized in Table 6.9. Ruppert and Carroll in their original presentation do not give the units for the response, which is salinity. If the measurement is of concentration of salt, then a log transformation is not unreasonable. Half-normal plots of $C_i(w_p)$ for $\lambda = 1$ and $\lambda = 0$, not shown here, are quite different from Fig. 6.27. They show that, after deletion of observation 3, the evidence for a power transformation is spread throughout

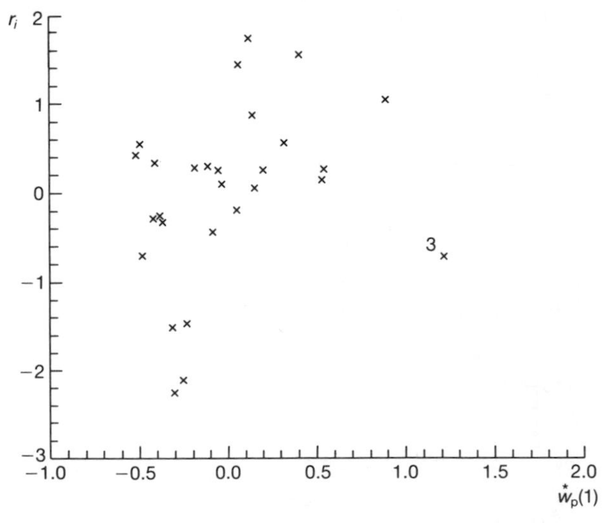


**Fig. 6.26** Example 4: Salinity data 'corrected': added variable plot for constructed variable $w_p(1)$

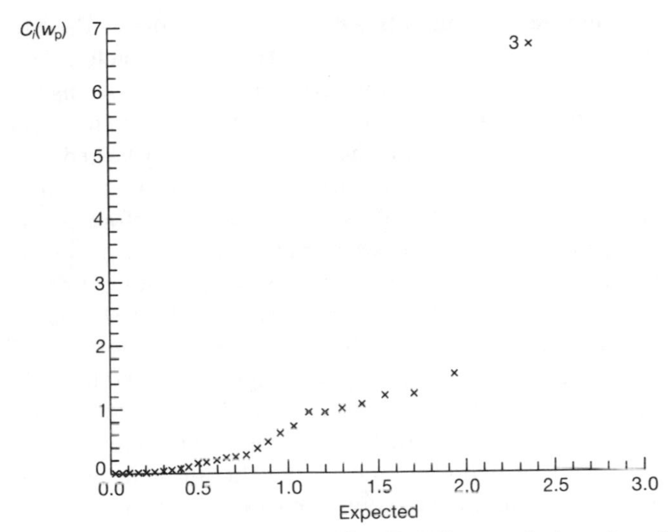


**Fig. 6.27** Example 4: Salinity data 'corrected': half-normal plot of modified Cook statistic $C_i(w_p)$ for Fig. 6.26

the data. As a result of the whole process of data manipulation, consisting of correction, deletion and transformation, the residual sum of squares has been reduced from 42.47 to 20.60. The details are in Table 6.9. Interpretation of this process, including judgements about its plausibility and relevance, requires the collaboration of the scientists who collected and understand the data. Nevertheless, this example again serves to show the use of graphical methods in detecting how the evidence for a transformation can depend on a single observation. □

**Table 6.9** Example 4: Salinity data. Tests for transformations and measures of influence

| | All observations | Observation 16 'corrected' | Observation 16 'corrected' and 3 deleted |
|---|---|---|---|
| $T_p(1)$ | −0.08 | −1.61 | −2.50 |
| $C_{i\,\max}(w_p)$ | 3.24 | 6.71 | 1.37 |
| $i_{\max}$ | 3 | 3 | 7 |
| Residual sum of squares $R(1)$ | 42.47 | 26.24 | 25.56 |
| $T_p(0)$ | 2.66 | 1.54 | −0.35 |
| Residual sum of squares $R(0)$ | 52.18 | 26.24 | 20.60 |

**Example 6 (continued) Minitab tree data 5** In Chapter 5 the Minitab tree data served to illustrate the use of a variety of plots, including the added variable plot, in establishing a satisfactory linear model. In this further look at the data we consider transformation of the response, which is volume of the tree. Some of the general points which arise have already been discussed, such as the possibility that a transformation may provide an alternative to elaboration of the linear model. These points will therefore be passed over rapidly. But a new point which does arise is the use of dimensional arguments in suggesting and interpreting models. In many branches of engineering, for example in the study of fluid flow and heat transfer, dimensionless quantities such as Reynolds' and Prandtl's numbers, are of great importance. The use of such guides to meaningful models is often ignored in statistical work, an omission which has been lamented by Finney (1977).

In Chapter 5 the models studied for the tree data were the regression of volume on $x_1$, girth or diameter, and $x_2$, height. Two suggestions of inadequacy for the model including $x_1$ and $x_2$ came from the scatter plot of $y$ against $x_2$ (Fig. 5.2), which suggests that variance may increase with the response, and from the index plots of $r_i^*$ and $C_i$ (Figs. 5.9 and 5.10). These show evidence of curvature, although it is not clear from the index plots how this is best modelled. As in earlier examples, we follow two routes in finding an improved model. First we consider transformations, then elaboration of the linear model.

For the model with terms in only $x_1$ and $x_2$, which will be called Model 1, the score statistic $T_p(1)$ is $-7.41$, strong evidence of the advantages of a transformation. The maximum value of $C_i(w_p)$ is 6.29 for observation 31. The added variable plot for the constructed variable $w_p(1)$ (Fig. 6.28) shows how influential observation 31 is in determining the need for a transformation, which is to be expected of appreciably the largest tree. If the observation is deleted, $T_p(1)$ is reduced in magnitude to $-6.84$, a reduction which does not markedly change the significance of the evidence for a transformation. For all 31 observations the maximum likelihood estimate of $\lambda$ is $\hat{\lambda} = 0.3066$, with 95 per cent confidence interval calculated from the likelihood as (0.118, 0.492). The effect of the transformation is to reduce the residual sum of squares of $z(\lambda)$ from 421.9 to 135.4. The results are summarized in Table 6.10.

This value of $\hat{\lambda}$ suggests that the transformation $\lambda = \frac{1}{3}$ should be employed, namely that $(\text{volume})^{1/3}$ be treated as a linear function of diameter and height. The technique of parametric transformation has thus yielded a dimensionally homogeneous equation, both sides of which have the dimension of length.

An interesting feature of the data is that there are several other models which fit equally well. One comes from elaboration of the linear model. On the untransformed scale the standard regression technique of adding a full second-order model, followed by the sequential deletion of terms judged non-

**Table 6.10** Example 6: Minitab tree data. Tests for transformations and measures of influence

| Carriers in model | $\lambda$ | $R(\lambda)$ | $T_p(\lambda)$ | $i_{max}$ | $C_{i\,max}$ |
|---|---|---|---|---|---|
| $x_1, x_2$ (model 1) | 1 | 421.9 | −7.41 | 31 | 6.29 |
| | 1/3 | 135.8 | — | — | — |
| | 0.3066 ($\hat{\lambda}$) | 135.4 | 0 | (0.118, 0.492)† | |
| | 0 | 182.5 | 3.08 | 31 | 2.16 |
| $x_1, x_2, x_1^2$ (model 2) | 1 | 186.0 | −3.18 | 31 | 2.17 |
| | 1/3 | 135.7 | — | — | — |
| | 0 | 126.6 | −0.29 | 18 | 2.19 |
| | −0.0663 ($\hat{\lambda}$) | 126.3 | 0 | (−0.489, 0.482)† | |
| Cone: $x_1^2 x_2$ no constant (model 3) | 1 | 180.8 | 0.52 | 31 | 2.46 |
| | 0.9499 ($\hat{\lambda}$) | 177.0 | 0 | (0.889, 1.014)† | |
| | 0 | 25 984.2 | 51.73 | 31 | 7.02 |
| $\log x_1, \log x_2$ (model 4) | 1 | 843.0 | −11.40 | 31 | 9.63 |
| | 0 | 129.1 | −0.72 | 31 | 1.76 |
| | −0.0672 ($\hat{\lambda}$) | 126.7 | 0 | (−0.242, 0.110)† | |

† Approximate 95 per cent confidence interval for $\lambda$ calculated from the change in log-likelihood (6.2.9).

significant by $t$ tests, leads to a linear model with terms in only $x_1$, $x_2$ and $x_1^2$. Inclusion of the term in $x_1^2$ is also indicated by the plot of residuals against $x_1$ (Fig. 6.29), which identifies the cause of the curvature noted in Figs. 5.9 and 5.10. The residual sum of squares for the model with a quadratic term, Model 2, is 186.0, which is not as good a fit as Model 1 with the cube root transformation for $y$. That there is still some systematic departure from the model after introduction of the quadratic term is shown by the plot of residuals against $x_2$ (Fig. 6.30). The suggestion of non-constancy of variance, which was detected in the plot of $y$ vs $x_2$ (Fig. 5.2) shows in this picture as a megaphone shape of increasing residuals. This shape is one of the standard indications of inhomogeneous variance, one cure for which is to transform the response.

For the first-order model the score statistic $T_p(1)$ was −7.41. Introduction of the term in $x_1^2$ reduces this value in magnitude to −3.18, which shows appreciably less evidence of the need for a transformation. However, the value is significant at the 1 per cent level. In this case the log transformation is indicated by $\hat{\lambda} = -0.0663$, with approximate 95 per cent confidence interval calculated from the likelihood of (−0.489, 0.482). For the log transformation the residual sum of squares is 126.6.

An interesting feature of the transformation applied to the second-order model is that the log transformation is indicated. Since the model includes a

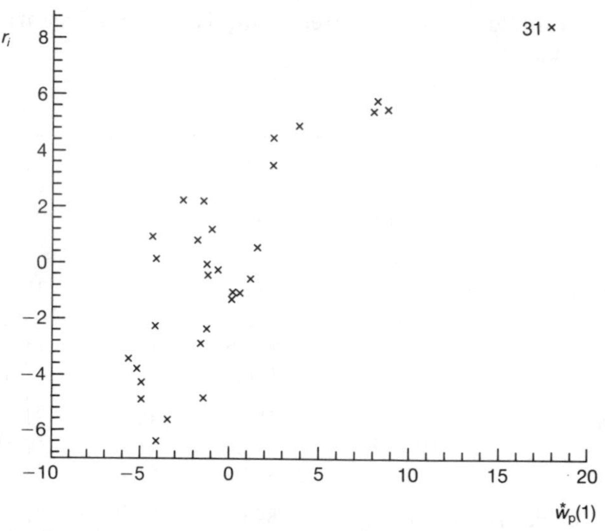


**Fig. 6.28** Example 6: Minitab tree data, first-order model: added variable plot for constructed variable $w_p(1)$

quadratic term, the right-hand side is no longer dimensionally homogeneous in length, so that dimensional considerations no longer suggest the one third transformation. Although the 95 per cent confidence interval for the quadratic model includes the value $\frac{1}{3}$, that for the first-order model does not include 0. As the values of the score statistics demonstrate, addition of the

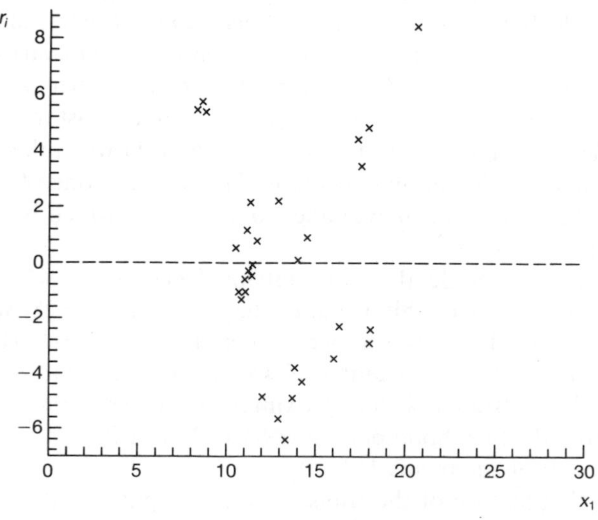


**Fig. 6.29** Example 6: Minitab tree data, first-order model: residuals $r_i$ against $x_1$

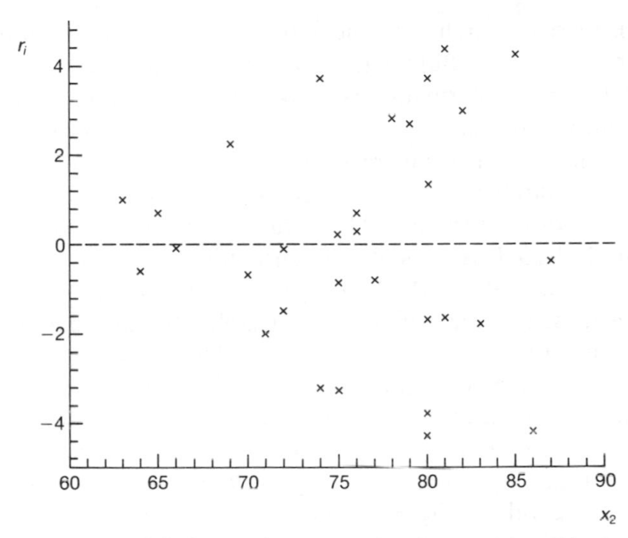


**Fig. 6.30** Example 6: Minitab tree data, second-order model: residuals $r_i$ against $x_2$

quadratic term makes the transformation less important. This can be represented graphically by a plot of $L_{max}(\lambda)$ for the two models. The plot, given by Cook and Weisberg (1982, Fig. 2.4.3), is similar to the plot of $2L_{max}(\lambda)$ for the textile data given in Fig. 6.2.

Several authors, including Sprent (1982), have commented that the shape of a tree trunk is pretty much that of a cone. It might therefore be sensible to consider models of the form $V = \beta hd^2$. For this model, which is called Model 3 in Table 6.10, there is, gratifyingly, no evidence of the need for a constant term. Regression on the single carrier $x_1^2 x_2$ leads to a residual sum of squares of 180.8. Further, there is no evidence of the need for a transformation with $T_p(1) = -0.52$.

Comparison of Model 3 with the two polynomial models with transformed response is not entirely straightforward since the models are not nested. That is, Model 3 is not a special case of either. A rough guide to comparison of the models which allows for the difference in the number of parameters is provided by an analysis of variance based on Model 1 at $\hat{\lambda}$. This yields an estimate for $\sigma^2$ of $135.4/(31 - 4) = 5.02$ on twenty-seven degrees of freedom when allowance is made in the degrees of freedom for the effect of estimating $\lambda$. The difference in residual sum of squares between this model and the one-parameter conical model is 45.4, yielding an approximate $F$ value of 3.02 on three and twenty-seven degrees of freedom. The 5 per cent point of the $F$ distribution with these degrees of freedom is 2.96, so that the observed value is just significant at this level. The separate nature of the models, combined with the approximate nature of the distributional results for sums of squares of the $z(\lambda)$, render it foolish to yield to a dogmatic interpretation of the significance of this result. But the test does afford some evidence that the first-order model with transformation is to be preferred to the cone.

An alternative, stronger, line of argument is that it is physically plausible that measurement error will increase with volume. The plots and score tests so far have confirmed this. Therefore we consider further models with transformation of the response. The logarithmic transformation of just the response in Model 3 does not make dimensional sense and is strongly rejected by the data with $T_p(0) = 51.73$. Another set of models is that in which both the response and the explanatory variables are logged. One example is the model which Box and Cox (1964) proposed for the wool data, Example 7. In the present case a model with log $y$ regressed on log $x_1$ and log $x_2$ yields a residual sum of squares of 129.1, close to that for Model 2, when the carriers for log $y$ were $x_1$, $x_2$ and $x_1^2$. This model seems to deal adequately with the variance heterogeneity observed in Fig. 6.29. The scatter of residual log $y$ against predicted values, Fig. 6.31, shows no systematic trend.

The result of these analyses is to suggest several non-nested models. Of these, two of the most satisfactory are those which not only fit well but make dimensional sense. The alternatives are to regress:

1. $(\text{volume})^{1/3}$ on height and diameter, or
2. log volume on log height and log diameter.

The data give no evidence for preferring one form to the other. If several sets

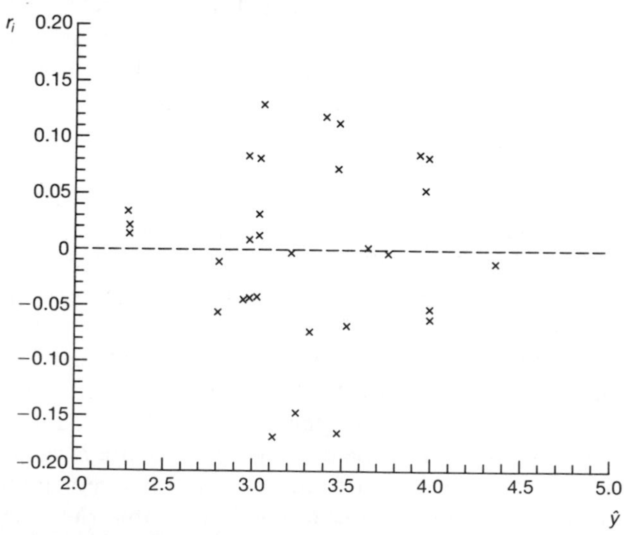


**Fig. 6.31** Example 6: Minitab tree data, logged response and logged carriers: residuals $r_i$ against predicted values.

of data, for example on different locations and species, were available, constancy of regression coefficients over the sets of data would be a strong recommendation. For the second of the two models in the list above the coefficients of log height and log diameter are respectively 1.12 and 1.98, which are not far from the theoretical values of 1 and 2 for the cone. We return to these data in Chapter 8 for the discussion of parametric transformations of both response and explanatory variables. □

**Example 10 Brownlee's stack loss data 1** The last example of the use of score statistics and constructed variables for the power transformation illustrates the reconciliation of a suspected outlier with a simple model once the data have been transformed. However, there is the possibility that the fitted model has been unduly influenced by this one observation. We investigate this with the help of index plots of the changes in the parameter estimates on deletion of individual observations.

The data, taken from Brownlee (1965, p. 454), are given in Table 6.11. There

**Table 6.11** Example 10: Brownlee's stack loss data

| Observation number | Air flow $x_1$ | Cooling water inlet temperature $x_2$ | Acid concentration $x_3$ | Stack loss $y$ |
|---|---|---|---|---|
| 1 | 80 | 27 | 89 | 42 |
| 2 | 80 | 27 | 88 | 37 |
| 3 | 75 | 25 | 90 | 37 |
| 4 | 62 | 24 | 87 | 28 |
| 5 | 62 | 22 | 87 | 18 |
| 6 | 62 | 23 | 87 | 18 |
| 7 | 62 | 24 | 93 | 19 |
| 8 | 62 | 24 | 93 | 20 |
| 9 | 58 | 23 | 87 | 15 |
| 10 | 58 | 18 | 80 | 14 |
| 11 | 58 | 18 | 89 | 14 |
| 12 | 58 | 17 | 88 | 13 |
| 13 | 58 | 18 | 82 | 11 |
| 14 | 58 | 19 | 93 | 12 |
| 15 | 50 | 18 | 89 | 8 |
| 16 | 50 | 18 | 86 | 7 |
| 17 | 50 | 19 | 72 | 8 |
| 18 | 50 | 19 | 79 | 8 |
| 19 | 50 | 20 | 80 | 9 |
| 20 | 56 | 20 | 82 | 15 |
| 21 | 70 | 20 | 91 | 15 |

are observations from 21 days operation of a plant for the oxidation of ammonia as a stage in the production of nitric acid. The variables are:

$x_1$: air flow
$x_2$: cooling water inlet temperature
$x_3$: 10 × (acid concentration − 50)
$y$: stack loss; ten times the percentage of ingoing ammonia escaping unconverted.

The air flow measures the rate of operation of the plant. The nitric oxides produced are absorbed in a counter-current absorption tower: $x_2$ is the inlet temperature of cooling water circulating through coils in this tower and $x_3$ is proportional to the concentration of acid in the tower. Small values of the response correspond to efficient absorption of the nitric oxides.

These data have been extensively analysed in the statistical literature. Some references are given in Chapter 12. One general conclusion is that it is a paradigm of robust regression that observations 1, 3, 4 and 21 are outliers. Instead of down-weighting these four observations, which is the consequence of the robust analysis, we find a model which fits all 21 observations.

An index plot of the response $y$ is given in Fig. 6.32. Although Brownlee does not explicitly state that the observations are ordered in time, they have sometimes been so interpreted. Looked at in this way, Fig. 6.32 shows a successful adjustment of the operating conditions, leading to a reduction of the stack loss from 42 to a minimum value of 7. Since the loss is necessarily

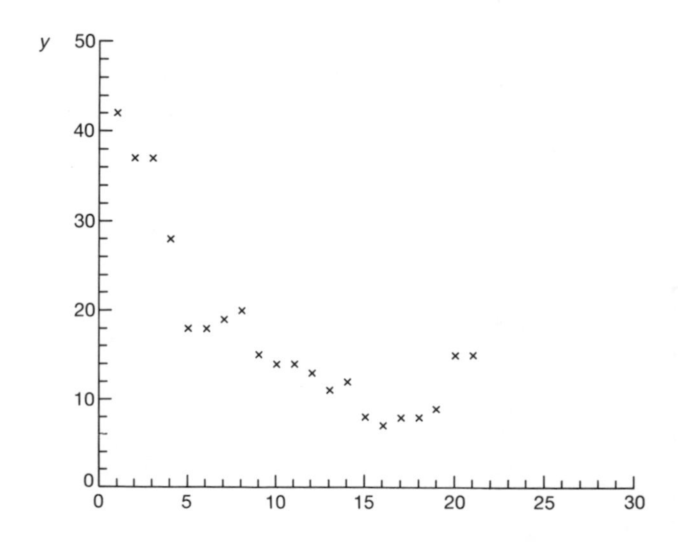


**Fig. 6.32** Example 10: Brownlee's stack loss data: index plot of response.

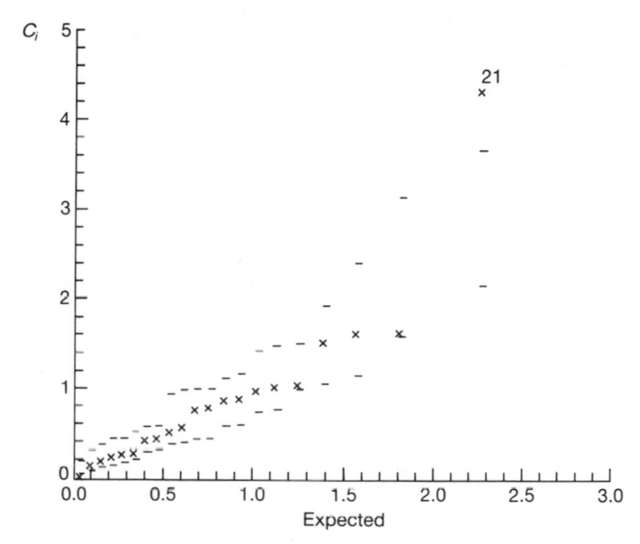


**Fig. 6.33** Example 10: Brownlee's stack loss data: half-normal plot of modified Cook statistic $C_i$

non-negative with range of almost a cycle, it is quite likely *a priori* that a transformation of $y$ will be desirable.

When a first-order model in the three explanatory variables is fitted, the half-normal plot of the modified Cook statistic, Fig. 6.33, shows observation 21 as an outlier. Techniques of transformation similar to those amply exemplified in this chapter lead to the conclusion that a log transformation should be taken and that the linear model should contain terms in $x_1$, $x_2$, $x_1x_2$ and $x_1^2$. The third explanatory variable, $x_3$, is not relevant. A summary of the evidence for the transformation is given in Table 6.12 which lists score

**Table 6.12** Example 10: Brownlee's stack loss data. Tests for transformations and residual sums of squares

| Carriers in model | $\lambda$ | $R(\lambda)$ | $T_p(\lambda)$ |
|---|---|---|---|
| $x_1, x_2, x_3$ | 1 | 178.8 | $-3.29$ |
| | 0 | 123.6 | 1.25 |
| $x_1, x_2, x_3$ with | 1 | 105.6 | $-2.53$ |
| observation 21 deleted | 0 | 95.9 | 2.15 |
| $x_1, x_2, x_1x_2, x_1^2$ | 1 | 120.8 | $-5.72$ |
| | 0 | 64.4 | $-1.12$ |
| $x_1, x_2, x_1x_2, x_1^2$ with | 1 | 103.9 | $-5.14$ |
| observation 21 deleted | 0 | 53.9 | $-1.19$ |
| $x_1, x_2, x_1x_2$ with | 1 | 104.9 | $-4.91$ |
| observation 21 deleted | 0 | 54.2 | $-1.13$ |

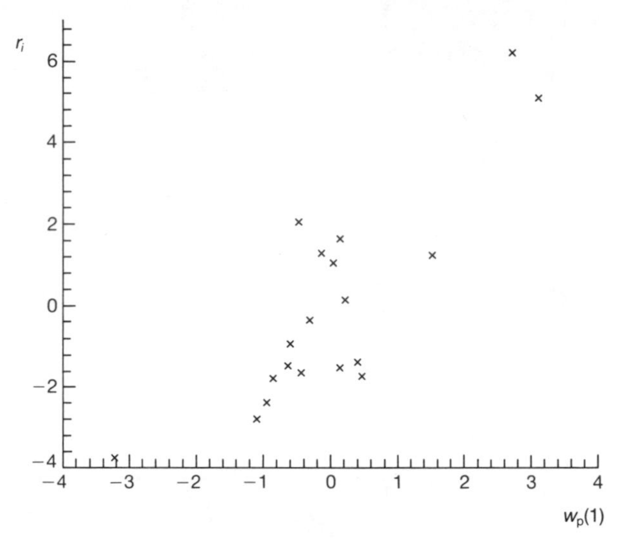


**Fig. 6.34** Example 10: Brownlee's stack loss data, second-order model: added variable plot for constructed variable $w_p(1)$

statistics and residual sums of squares. As judged by $t$ tests of coefficients, the five-carrier second-order model is justified on the log scale, whereas it is not so on the untransformed scale. That the evidence for a transformation is not being unduly influenced by a few observations is shown by the added variable plot of $w_p(1)$ for the second-order model, Fig. 6.34.

For the original first-order model the residual sum of squares is 178.8. The combination of the second-order terms and a log transformation reduces this to 64.4. The half-normal plot of the modified Cook statistic $C_i$, Fig. 6.35, when compared with Fig. 6.33 shows how observation 21 has been reconciled with the rest of the data through the use of a more complicated model.

It however remains a possibility that the more complicated model has been fitted solely to accommodate one outlier. The situation would then be comparable to that which arose in fitting a quadratic to Huber's 'data'. Indeed the curved envelope of Fig. 6.35 suggests the presence of a point of high leverage as did Fig. 4.10 for the simulated factorial and Fig. 4.16 for Huber's 'data'.

To check the influence of individual observations on the second-order model we first look at the index plot of the $h_i$, Fig. 6.36. The highest value by far is 0.869 for observation 21, followed by 0.434 for observations 1 and 2. It is clear from this plot that observation 21 is indeed a point of appreciable leverage for the second-order model. The physical interpretation can be found from Fig. 6.37 which shows a plot of $x_1$ against $x_2$. For all observations except 21 there is a roughly linear relationship between the two explanatory

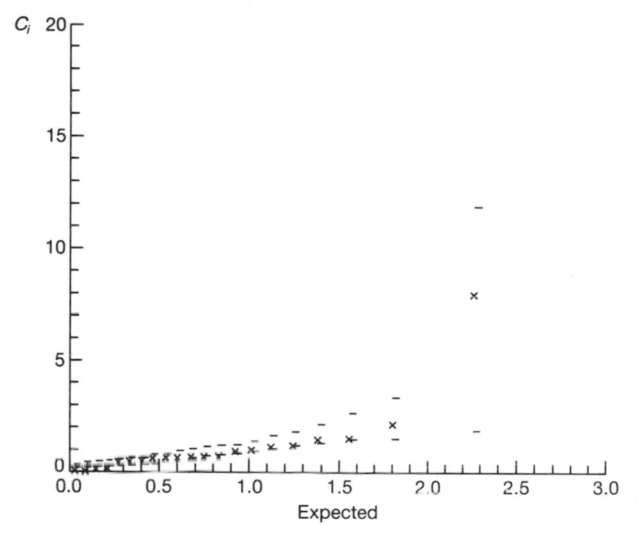


**Fig. 6.35** Example 10: Brownlee's stack loss data, second-order model, logged response: half-normal plot of modified Cook statistic $C_i$

variables, with higher values of one variable tending to be associated with higher values of the other. Observation 21 stands apart from this trend and so corresponds to the exploration of a rather different region in factor space.

The influence of individual observations is clearly shown by index plots of the changes in the components of $\hat{\beta}$ when observations are deleted. If we

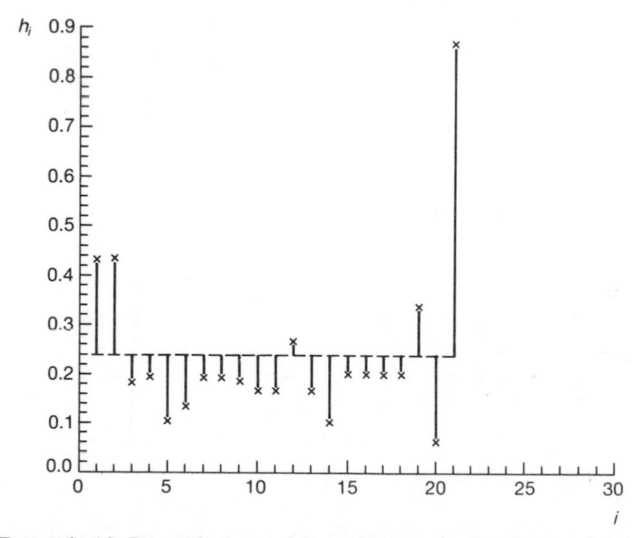


**Fig. 6.36** Example 10: Brownlee's stack loss data, second-order model: index plot of leverage measure $h_i$

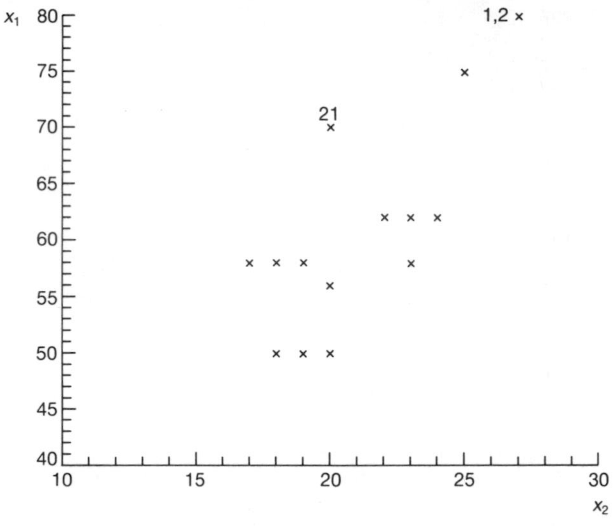


**Fig. 6.37** Example 10: Brownlee's stack loss data: scatter plot of $x_1$ against $x_2$

define $\Delta\hat{\beta}_{j(i)}$ as the change in the $j$th component of $\hat{\beta}$ on deletion of the $i$th observation,

$$\Delta\hat{\beta}_{j(i)} = \hat{\beta}_j - \hat{\beta}_{j(i)},$$

with the right-hand side given explicitly by (2.2.8). Figure 6.38 shows the

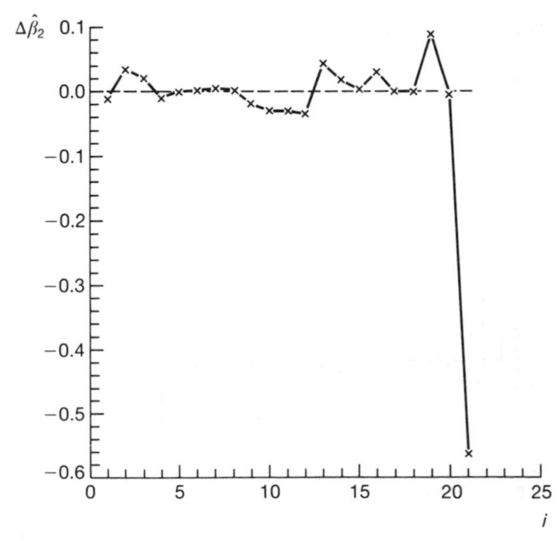


**Fig. 6.38** Example 10: Brownlee's stack loss data, second-order model, logged response: index plot of $\Delta\hat{\beta}_2$

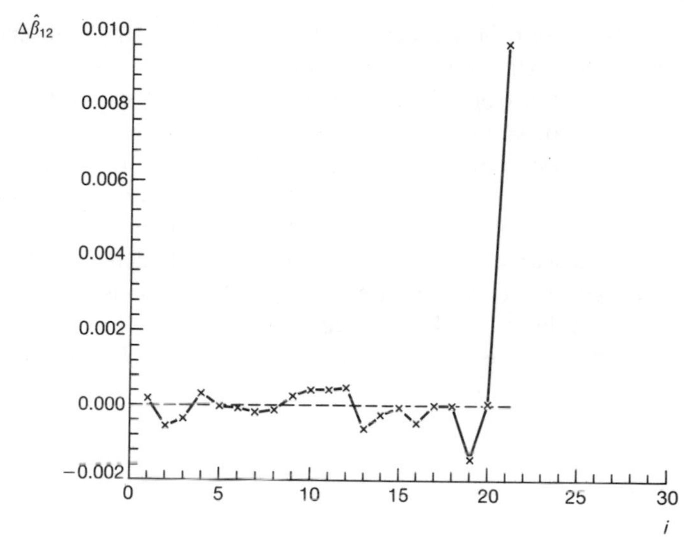


**Fig. 6.39** Example 10: Brownlee's stack loss data, second-order model, logged response: index plot of $\Delta\hat{\beta}_{12}$

index plot of $\Delta\hat{\beta}_2$. The effect of observation 21 is unmistakable, as it is in Fig. 6.39 for $\Delta\hat{\beta}_{12}$. The plot for $\Delta\hat{\beta}_{11}$ is virtually identical, while that for $\Delta\hat{\beta}_1$ is rather less dramatic, although observation 21 is still clearly the most influential observation.

The conclusion from these plots is that deletion of observation 21 would drastically change the values of the coefficients in the fitted model. For this example deletion of observation 21 also causes the form of the fitted model to change. If the second-order model with logged response is fitted to only the first 20 observations, the coefficients of all five carriers are no longer significant. The smallest absolute $t$ value is $-0.25$ for the coefficient of $x_1^2$. If the corresponding term is dropped from the model, the remaining terms are all significant, the smallest $t$ value now being $-3.60$ for the interaction $x_1x_2$. This change in the form of the fitted model is important because it will have physical significance. The interpretation of dramatic plots such as Figs. 6.38 and 6.39 is sometimes less clear, because the changes in the parameter estimates may virtually cancel out in the fitted model. For example, Figs. 6.38 and 6.39 show changes of opposite sign in a model with all explanatory variables positive. As a general rule diagnostic measures should be sought which are independent of parameterization as, for example, are residuals and Cook's distance measure.

For the five carrier model with 21 observations the residual sum of squares is 53.94. For the four-carrier model without $x_1^2$ the residual sum of squares, after deletion of observation 21, is 54.16. As Figs. 6.38 and 6.39 suggest, the presence of observation 21 is crucial to the model. This single observation is in fact causing one extra term to be fitted.

From the standpoint of statistical methodology this example shows the effect one observation can have on a fitted model and how the graphical tools of diagnostic regression analysis can be used to detect that influence. The physically important question is whether observation 21 is genuinely extending the range of observation in an important manner, or whether it is an outlier to be discarded. Further statistical analysis cannot elucidate this point, which requires new measurements of the response in this under-explored region of the space of the explanatory variables. The achievement of the diagnostic methods is to pinpoint the importance of this one observation. □

# 7
# Transformations for percentages and proportions

## 7.1 A general class of transformations

The power transformation so extensively studied in Chapter 6 is appropriate for responses constrained to be non-negative. But in some cases the response may also be subject to an upper constraint. For example the response might be the percentage yield of a chemical, the true value of which must lie between 0 and 100. Unless the values lie near zero, when the upper constraint is of little importance, a power transformation of the observations will not, in general, provide a simple model. If the percentages all lie near one hundred, for example in the range 90–100, then a transformation from the family $(100 - y)^{\lambda}$ will sometimes be useful. But for the complete range of values of $y$, a more general class of transformations needs to be studied.

In this chapter we develop analogues of the power transformation for data which are proportions, so that the true values are constrained to lie between 0 and 1. The closely related procedures for percentages also receive some mention. In this section the theory is developed for a general family of parametric transformations, which includes the results of the previous chapter on the power transformation. The central three sections of the chapter cover three different transformations which have been suggested for data between 0 and 1. These are respectively the folded power transformation and the proposals of Guerrero and Johnson (1982) and of Aranda-Ordaz (1981). In the concluding section the three transformations are compared and some comments are made on the purpose of the diagnostic analysis of data.

For the case of the general transformation we assume that there is some transformation of the data $u = u(\lambda)$ for which the transformed observations satisfy the linear model $E(U) = X\beta$ when the errors of observation will be at least approximately normally distributed with constant variance. The parameter $\lambda$ could, in principle, be a vector as it is in Chapter 9 when we come to consider power transformations with a shift in location. But, in this chapter, as it was in the preceding chapter, $\lambda$ remains a scalar.

The development of the normalized transformation and of the score statistic for testing hypotheses about $\lambda$ parallel those of Chapter 6. Thus, to correct the likelihood of the observations for the change of scale consequent upon transformation, the likelihood has to be multiplied by the Jacobian $J$

given by

$$J = \prod_{i=1}^{n} \left| \frac{\partial u_i}{\partial y_i} \right|.$$

If we let

$$\lambda v = |\partial u / \partial y|$$

an alternative form for the Jacobian is

$$J = \prod_{i=1}^{n} \lambda v_i.$$

For independent observations it is convenient, as we saw before, to work with the normalized variables

$$z(\lambda) = u / \lambda \dot{v} \tag{7.1.1}$$

where

$$\dot{v} = G(v) = \left( \prod_{i=1}^{n} v_i \right)^{1/n}, \tag{7.1.2}$$

the geometric mean of the $v_i$.

With this more general definition of $z(\lambda)$ the results of Section 6.2 go through unchanged. Thus $R(\lambda)$, the residual sum of squares of $z(\lambda)$, is given by (6.2.8) and the partially maximized log-likelihood by (6.2.7). The score statistic for testing hypotheses about $\lambda$ is given by (6.4.3). The only difference is that a more general definition of the constructed variable $w(\lambda)$ is required, namely,

$$w(\lambda) = \frac{\partial z}{\partial \lambda} = z \left( \frac{1}{u} \frac{\partial u}{\partial \lambda} - \frac{1}{\lambda} - \frac{Q}{n} \right) \tag{7.1.3}$$

where

$$Q = \sum_{i=1}^{n} \frac{1}{v_i} \frac{\partial v_i}{\partial \lambda}. \tag{7.1.4}$$

In many cases more compact forms can be found for $w(\lambda)$. The special case of the power transformation is recovered by putting

$$u(\lambda) = (y^{\lambda} - 1),$$

when, for example, the constructed variable $w(\lambda)$ given by (7.1.3) and (7.1.4) reduces to the form for $w_{\mathrm{p}}(\lambda)$ given by (6.4.5).

## 7.2 The folded power transformation

For the power transformation the limit of $y(\lambda)$ as $\lambda$ approaches zero is the log transformation. For proportions the analogue of the log is the logit, where

$$\operatorname{logit}(y) = \log \{ y/(1-y) \}.$$

In this chapter several parametric families of transformations for proportions will be described which yield the logit as $\lambda$ approaches zero. In many ways the simplest family with this property is obtained by setting

$$u(\lambda) = y^\lambda - (1 - y)^\lambda \qquad (0 \leqslant y \leqslant 1). \tag{7.2.1}$$

In the nomenclature of Mosteller and Tukey (1977, p. 92) this is called the folded power transformation. For values of $y$ near zero the transformation behaves like the power transformation $y^\lambda$, whereas for $y$ near one it behaves like $(1 - y)^\lambda$.

The normalized version of this transformation, analogous to (6.2.6), will be called $z_f(\lambda)$, to indicate that we are dealing with the folded power transformation. From the definition of the Jacobian of the transformation in Section 7.1

$$v_f(\lambda) = y^{\lambda-1} + (1 - y)^{\lambda-1}$$

and

$$G_f(\lambda) = G\{y^{\lambda-1} + (1 - y)^{\lambda-1}\} \tag{7.2.2}$$

where $G(\cdot)$ is the geometric mean function defined in (7.1.2). It therefore follows from (7.1.1) that the appropriate normalized response variable is

$$z_f(\lambda) = \begin{cases} \dfrac{y^\lambda - (1 - y)^\lambda}{\lambda G\{y^{\lambda-1} + (1 - y)^{\lambda-1}\}} & (\lambda \neq 0) \\ \log\{y/(1 - y)\}G\{y(1 - y)\} & (\lambda = 0). \end{cases} \tag{7.2.3}$$

Thus a multiple of the desired logit form is obtained for $\lambda = 0$. When $\lambda = 1$, the response $z_f(1) = y - \frac{1}{2}$. A slight computational disadvantage of (7.2.3) is that the geometric mean has to be calculated afresh from all $n$ observations for each value of $\lambda$. By comparison, the normalized power transformation (6.2.6) requires only the geometric mean of the observations, raised to the appropriate power.

To obtain the constructed variable $w_f(\lambda)$, let

$$p = y/(1 - y);$$

then $Q_f(\lambda)$, defined for the general transformation by (7.1.4), can be written as

$$Q_f(\lambda) = \sum_{i=1}^{n} \frac{p_i^{\lambda-1} \log y_i + \log(1 - y_i)}{p_i^{\lambda-1} + 1}. \tag{7.2.4}$$

The constructed variable $w_f(\lambda)$ is accordingly given by

$$w_f(\lambda) = \frac{(1 - y)^\lambda}{\lambda G_f(\lambda)} \{p^\lambda \log y - \log(1 - y) - (p^\lambda - 1)(1/\lambda + Q_f(\lambda)/n)\}.$$

For the special cases of $\lambda = 1$, no transformation, and $\lambda = 0$, the logit transformation, the constructed variable becomes

$$w_f(1) = \tfrac{1}{2}\{y \log y - (1-y)\log(1-y)\} - (y - \tfrac{1}{2})[1 + \tfrac{1}{2}\log G\{y(1-y)\}]$$

and

$$w_f(0) = G\{y(1-y)\}[\{\log^2 y - \log^2(1-y)\}/2 - (1/n)\log\{y/(1-y)\}Q_f(0)].$$

Given these definitions of $z_f(\lambda)$ and $w_f(\lambda)$ the analysis of the folded power transformation is entirely analogous to that of the power transformation of Chapter 6. For example, the partially maximized log-likelihood is a function of the residual sum of squares of $z_f(\lambda)$. To test hypotheses about the value of the transformation parameter, the likelihood ratio test can be used or, alternatively, the $F$ test on the residual sum of squares of $z_f(\lambda)$ can be employed. The score test $T_f(\lambda)$ is, by analogy with (6.4.3), given by

$$T_f(\lambda) = -\frac{z_f(\lambda)^T A w_f(\lambda)}{s_z\{w_f(\lambda)^T A w_f(\lambda)\}^{1/2}} = -\frac{\overset{*}{z}_f(\lambda)^T \overset{*}{w}_f(\lambda)}{s_z\{\overset{*}{w}_f(\lambda)^T \overset{*}{w}_f(\lambda)\}^{1/2}}. \tag{7.2.5}$$

The influence of individual observations on the evidence for a transformation can be assessed from an added variable plot of $\overset{*}{z}_f(\lambda)$ against $\overset{*}{w}_f(\lambda)$. A more formal procedure is given by looking at a half-normal plot of the modified Cook statistic $C_i(w_f)$ for the regression line in the added variable plot.

The procedures developed so far in this section are for transformations of responses $y$ which are proportions so that $0 \leqslant y \leqslant 1$. For the analysis of data which arise as percentages there are two possibilities. One is to divide all responses by one hundred and to use the analysis for proportions. If so desired, residual sums of squares can be rescaled through multiplication by $10^4$. The other, entirely equivalent, possibility is to modify the expressions for $z_f(\lambda)$ and $w_f(\lambda)$, as did Atkinson (1982a), so that the permitted range of the response is $0 \leqslant y \leqslant 100$. In this case one obtains

$$z_f(\lambda) = \frac{y^\lambda - (100-y)^\lambda}{\lambda G\{y^{\lambda-1} + (100-y)^{\lambda-1}\}}$$

with

$$z_f(1) = y - 50$$

and

$$z_f(0) = \frac{G\{y(100-y)\}}{100}\log\{y/(100-y)\}.$$

Similar modifications, ascertainable from dimensional considerations, need to be made to the expressions for the constructed variables.

**Example 5 (continued) Prater's gasoline data 2** The analysis of Prater's gasoline data in Section 4.3 led to the detection of a transcription error in Table 4.5. The analysis of the corrected data showed that the data were not multiple regression data with four explanatory variables, but that there was a nested structure. The final analysis in the section used a model with a different intercept for each of the ten crude oils identified, but with the same regression on the single explanatory variable $x_4$, the end point of the gasoline or petrol. The corrected data are given in this new form in Table 7.1.

The analysis of Section 4.3 concluded with the suggestion, based on the half-normal plots of Figs. 4.29 and 4.30, that a transformation might be appropriate. One possibility is to try the power transformation of Chapter 6. Although we have argued that special transformations are needed for proportions, the maximum yield in Table 7.1 is only 45.7 per cent for observation 4: the next largest response is 34.9 per cent for observation 14. It is therefore to be anticipated that, for these data, there will not be much difference between the power and folded power transformations.

Some results for the power transformation are given in Table 7.2. The score statistic $T_p(1)$ has the value $-3.31$ indicating that a transformation should be considered. The maximum likelihood estimate of $\lambda$ is given by $\hat{\lambda} = 0.6862$. At this point the residual sum of squares of $z(\lambda)$ has dropped to 47.50 from 74.13 for the untransformed data. Possible values of $\lambda$ include $\frac{1}{2}$, $\frac{2}{3}$ and $\frac{3}{4}$. But the log transformation, $\lambda = 0$, is firmly rejected by the data with $T_p(0) = 7.67$.

Closely related results for the folded power transformation are given in Table 7.3. In calculating these quantities the response was divided by 100 and use made of the transformation (7.2.3). The resulting residual sums of squares were therefore multiplied by $10^4$ for comparability with the results from the power transformation. The scaling of the observations obviously does not affect the values of $F$ statistics calculated from ratios of sums of squares, nor can it affect the values of the score statistics.

From Table 7.3 the score statistic $T_f(1)$ is seen to have the value $-3.57$, indicating that the data should be transformed. At the maximum likelihood estimate of $\lambda$, 0.4818, the residual sum of squares is 45.01, close to the minimum value of 47.50 for the power transformation. Possible simple values of $\lambda$ include $\frac{1}{3}$, $\frac{1}{2}$ and $\frac{2}{3}$, but 0 is again rejected by the data with $T_f(0) = 5.78$. The residual sum of squares for the logit transformation is 120.10, an appreciably better fit than the log transformation for which the corresponding value is 212.74. Otherwise there is little, apart from the value of $\hat{\lambda}$, to choose between the two transformations.

The similarity of results for the two transformations extends to measures of influence of the individual observations. Figure 7.1 shows the added variable

**Table 7.1** Example 5: Prater's gasoline data divided according to batch of crude oil†

| | | Crude oil | | | | |
|---|---|---|---|---|---|---|
| Observation number | Batch number | Gravity °API $x_1$ | Vapour pressure, (lbf/in$^2$) $x_2$ | ASTM 10% point $x_3$ | Gasoline end point, °F $x_4$ | Gasoline yield, per cent $y$ |
| 1 | 1 | 50.8 | 8.6 | 190 | 205 | 12.2 |
| 2 | | | | | 275 | 22.3 |
| 3 | | | | | 345 | 34.7 |
| 4 | | | | | 407 | 45.7 |
| 5 | 2 | 41.3 | 1.8 | 267 | 235 | 2.8 |
| 6 | | | | | 275 | 6.4 |
| 7 | | | | | 358 | 16.1 |
| 8 | | | | | 416 | 27.8 |
| 9 | 3 | 40.8 | 3.5 | 210 | 218 | 8.0 |
| 10 | | | | | 273 | 13.1 |
| 11 | | | | | 347 | 26.6 |
| 12 | 4 | 40.3 | 4.8 | 231 | 307 | 14.4 |
| 13 | | | | | 367 | 26.8 |
| 14 | | | | | 395 | 34.9 |
| 15 | 5 | 40.0 | 6.1 | 217 | 212 | 7.4 |
| 16 | | | | | 272 | 18.2 |
| 19 | | | | | 340 | 30.4 |
| 20 | 6 | 38.4 | 6.1 | 220 | 235 | 6.9 |
| 21 | | | | | 300 | 15.2 |
| 22 | | | | | 365 | 26.0 |
| 23 | | | | | 410 | 33.6 |
| 24 | 7 | 38.1 | 1.2 | 274 | 285 | 5.0 |
| 25 | | | | | 365 | 17.6 |
| 26 | | | | | 444 | 32.1 |
| 27 | 8 | 32.2 | 5.2 | 236 | 267 | 10.0 |
| 28 | | | | | 360 | 24.8 |
| 29 | | | | | 402 | 31.7 |
| 30 | 9 | 32.2 | 2.4 | 284 | 351 | 14.0 |
| 31 | | | | | 424 | 23.2 |
| 33 | 10 | 31.8 | 0.2 | 316 | 365 | 8.5 |
| 34 | | | | | 379 | 14.7 |
| 35 | | | | | 428 | 18.0 |

† The arbitrary batch numbers, taken from Daniel and Wood (1971, p. 237), are in order of decreasing $x_1$.

**Table 7.2** Example 5: Prater's gasoline data. Summary statistics for power transformations

| $\lambda$ | Score statistic $T_p(\lambda)$ | Residual sum of squares of $z(\lambda)$ | $C_{i\max}$ | $i_{\max}$ |
|---|---|---|---|---|
| 1 | −3.308 | 74.13 | 3.54 | 4 |
| 0.75 | — | 48.61 | — | — |
| 0.6862 | 0 | 47.50 | — | — |
| 0.6667 | — | 47.61 | — | — |
| 0 | 7.668 | 212.74 | 6.69 | 5 |

plot of $\overset{*}{z}_f(1)$ against $\overset{*}{w}_f(1)$. Observation 4 is most influential as assessed by the value of the modified Cook statistic. But the half-normal plot of these values, which is not shown, does not suggest that 2.70 is an unduly large value for the maximum of these 32 quantities. When observation 4 is deleted, both the details and the conclusions of the analysis remain virtually unchanged. For example $T_f(1)$ increases slightly in magnitude to $-3.79$, so that a transformation is still indicated. For the power transformation the added variable plot for the constructed variable $w_p(1)$ is practically indistinguishable from Fig. 7.1.

For both analyses the largest observation, the fourth, is most influential. When logs are taken the most influential observation, again in both cases, is observation 5 for which the response has the minimum value of 2.8 per cent. As this is the value of the response closest to zero, it is to be expected that its importance will increase when the observations are logged. □

**Table 7.3** Example 5: Prater's gasoline data. Summary statistics for folded power transformations

| $\lambda$ | Score statistic $T_f(\lambda)$ | Residual sum of squares of $z_f(\lambda)$ | $C_{i\max}$ | $i_{\max}$ |
|---|---|---|---|---|
| | | (a) all observations | | |
| 1 | −3.575 | 74.13 | 2.70 | 4 |
| 0.6667 | −1.614 | 50.76 | — | — |
| 0.5 | −0.174 | 45.08 | — | — |
| 0.4818 | 0 | 45.01 | — | — |
| 0.3333 | 1.539 | 50.25 | — | — |
| 0 | 5.780 | 120.10 | 4.08 | 5 |
| | | (b) observation 4 deleted | | |
| 1 | −3.786 | 71.72 | 2.25 | 14 |
| 0 | 4.804 | 90.63 | 2.88 | 5 |

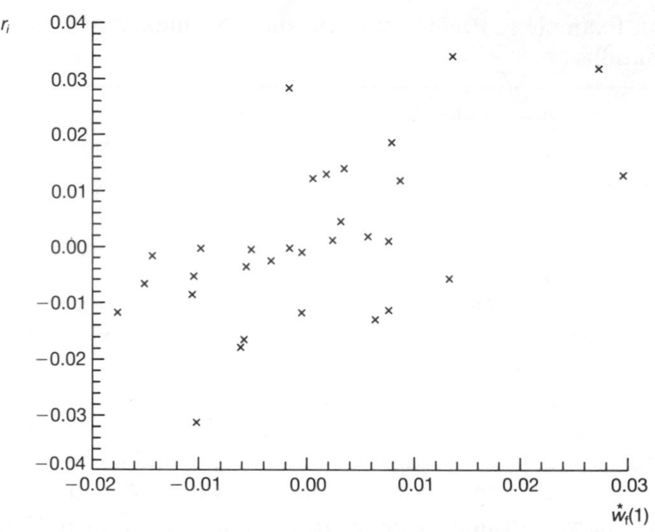


**Fig. 7.1** Example 5: Prater's gasoline data: added variable plot of the constructed variable $w_f(1)$ for the folded power transformation

In this example, due to the range of the data, there is little to choose between the power and folded power transformations. The former suggests analysis in terms of $y^{2/3}$, the latter that the appropriate response is $y^{0\cdot5} - (1 - y)^{0\cdot5}$. On grounds of simplicity, a square root transformation is slightly to be preferred to one in which the power is two thirds. There is however a serious drawback to the folded power transformation. Suppose that, once the model has been fitted, the predicted response on the transformed scale is $\hat{\mu}$. Then $\hat{y}$, the predicted response on the original scale, satisfies the relationship

$$\hat{y}^{\lambda} - (1 - \hat{y})^{\lambda} = \lambda\hat{\mu}. \tag{7.2.6}$$

Except for a few special values of $\lambda$, such as zero and one, analytical inversion of the transformation (7.2.6) is not possible. Thus to obtain predictions on the original scale of the observations the iterative solution of relationships like (7.2.6) is required.

This serious drawback is not necessarily an argument in favour of the power transformation. In the remainder of this chapter we consider two further transformations, both of which can be readily inverted and which both give the logit transformation for $\lambda = 0$. The two transformations were originally proposed for the analysis of binary data but can readily be adapted for the analysis of continuous proportions.

### 7.3 The Guerrero and Johnson transformation

To obtain an invertible family of transformations which includes the logit, Guerrero and Johnson (1982) suggested application of the Box and Cox power transformation not to $y$, but to the odds $p = y/(1-y)$. The transformation was employed in the analysis of binary data to model the relationship between probabilities and carriers. We use it to model responses which, like probabilities, are constrained to lie between zero and one.

In the notation of this chapter the proposal is to let

$$u_{GJ}(\lambda) = p^{\lambda} - 1 = \{y/(1-y)\}^{\lambda} - 1. \qquad (7.3.1)$$

The normalized transformation is found, in a manner similar to that of Section 7.2, by letting

$$v_{GJ}(\lambda) = p^{\lambda-1}(1+p)^2 \qquad \text{and} \qquad G_{GJ}(\lambda) = G\{y^{\lambda-1}/(1-y)^{\lambda+1}\}, \qquad (7.3.2)$$

where, as before, $G(\cdot)$ is the geometric mean function. From (7.3.2) the expressions for the standardized transformation are

$$z_{GJ}(\lambda) = \begin{cases} \{(y/(1-y))^{\lambda} - 1\}/\lambda G_{GJ}(\lambda) & (\lambda \neq 0) \\ \log\{y/(1-y)\}G\{y(1-y)\} & (\lambda = 0). \end{cases} \qquad (7.3.3)$$

The transformation thus yields the logit for $\lambda = 0$. But for $\lambda = 1$ we obtain

$$z_{GJ}(1) = \{(2y-1)/(1-y)\}G^2(1-y), \qquad (7.3.4)$$

where $G^2(1-y) = \{G(1-y)\}^2$. The family of transformations thus does not yield the untransformed data for $\lambda = 1$. In this the transformation is different from the power transformation and the folded power transformation for both of which $\lambda = 1$ corresponds to no transformation of the response.

The advantage of the transformation (7.3.3) is that it is readily inverted. For a given $\lambda$ the constants to allow for change of scale are immaterial and we can write

$$\{y/(1-y)\}^{\lambda} - 1 = \mu$$

which can be inverted straightforwardly to yield

$$y = \frac{(\mu+1)^{1/\lambda}}{1+(\mu+1)^{1/\lambda}}. \qquad (7.3.5)$$

Thus the fitted model can readily be used to provide predictions and confidence intervals on the original scale of the observations once these have been calculated on the transformed scale.

The score statistic for testing hypotheses about values of $\lambda$ is similar to that for the power transformation of Chapter 6, but with the odds $p$ replacing $y$. The sum $Q$ defined by (7.1.4) becomes

$$Q_{GJ} = \log G\{y/(1-y)\} = n \log \dot{p},$$

where $\dot{p}$, the geometric mean of the $p_i$, is the analogue of $\dot{y}$. The constructed variable in terms of $p$ is

$$w_{GJ}(\lambda) = \frac{1}{\lambda G_{GJ}(\lambda)} \{p^\lambda \log p - (p^\lambda - 1)(1/\lambda + \log \dot{p})\} \tag{7.3.6}$$

which, except for the scaling introduced by the geometric mean functions, is the constructed variable for the power transformation (6.4.5) with $p$ substituted for $y$. This scaling has, of course, no effect on the value of the score statistic, although it does affect the slope of the added variable plot.

As the constructed variable for the power transformation can be simplified, so can (7.3.6). For a model with a constant, since interest is only in the residual variable $\overset{*}{w}_{GJ}(\lambda)$, (7.3.6) becomes

$$w_{GJ}(\lambda) = \frac{p^\lambda}{\lambda G_{GJ}(\lambda)} \{\log (p/\dot{p}) - 1/\lambda\}$$

with special cases

$$w_{GJ}(1) = p\{\log (p/\dot{p}) - 1\}G^2(1 - y)$$

and

$$w_{GJ}(0) = \log p\{\log p - 2 \log \dot{p}\}G\{y(1 - y)\}. \tag{7.3.7}$$

The score statistic $T_{GJ}(\lambda)$ follows with a definition analogous to that of $T_f(\lambda)$ in (7.2.5).

**Example 5 (continued) Prater's gasoline data 3** Results from the application of the Guerrero–Johnson transformation to the gasoline data are given in Table 7.4. These results are close to those for the folded power transformation in Table 7.3. The point of common comparison is the logistic transformation, $\lambda = 0$, for which the residual sums of squares of $z_f(\lambda)$ and $z_{GJ}(\lambda)$ are both 120.10. For the Guerrero–Johnson transformation the maximum likelihood estimate of $\lambda$ is at $\hat{\lambda} = 0.3834$ for which the residual sum of squares is 42.05, compared with 45.01 for the folded transformation at its maximum likelihood estimate of $\lambda$. For the power transformation the corresponding value is 47.50. The results of Table 7.4 show that both 0 and 1 are rejected at the 5 per cent level as possible values of $\lambda$, whereas $\frac{1}{3}$ and $\frac{1}{2}$ are acceptable.

The value of $T_{GJ}(1)$ is $-9.78$, implying much stronger evidence of the need for a transformation than is given by the other two transformations. This is an indication that the straight response $y$ is much more acceptable than the odds $y/(1 - y)$. The strength of this evidence is shown in Fig. 7.2 by the definite slope of the added variable plot for $w_{GJ}(\lambda)$. It is clear from this figure that observation 4 has high leverage and so could have appreciable influence

**Table 7.4** Example 5: Prater's gasoline data. Summary statistics for Guerrero and Johnson transformation

| $\lambda$ | Score statistic $T_{GJ}(\lambda)$ | Residual sum of square of $z_{GJ}(\lambda)$ | $C_{i\,max}$ | $i_{max}$ |
|---|---|---|---|---|
| | | (a) all observations | | |
| 1 | −9.775 | 254.40 | 5.68 | 4 |
| 0.5 | −1.793 | 48.61 | — | — |
| 0.3834 | 0 | 42.05 | — | — |
| 0.3333 | 0.771 | 43.27 | — | — |
| 0 | 6.157 | 120.10 | 3.12 | 5 |
| | | (b) observation 4 deleted | | |
| 1 | −7.975 | 157.74 | 2.50 | 14 |
| 0 | 4.918 | 90.63 | 2.28 | 5 |

on the transformation. However, since the observation continues the linear relationship of the other observations, its deletion does not appreciably change inferences about the transformation. The half-normal plot of the modified Cook statistic for this transformation is shown in Fig. 7.3. Observation 4 is indeed appreciably more influential than the other observations. But the results of Table 7.4 show that, when observation 4 is

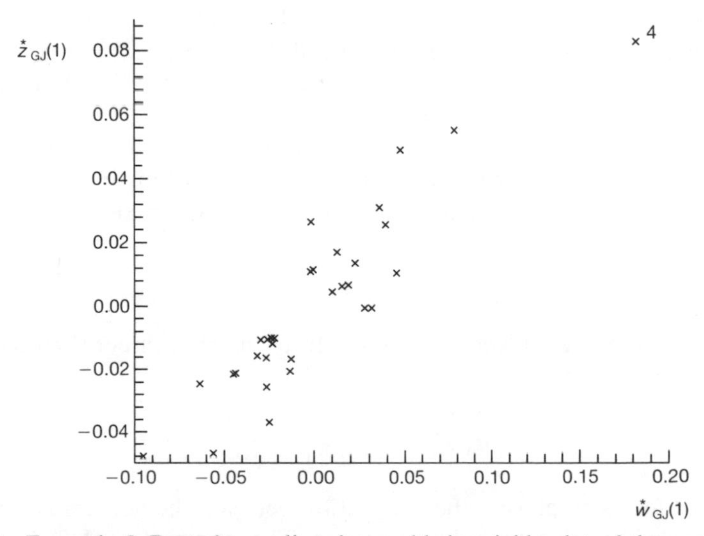


**Fig. 7.2** Example 5: Prater's gasoline data: added variable plot of the constructed variable $w_{GJ}(1)$ for the Guerrero and Johnson transformation

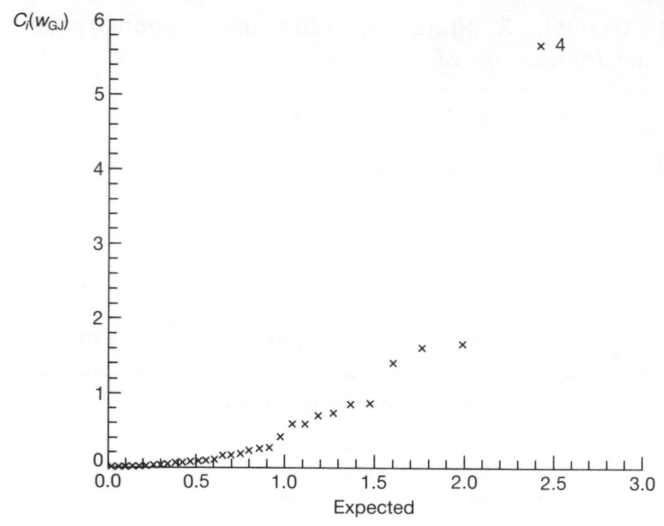


**Fig. 7.3** Example 5: Prater's gasoline data: half-normal plot of modified Cook statistic $C_i(w_{GJ})$ for Fig. 7.2

deleted, the hypotheses $\lambda = 0$ and $\lambda = 1$ are still firmly rejected by the data, although with some reduction in the absolute values of the statistics. □

## 7.4 The Aranda-Ordaz transformation

The last transformation for proportions which we shall consider is more complicated than those of earlier sections. But it has the advantages of being invertible and of yielding the untransformed data for $\lambda = 1$ as well as reducing to the logit transformation for $\lambda = 0$.

The suggestion of Aranda-Ordaz (1981) was, like that of Guerrero and Johnson, originally for probabilities in the analysis of binary data. For the analysis of proportions we ignore a scaling factor and take

$$u_{AO}(\lambda) = \frac{p^\lambda - 1}{p^\lambda + 1} \qquad (\lambda \neq 0), \tag{7.4.1}$$

where, as before, the odds $p = y/(1 - y)$. In terms of $y$, rather than the odds, (7.4.1) is

$$u_{AO}(\lambda) = \frac{y^\lambda - (1 - y)^\lambda}{y^\lambda + (1 - y)^\lambda}. \tag{7.4.2}$$

These two forms emphasize the connection between the new transformation and those of the two preceding sections. The numerator of (7.4.1) is the Guerrero and Johnson transformation, while the numerator of (7.4.2) yields the folded power transformation. The inverse of the transformation is found, by equating (7.4.1) to $\mu$, to satisfy

$$p^{\lambda} = (1 + \mu)/(1 - \mu),$$

which can readily be further re-arranged to give a slightly cumbersome expression for $y$ as a function of $\mu$.

A strange feature of the transformation given by (7.4.1) is that

$$u_{AO}(\lambda) = -u_{AO}(-\lambda), \tag{7.4.3}$$

which is easily shown since $p^{-\lambda} = 1/p^{\lambda}$. The partially maximized log-likelihood is therefore symmetrical about $\lambda = 0$. The inferential implications of this relationship become apparent when we derive the score statistic $T_{AO}(\lambda)$.

As before, we require the components of the Jacobian of the transformation. From the definitions in Section 7.1 we set

$$v_{AO}(\lambda) = \frac{2p^{\lambda-1}(1+p)^2}{(p^{\lambda}+1)^2} \quad \text{and} \quad G_{AO}(\lambda) = G\{v_{AO}(\lambda)\}. \tag{7.4.4}$$

The combination of (7.4.2) and (7.4.4) then yields the normalized transformation

$$z_{AO}(\lambda) = \begin{cases} \dfrac{p^{\lambda}-1}{\lambda(p^{\lambda}+1)}\{G_{AO}(\lambda)\}^{-1} & (\lambda \neq 0) \\ \log\{y/(1-y)\}G\{y(1-y)\} & (\lambda = 0). \end{cases} \tag{7.4.5}$$

For $\lambda = 1$ (7.4.5) reduces to $z_{AO}(1) = y - \frac{1}{2}$ which confirms that the transformation does, as claimed, combine both the logistic transformation and the untransformed data in a single parametric family.

To find constructed variables and the score statistic we require the sum $Q$ given by (7.1.4). For this transformation

$$Q_{AO}(\lambda) = \sum_{i=1}^{n} \{(1 - p_i^{\lambda})/(1 + p_i^{\lambda})\} \log p_i. \tag{7.4.6}$$

As with the other transformations, the special cases $\lambda = 0$ and $\lambda = 1$ are of particular interest. For these values (7.4.6) yields

$$Q_{AO}(1) = \sum_{i=1}^{n} (1 - 2y_i) \log y_i$$

and

$$Q_{AO}(0) = 0.$$

The general expression for the constructed variable, written as a function of $p$

or $y$ is lengthy, although easy to compute. In terms of $p$

$$w_{AO}(\lambda) = \frac{1}{\lambda(p^\lambda + 1)G_{AO}(\lambda)} \times \left[\frac{2p^\lambda \log p}{p^\lambda + 1} - (p^\lambda - 1)\{1/\lambda + Q_{AO}(\lambda)/n\}\right] \quad (\lambda \neq 0)$$

with $w_{AO}(0) = 0$.

The result that the constructed variable $w_{AO}(0)$ is identically zero is, at first sight, surprising. However, it reflects the symmetry of the transformation, and hence of the likelihood, about $\lambda = 0$. In the absence of a discontinuity in the likelihood and its derivatives, the symmetry implies that the slope must be zero. It is therefore not possible to use the score test to explore departures from the logistic transformation. The likelihood ratio test does however remain available.

**Example 5 (continued) Prater's gasoline data 4** The results from the analysis of the gasoline data using the Aranda-Ordaz transformation are similar to those using the other transformations. Values of residual sums of squares and score statistics are given in Table 7.5, from which it can be seen that there is evidence of the need for a transformation with $T_{AO}(1) = -3.44$. The maximum likelihood estimate of the transformation parameter is $\hat{\lambda} = 0.7286$ for which the residual sum of squares of $z(\lambda)$, after multiplication by $10^4$, is 46.10. This value is similar to those for the other transformations which range from 42.05 to 47.50. The approximate 95 per cent confidence interval for $\lambda$ includes $\frac{2}{3}$ and $\frac{3}{4}$ but not $\frac{1}{2}$.

**Table 7.5** Example 5: Prater's gasoline data. Summary statistics for the Aranda-Ordaz transformation

| $\lambda$ | Score statistic $T_{AO}(\lambda)$ | Residual sum of squares of $z_{AO}(\lambda)$ | $C_{i\,max}$ | $i_{max}$ |
|---|---|---|---|---|
| | | (a) all observations | | |
| 1 | −3.437 | 74.13 | 2.99 | 4 |
| 0.75 | −0.259 | 46.26 | — | — |
| 0.7286 | 0 | 46.10 | — | — |
| 0.6667 | 0.731 | 47.35 | — | — |
| 0.5 | 2.549 | 61.61 | — | — |
| 0 | 0 | 120.10 | — | — |
| | | (b) observation 4 deleted | | |
| 1 | −3.698 | 71.72 | 2.26 | 14 |
| 0 | 0 | 90.63 | — | — |

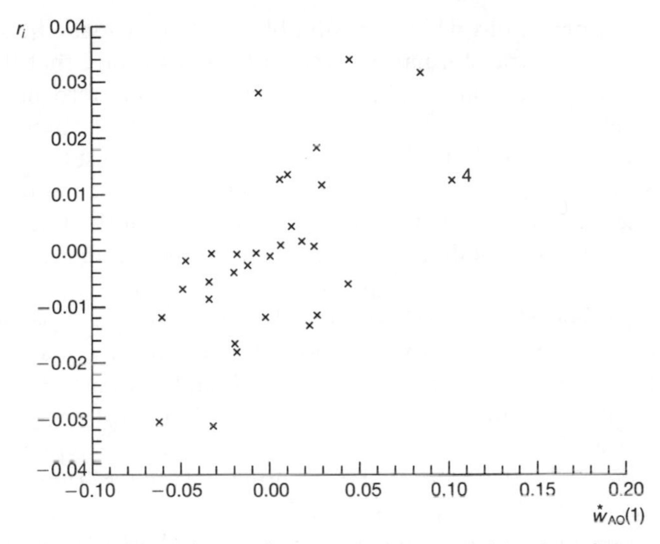


**Fig. 7.4** Example 5: Prater's gasoline data: added variable plot of the constructed variable $w_{AO}(1)$ for the Aranda-Ordaz transformation.

The most influential observation, when $\lambda = 1$, is again number 4. The added variable plot of $w_{AO}(1)$, Fig. 7.4, shows that the effect of deletion of observation 4 is unlikely to be great. In fact, deletion of this observation causes the score statistic to increase slightly in magnitude to $-3.70$. This slight increase in the evidence for a transformation does not affect the conclusions to be drawn from this analysis. □

## 7.5 Discussion

The results of this chapter indicate that Prater's data ought to be transformed, with values of the test statistic for the hypothesis of no transformation which are significant at least at the 1 per cent level. For the Guerrero and Johnson transformation a value of $\lambda$ near $\frac{1}{3}$ is indicated, whereas for the other transformations a value of $\lambda$ around $\frac{2}{3}$ is acceptable. This difference might be expected as the Guerrero and Johnson transformation is the only family for which $\lambda = 1$ does not correspond to the untransformed response. The question remains as to which of these transformations should be employed.

For proportions which range over the whole interval between 0 and 1 the power transformation is not appropriate. That it is so little to be distinguished from the other transformations in the present example is a reflection of the limited spread of the observations. As the discussion of influence has repeatedly emphasized, the maximum response is 45.7 per cent with 34.9 per cent the next largest observation.

Of the transformations which are suitable for proportions, the seemingly simple folded power transformation suffers the disadvantage that the inverse cannot be explicitly calculated. The transformation of Guerrero and Johnson, although readily inverted, fails to yield the untransformed observations for any value of $\lambda$. The Aranda-Ordaz transformation does not suffer from any of these drawbacks, being easy to invert and including both the untransformed data and the logit transformation. But the symmetry of the transformation about $\lambda = 0$ is a potential disadvantage which requires further investigation.

**Example 5 (concluded) Prater's gasoline data 5** As one comparison of the transformations, Fig. 7.5 shows a plot of the partially maximized log likelihood $L_{\max}(\lambda)$ as a function of $\lambda$ between 0 and 1. Figure 7.6 shows the related plot of the residual sum of squares $R(\lambda)$. For clarity of presentation the power transformation has not been included. The remaining three transformations all yield the logistic transformation for $\lambda = 0$ and so coincide at that point. The folded power transformation and the Aranda-Ordaz transformation also coincide at $\lambda = 1$. For $\lambda = \frac{1}{2}$ the folded power transformation and that due to Guerrero and Johnson seem similar although, as has been stressed in this chapter, they diverge as $\lambda$ increases towards 1. The behaviour of the Aranda-Ordaz transformation, however, is very different from these two for small $\lambda$. Because of the symmetry of $L_{\max}(\lambda)$ and $R(\lambda)$, the slope of the two functions is zero at $\lambda = 0$ which, as has already been stressed,

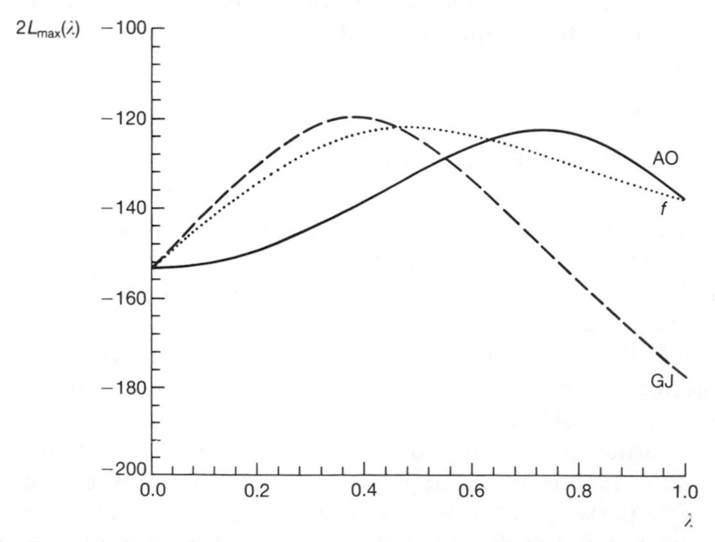


**Fig. 7.5** Example 5: Prater's gasoline data: comparison of log-likelihoods; $2L_{\max}(\lambda)$ as a function of $\lambda$ for three transformations; —— Aranda-Ordaz; ······ folded; --- Guerrero and Johnson

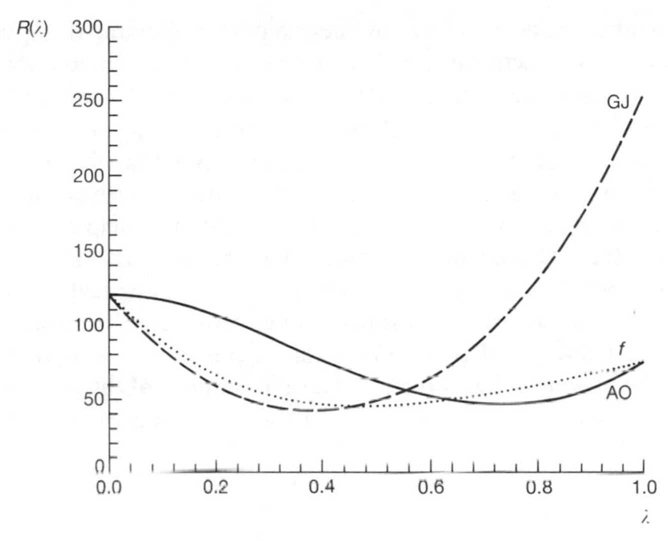


**Fig. 7.6** Example 5: Prater's gasoline data: residual sum of squares $R(\lambda)$ as a function of $\lambda$ for three transformations; Aranda-Ordaz; ······ folded; – – – Guerrero and Johnson

leads to the value of zero for the score statistic $T_{\mathrm{AO}}(0)$. For this example the functions have two stationary points and are not at all quadratic. It is therefore likely that asymptotic results for the distribution of $\hat{\lambda}$ will hold less well for the Aranda-Ordaz transformation in small samples than they will for the other transformations. In practice, if $\hat{\lambda}$ is near zero, the asymptotic results could, with advantage, be supported by a small-scale simulation to investigate the distribution of quantities of interest. □

With the numerical and graphical results of this chapter, our analysis of the gasoline data is complete. Steps of the analysis have included:

1. checks for the presence of outliers;
2. the detection and correction of a transcription error;
3. discovery of the nested structure of the data;
4. testing and estimation of the four transformations of the response and
5. checks on the transformations for the influence of individual observations.

A danger in a book like this which emphasizes diagnostic techniques is that these procedures can be taken to constitute a complete analysis of the data. It can hardly be over-emphasized that they do not. The primary interest must nearly always be in the structure of the means. Usually once outliers have been detected, influential observations identified and checked and, where necessary, a transformation established, the analysis moves on to the establishment of the correct form of linear model. Following this, predictions and associated confidence intervals can be established. There may be some iteration between these stages, as we saw in Chapter 6 where the addition of extra carriers was sometimes an alternative to transformation of the response. But in the analysis of the gasoline data the model is extremely simple and the process of building a model is straightforward. However, given more information on the ten different crude oils it would in principle be possible to examine contrasts between the crudes and perhaps to establish simple relationships between similar crudes. The purpose of statistical analysis is not to obtain random scatters of residuals and constructed variables together with straight half-normal plots. For didactic purposes the analyses in this book may be biased in that direction. But the purpose of the plots is to serve as indicators that one can now turn to the more important questions of structure.

# 8
# Other constructed variables

## 8.1 Constructed variables for power transformations

In the two preceding chapters constructed variables were derived for a variety of transformations. Values of test statistics coupled with plots of the related constructed variables were used to provide information about the influence of individual observations on the estimated transformation. In this section several constructed variables related to the power transformation are compared with the variable $w_p(\lambda)$. The most famous of these alternatives yields the test for one degree of freedom for non-additivity introduced by Tukey (1949). This, together with the other constructed variable to be investigated in detail, which is due to Andrews (1971), yields an exact test.

The remainder of the chapter describes a variety of techniques derived from the constructed variable for the power transformation. In Section 8.2 quick estimates of the transformation parameter are derived and calculated for a number of examples. In the next section transformation of the explanatory variables is discussed within the parametric framework originally described by Box and Tidwell (1962). The chapter concludes with methods for investigating transformations of both response and explanatory variables.

We begin with alternatives to the constructed variable for the power transformation $w_p(\lambda)$ given by (6.4.5). A disadvantage of the score test $T_p(\lambda)$ derived from this variable is that its distribution does not exactly follow a $t$ distribution. Even if the estimated variance $s_z^2$ is replaced by a known value, the distribution of the statistic is only asymptotically standard normal. Theoretical disadvantages of this are that, in a particular case, the null distribution of the test statistic may be skewed, it may be long or short tailed with a variance far from one, and it may not be centred on zero. The practical consequence is that interpretation of an observed numerical value of the test statistic has to be slightly hedged with doubt. We shall see, in an example at the end of this section, that a test of nominal size 5 per cent has a true size nearer 9 per cent. Thus evidence for a transformation would be found more often than was justified by the nominal significance level.

To overcome this difficulty and to provide an easily calculated test statistic, Andrews (1971) used the procedure suggested by Milliken and Graybill (1970) to define an exact test. The key result is that for the normal theory linear model the vectors of predicted values and of residuals are distributed independently of each other. It therefore follows that functions of the residuals will be distributed independently of the predicted values. Andrews, in effect, accordingly suggested replacement of $w(\lambda)$ in the definition of the test statistic (6.4.3) by a constructed variable in which all observed values are replaced by predictions from the fitted model. The resulting test statistic will therefore have exactly a $t$ distribution on $n - p - 1$ degrees of freedom.

The constructed variable for the power transformation (6.4.5) can be written as

$$w_p(\lambda) \propto y^\lambda \log y - (y^\lambda - 1)(1/\lambda + \log \dot{y}), \tag{8.1.1}$$

which ignores a constant divisor. The constructed variable for the exact test is found by replacing the observed values of $y$ in (8.1.1) by the predictions $\hat{y}$ which yields

$$w(\lambda) = \hat{y}^\lambda \log \hat{y} - (\hat{y}^\lambda - 1)(1/\lambda + \log \dot{y}). \tag{8.1.2}$$

The score statistic and added variable plots require the residual constructed variables

$$\overset{*}{w}(\lambda) = Aw(\lambda)$$

where, as in (5.2.5), the projection matrix $A = I - H$. Because interest is in residuals, the constructed variable (8.1.2) can be simplified. For fixed $\lambda$, fitting the model with response $z(\lambda)$ differs only by a scaling factor from fitting the model with response $y(\lambda)$. The predicted values and residuals from one model are a constant multiple of those from the other and the properties of the residuals are the same for the two models. Accordingly

$$\hat{y}^\lambda - 1 = H(y^\lambda - 1)$$

so that

$$A(\hat{y}^\lambda - 1) = AH(y^\lambda - 1) = 0.$$

It therefore follows from (8.1.2) that replacement of the observed values of $y$ by the predicted values leads, after consideration of the properties of the residual variable, to the constructed variable

$$w_A(\lambda) = \hat{y}^\lambda \log \hat{y} \tag{8.1.3}$$

proposed by Andrews (1971). A noteworthy feature of (8.1.3) is that the constructed variable does not depend on $\dot{y}$.

An alternative and informative derivation of this variable, and so of the related test statistic, which we shall call $T_A(\lambda)$, is given by Andrews. He starts from the model for $y(\lambda)$, that is from the model which does not include the Jacobian and so does not allow for the effect of the transformation on the scale of the observations. By analogy with (6.4.1) we expand the model in a

Taylor series about $\lambda_0$ to obtain

$$y(\lambda) \simeq y(\lambda_0) + (\lambda - \lambda_0)y'(\lambda_0) = X\beta,$$

where $y'(\lambda_0) = \partial y(\lambda)/\partial\lambda|_{\lambda=\lambda_0}$. The linearized model is thus

$$y(\lambda_0) = X\beta - (\lambda - \lambda_0)y'(\lambda_0) + \varepsilon \tag{8.1.4}$$

the direct analogue of (6.4.1). The constructed variable $y'(\lambda_0)$, similar to (8.1.1), is given by

$$w(\lambda) = y'(\lambda) = y^{\lambda}\log y/\lambda - (y^{\lambda} - 1)/\lambda^2. \tag{8.1.5}$$

Comparison of (8.1.5) with (8.1.1) shows that the exact version of this new constructed variable formed on replacement of $y$ with $\hat{y}$ will, apart from a scaling factor, be identical with Andrews' constructed variable (8.1.3).

This alternative derivation makes explicit the implication of the absence of $\dot{y}$ in (8.1.3). That is, regression on $w_A(\lambda)$ points towards the minimum of $S(\lambda)$ given by (6.2.4), that is towards the minimum of the residual sum of squares of $y(\lambda)$, rather than towards the minimum of the residual sum of squares of $z(\lambda)$. Whether, in a particular case, this is misleading remains to be seen. The test statistic based on $w_A(\lambda)$ is explicitly

$$T_A(\lambda) = \frac{y(\lambda)^T A w_A(\lambda)}{s_y(\lambda)\sqrt{\{w_A(\lambda)^T A w_A(\lambda)\}}}, \tag{8.1.6}$$

where $s_y(\lambda)$ is an estimate of the variance of $y(\lambda)$. Unlike $T_p(\lambda)$ this statistic has an exact $t$ distribution under the null hypothesis. Often the hypothesis of interest is $\lambda_0 = 1$, corresponding to no transformation, when the constructed variable is

$$w_A(1) = \hat{y}\log\hat{y}. \tag{8.1.7}$$

A famous related test derived from a constructed variable is Tukey's one degree of freedom for non-additivity. This provides a test for departures from an assumed linear model, but is not explicitly related to a parametric family of transformations. The test and related constructed variable may be derived either from the expression for $w_p(1)$ (6.4.8) or from the expression for $w_A(1)$ given above. If we start from

$$w_p(1) = y\{\log(y/\dot{y}) - 1\}, \tag{8.1.8}$$

expansion in a Taylor series about $\dot{y}$ yields the constructed variable

$$w_{TS}(1) = (y - \dot{y})^2/2\dot{y}. \tag{8.1.9}$$

A variable yielding an exact test is obtained upon replacement of $y$ by $\hat{y}$. If further $\dot{y}$ in (8.1.9) is replaced by $\bar{y}$, the resulting constructed variable

$$w_T(1) = (\hat{y} - \bar{y})^2/2\bar{y} \tag{8.1.10}$$

is the variable used in the calculation of Tukey's one degree of freedom for non-additivity. The same variable results from Taylor series expansion of $w_A(1)$ about $\bar{y}$, after allowance has been made for the dependence of the test statistic on residual variables.

The test for non-additivity is the exact $t$ test for regression on $w_T(1)$, directly analogous to (8.1.6). A simpler form for the constructed variable which gives the same value of the test statistic is obtained by ignoring the factor of $2\bar{y}$ in the denominator of (8.1.10). A further simplification for models which contain a constant is that the residual version of $(\hat{y} - \bar{y})^2$ is equivalent to regression on the residual of $\hat{y}^2$.

The three constructed variables $w_A(1)$, $w_{TS}(1)$ and $w_T(1)$ can, as has been shown, all be viewed as approximations to the constructed variable $w_p(1)$ for testing the hypothesis of no power transformation. Which of the four variables is to be preferred? Although ease of calculation was of importance even as recently as the introduction of Tukey's test, this feature is presumably by now only of marginal interest. The problem of preference for one test against another therefore comes down to the advantage of an exact test against any losses that are introduced as a result of the series of successive approximations to the score test. This test itself is, of course, only an approximation to the likelihood ratio test. There is, as we shall shortly see, some evidence that the use of exact tests and the related constructed variables does result in some loss of power. This may in particular be the case in replicated data, when the dependence of the variance within each cell on the mean value in the cell is one indication of the need to transform. With the score tests $T_p(1)$ and $T_{TS}(1)$, which are calculated from residuals and constructed variables which are direct functions of the individual observations, each observation contributes to both the residual $r$ and to $w$ in the numerator of the test statistic. But, for the exact tests, all observations in a cell have the same value of the constructed variable, since it is a function of $\hat{y}$, so that the variance within a cell will only affect the test statistic through the residuals.

The power transformation introduced in Chapter 6 was intended to achieve an additive model of simple structure, homogeneity of variance and also normality. The score statistics should therefore respond to the absence of any of these characteristics. Because of the close relationship which has been demonstrated between the score statistics and Tukey's test, it is to be expected that Tukey's test, which was originally introduced for lack of additivity in a two-way table, will also respond to other departures from an additive model. The simulation results of Yates (1972) do indeed show Tukey's non-additivity test responding to other features of the data, although in some cases with low power.

There are two small final theoretical points about Tukey's procedure before we illustrate the methods on an example. The first is that an alternative to the constructed variable (8.1.10) and the analysis of covariance procedure for calculating the score statistic is to perform the regression analysis twice, the second time with the inclusion of the extra carrier $\hat{y}^2$. Since virtually any regression program will permit operations on the vector of predicted values, Tukey's test requires a lower level of sophistication in software than that required by the other tests. The other point is that interpretation of the score statistic as a test for the significance of regression on a covariate which is a constructed variable generalizes and clarifies the results of Rojas (1973) on the relationship between Tukey's one degree of freedom and the analysis of covariance.

**Example 11 Box and Cox poison data 1** In Chapter 6 a textile example from Box and Cox (1964) was used to introduce the idea of transformation of the response. We now use the second of the two examples in their paper to compare the score test for transformations with the related exact tests. In this example the data are survival times of animals in a $3 \times 4$ factorial experiment, the factors being (a) three poisons and (b) four treatments. Each combination of the two factors is used for four animals, the allocation to animals being completely randomized.

The data are given in Table 8.1, where the response is survival time in units of 10 hours. We follow Box and Cox and fit an additive model without interactions in the two qualitative factors. The model therefore contains 6 parameters.

**Table 8.1** Example 11: Box and Cox poison data. Survival times in 10 hour units of animals in a $3 \times 4$ factorial experiment

| Poison | Treatment | | | |
|---|---|---|---|---|
| | A | B | C | D |
| I | 0.31 | 0.82 | 0.43 | 0.45 |
| | 0.45 | 1.10 | 0.45 | 0.71 |
| | 0.46 | 0.88 | 0.63 | 0.66 |
| | 0.43 | 0.72 | 0.76 | 0.62 |
| II | 0.36 | 0.92 | 0.44 | 0.56 |
| | 0.29 | 0.61 | 0.35 | 1.02 |
| | 0.40 | 0.49 | 0.31 | 0.71 |
| | 0.23 | 1.24 | 0.40 | 0.38 |
| III | 0.22 | 0.30 | 0.23 | 0.30 |
| | 0.21 | 0.37 | 0.25 | 0.36 |
| | 0.18 | 0.38 | 0.24 | 0.31 |
| | 0.23 | 0.29 | 0.22 | 0.33 |

**Table 8.2** Example 11: Box and Cox poison data. Comparison of score statistics from various constructed variables

| Constructed variable | Original data ($y_{20} = 0.23$) Score statistic | Maximum modified Cook statistic | | Modified data ($y_{20} = 0.13$) Score statistic | Maximum modified Cook statistic | |
|---|---|---|---|---|---|---|
| | | $i_{\max}$ | $C_{i\max}(w)$ | | $i_{\max}$ | $C_{i\max}(w)$ |
| $w_p(1)$ | −13.54 | 24 | 4.16 | −10.42 | 20 | 4.34 |
| $w_p(0)$ | −4.75 | 30 | 3.86 | −1.15 | 20 | 23.28 |
| $w_A(1)$ | −2.66 | 24 | 2.38 | −2.65 | 24 | 2.24 |
| $w_{TS}(1)$ | −13.30 | 24 | 9.45 | −11.52 | 24 | 6.64 |
| $w_T(1)$ | −2.65 | 24 | 2.23 | −2.62 | 24 | 2.07 |

Values of score statistics calculated from the set of constructed variables derived in this section are given in Table 8.2. It is clear from the value of −13.54 for $T_p(1)$ that the data ought to be transformed. This conclusion is to be expected with a non-negative response, in this case survival time, in which the range of the observations is 0.18 to 1.24. The value of the score statistic for the hypothesis $\lambda = 0$ is −4.75, so that the log transformation does not go far enough. The maximum likelihood estimate of $\lambda$ is −0.7502 with approximate 95 per cent confidence interval calculated from the likelihood ratio −1.13 to −0.37. The reciprocal transformation, $\lambda = -1$, is therefore indicated. The implication is thus that death rate, rather than survival time, has a simple additive structure.

This straightforward analysis parallels that of Box and Cox (1964). The

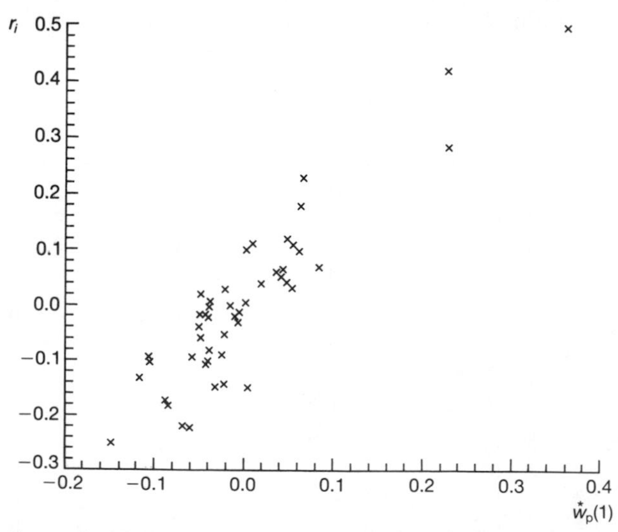


**Fig. 8.1** Example 11: Box and Cox poison data: added variable plot of $w_p(1)$

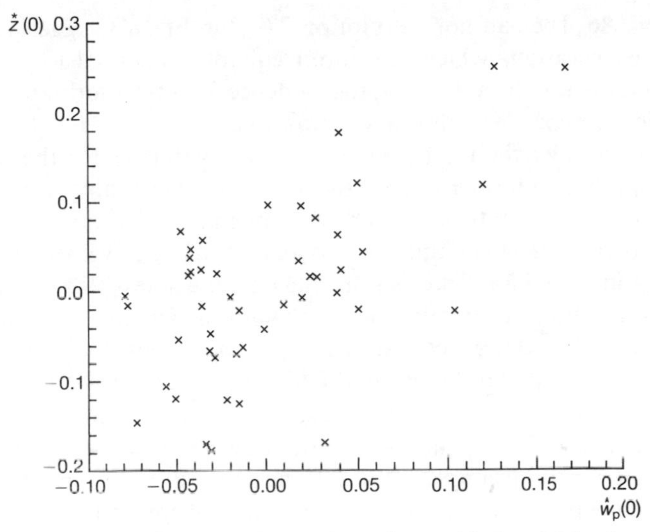


**Fig. 8.2** Example 11: Box and Cox poison data: added variable plot of $w_p(0)$

absence of observations unduly influential for the transformation can be checked by the graphical methods already developed. Figure 8.1 is the added variable plot for $w_p(1)$. This plot shows that the evidence for the transformation is spread throughout the data. Figure 8.2 is the same plot for $w_p(0)$, that is for evidence away from the log transformation. From Table 8.2 the most influential observation is number 30, for which the modified Cook statistic

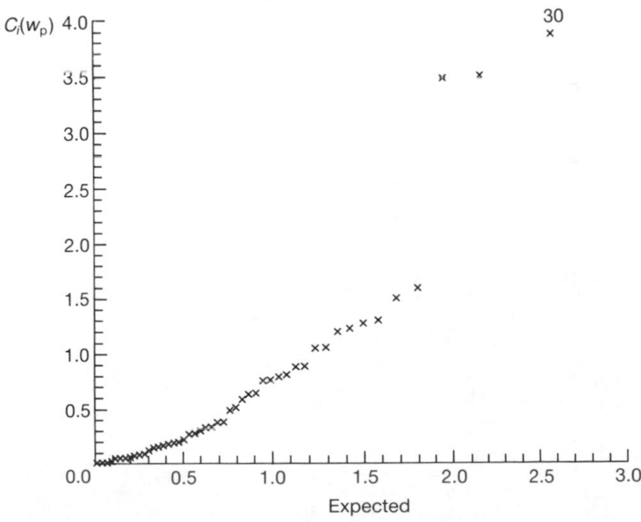


**Fig. 8.3** Example 11: Box and Cox poison data: half-normal plot of modified Cook statistic $C_i(w_p)$ for Fig. 8.2

$C_{30}(w_p) = 3.86$. The half-normal plot of $C_i(w_p)$ in Fig. 8.3 suggests that there are three observations which are about equally influential for this further transformation. But, from Fig. 8.2, the evidence for a transformation does not seem critically dependent on one observation.

We now consider the three new constructed variables for the hypothesis $\lambda = 1$. From the results of Table 8.2 the Taylor series variable, with a value of $-13.30$ for the score statistic $T_{TS}(1)$, gives much the same information as the constructed variable $w_p(1)$. But the two exact tests, with values of $-2.66$ for Andrews' and $-2.65$ for Tukey's tests, indicate much less strong evidence for the transformation. Two possible explanations for this, not necessarily exclusive, are that the exact tests are appreciably less powerful than the asymptotic tests and that the size of the likelihood ratio based test is far from its nominal value. We of course know that the exact tests must be of precisely their nominal size. The results of a simulation reported later in this section indicate that, for this example, although the nominal 5 per cent test $T_p(\lambda)$ has a size nearer 9 per cent, the main cause of the difference is loss of power.

It is informative to compare plots of the constructed variables. Figure 8.4 shows the added variable plot for Andrews' constructed variable $w_A(1)$. By comparison with Fig. 8.1 there is much more scatter in the plot and, if the strong visual effect of the upper edge of the plot is discounted, rather weak evidence for a transformation. The plot is clustered in fours on the value of the constructed variable because each of the four observations in a cell has the same value of $w_A(1)$. This effect has already been mentioned as one source

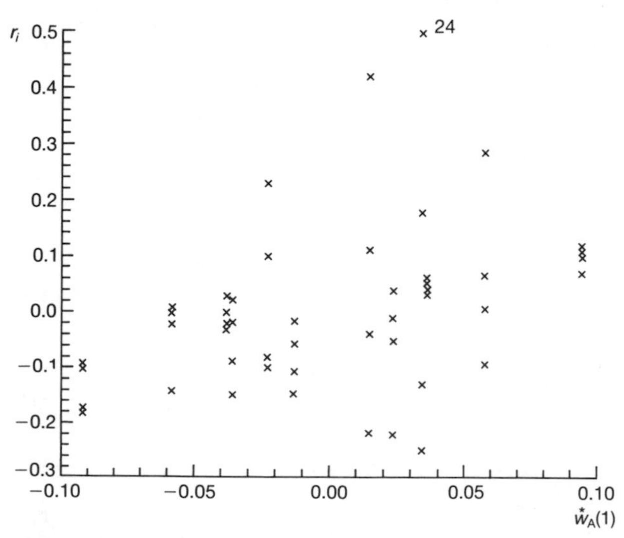


**Fig. 8.4** Example 11: Box and Cox poison data: added variable plot of Andrews' constructed variable $w_A(1)$

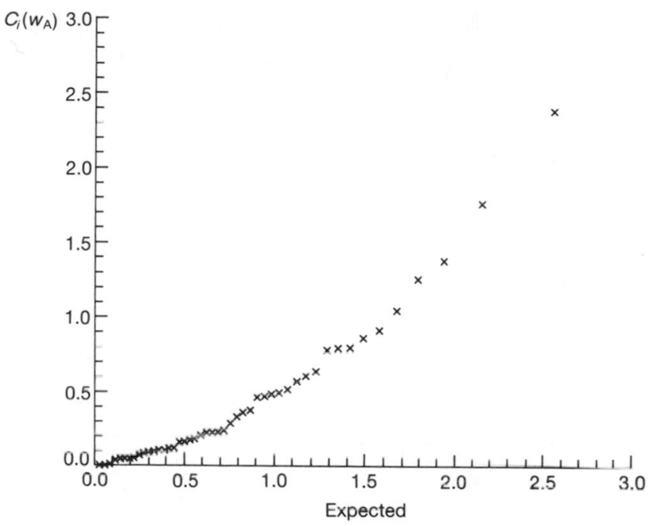


**Fig. 8.5** Example 11: Box and Cox poison data: half-normal plot of modified Cook statistic $C_i(w_A)$ for Fig. 8.4

of reduced power for the test using this variable. The half-normal plot of the modified Cook statistic derived from Fig. 8.4, which is shown in Fig. 8.5, is nearly a straight line. The relatively low value of the score statistic is therefore not due to any specific subset of the observations.

The added variable plot for the Taylor series variable $w_{TS}(1)$, shown in Fig.

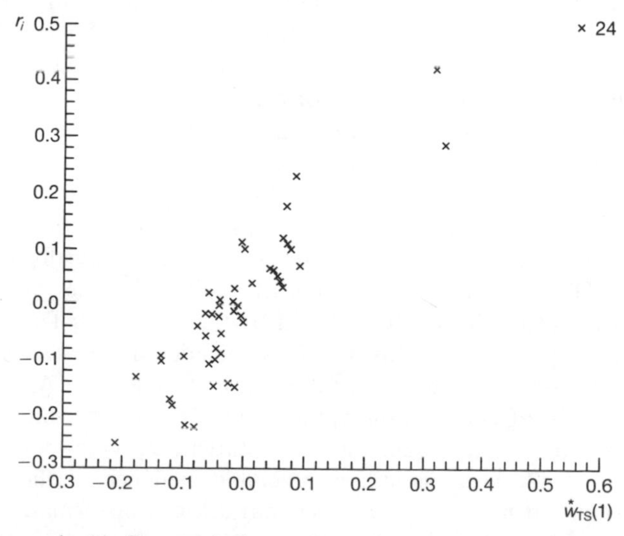


**Fig. 8.6** Example 11: Box and Cox poison data: added variable plot of Taylor series constructed variable $w_{TS}(1)$

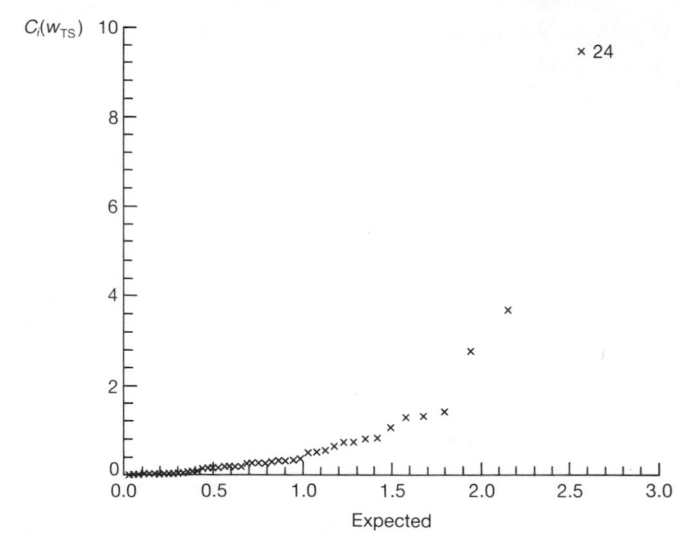


**Fig. 8.7** Example 11: Box and Cox poison data: half-normal plot of modified Cook statistic $C_i(w_{TS})$ for Fig. 8.6

8.6, is virtually indistinguishable from the plot for $w_p(1)$, Fig. 8.1. The half-normal plot of $C_i(w_{TS})$, Fig. 8.7, is markedly less straight than the corresponding plot for Andrews' variable in Fig. 8.4. It is, however, close to the plot for the constructed variable $w_p(1)$ which we do not reproduce.

Just as the plots for $w_{TS}(1)$ are virtually indistinguishable from those for $w_p(1)$, so the plots for Tukey's variable are virtually indistinguishable from those for Andrews' variable. Box (1980) gives the added variable plot for $w_T(1)$ for this example. The conclusion from these plots is that the step which leads to appreciable loss of power is not the approximation of $w_p(\lambda)$, but the introduction of the exact test. It was argued above that this loss is due to replacement of all values of $w$ within a cell for replicated data by a single value.

An advantage claimed by Andrews for his procedure is that it is robust, that is that the inference about $\lambda$ does not depend critically on a single observation. This effect is caused by the averaging of values of constructed variables within cells which is introduced by the exact test. To demonstrate the practical consequences of this Andrews considered the effect of changing the response for observation 20, that is Poison II, treatment A, from 0.23 to 0.13. As the results of Table 8.2 show, the effect of this change on the evidence for a transformation is to reduce the score statistic $T_p(1)$ in magnitude from $-13.54$ to $-10.42$. There is still appreciable evidence of the need for a transformation, but now $\hat{\lambda} = -0.15$, so that a log transformation might be considered, as opposed to the previous inverse transformation.

The effect of this change in the data on Andrews' procedure is much less apparent. For his exact test on the original data the 95 per cent confidence interval for $\lambda$ is $-1.18$ to 0.4 so that a log transformation would be considered and would be more plausible than the inverse transformation. The change in observation 20 alters Andrews' statistic only in the third figure and the suggested transformation is unchanged. Similar results are obtained by Carroll (1980) in a robust analysis of transformations which is considered at greater length in Chapter 12.

The changed data provide an example in which, due to the presence of a single outlier, methods based on a normal linear model give rather different answers from those obtained using robust methods, or methods claimed to be robust. However, the effect of the single altered observation can readily be detected using the diagnostic plots we have developed for transformations.

Figure 8.8 is the added variable plot of $w_p(1)$ when observation 20 has been altered. Comparison with Fig. 8.1 shows that the two plots are similar, both strongly indicating the need for a transformation, but in Fig. 8.8 observation 20 is revealed as slightly anomalous. For this altered data set the log transformation might be employed. As a check on this conclusion, Fig. 8.9 shows the added variable plot for $w_p(0)$. Observation 20 is now revealed as totally anomalous. The rest of the data indicate need for a further transformation, which information is almost totally negated by the single observation 20. The modified Cook statistic $C_{20}(w_p)$ is 20.38 which shows as extremely influential in the half-normal plot of Fig. 8.10. This is quite distinct from the plot of Fig. 8.3 for the unaltered data where there seemed to be three moderately influential observations.

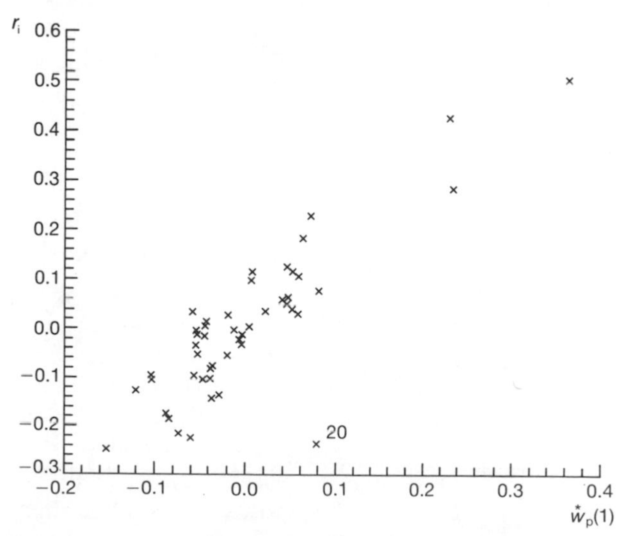


**Fig. 8.8** Example 11: Altered Box and Cox poison data with $y_{20} = 0.13$: added variable plot of $w_p(1)$

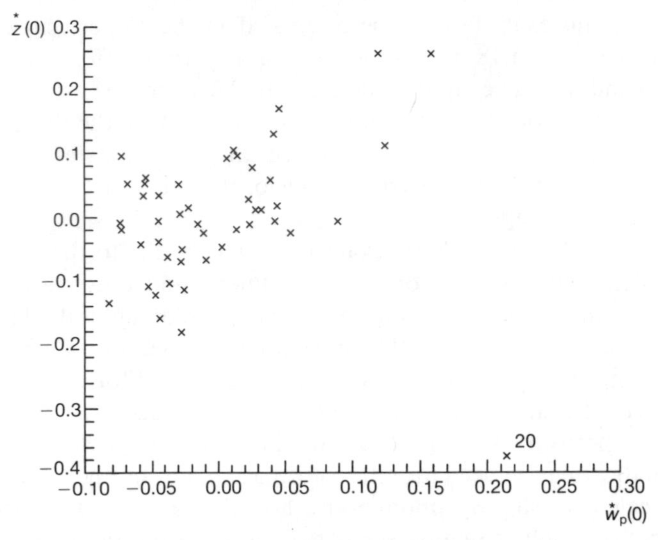


**Fig. 8.9** Example 11: Altered Box and Cox poison data with $y_{20} = 0.13$: added variable plot of $w_p(0)$

The results of Table 8.2 indicate that the alternative constructed variables leading to exact tests fail to diagnose the effect of observation 20. The Taylor's series variable gives a test which provides strong evidence of the need for a transformation, indeed slightly stronger than that from $T_p(1)$, with the

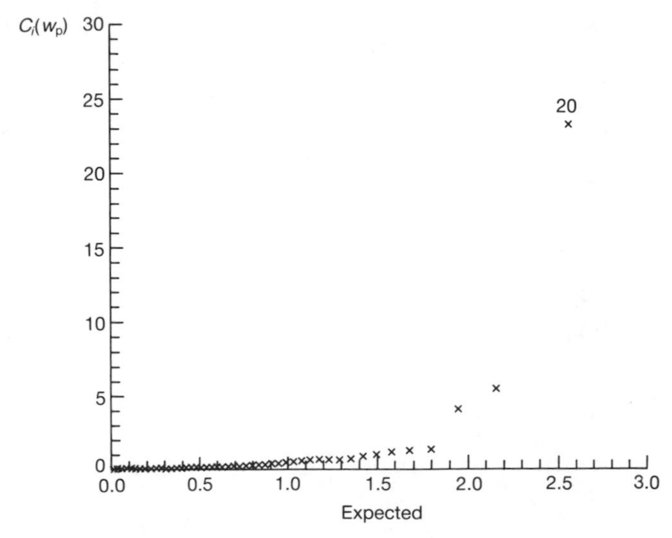


**Fig. 8.10** Example 11: Altered Box and Cox poison data with $y_{20} = 0.13$: half-normal plot of modified Cook statistic $C_i(w_p)$ for Fig. 8.9

maximum value of $C_i(w_{TS})$ a little greater than it is for $w_p(1)$. At $\lambda = 1$ the effect of the alteration in observation 20 on $w_{TS}(1)$ is slight, because of the strong information from all other observations of the need for a transformation. The effect of the alteration on the two variables leading to exact tests is negligible and cannot be discerned in the added variable plots. The effect is however just visible in the half-normal plot of $C_i(w)$ for Andrews' variable $w_A(1)$ in Fig. 8.11. Careful comparison with Fig. 8.5, which is the same plot for the unaltered data, shows that the third and fourth largest values are slightly changed.

The conclusion from the analysis of the altered data is that the robustness of the exact tests is achieved by lack of response to the information in the data. Use of diagnostic plots identifies the altered observation. There however remains the question of the comparative power of the two sets of tests for the unaltered data. To check the null distribution and power of the tests, Atkinson (1973) simulated 1000 samples of the 48 observations using the parameter estimates from the fitted model with $\lambda = -1$. Three tests were compared, Andrews' test $T_A(-1)$, the score test $T_p(-1)$ and the signed square root of the likelihood ratio test (6.3.1) which we shall call $T_l(-1)$: the sign is that of $T_p(-1)$, that is of the gradient of the likelihood at $\lambda = -1$. Asymptotically $T_p$ and $T_l$ should be equivalent and both should have standard normal distributions. For nominal 5 per cent tests we take the 95 per cent points of the $t$ distribution on forty-one degrees of freedom, that is $\pm 2.02$, as the limits for $T_p$ and $T_A$ and the normal limits of $+1.96$ for $T_l$.

Moments and cumulant ratios of the empirical distributions of the three

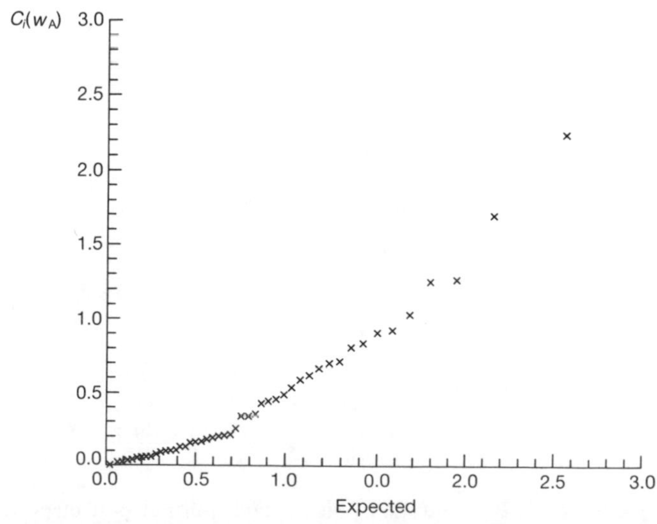


**Fig. 8.11** Example 11: Altered Box and Cox poison data with $y_{20} = 0.13$: half-normal plot of modified Cook statistic $C_i(w_A)$ for $\lambda_0 = 1$

**Table 8.3** Example 11: Box and Cox poison data. Moments and cumulant ratios of the empirical distributions of the three statistics for testing $\lambda = -1$. Data generated from $\lambda = -1$. One thousand simulations of 48 observations. Nominal test size 5 per cent

| | $T_A(-1)$ | $T_l(-1)$ | $T_p(-1)$ |
|---|---|---|---|
| Number significant† | | | |
| $+ (> C_{\alpha/2})$ | 23 | 47 | 64 |
| $- (\leqslant -C_{\alpha/2})$ | 26 | 13 | 25 |
| Total | 49 | 60 | 89 |
| Mean | 0.027 | 0.204 | 0.239 |
| Variance | 1.050 | 0.925 | 1.311 |
| $\gamma_1$ | −0.082 | 0.071 | 0.018 |
| $\gamma_2$ | 0.060 | 0.052 | 0.136 |

† For $T_l$, $C_{\alpha/2} = 1.96$; for the other statistics = 2.02.

statistics are given in Table 8.3. For an exactly normal distribution the skewness measure $\gamma_1$ should be zero, as should the kurtosis measure $\gamma_2$. The results show that Andrews' test has, as is to be expected, the required zero mean, unit variance and gives rise to a test with the nominal size of 5 per cent. Out of 1000 simulations, 49 values of the exact test statistic are significant at the 5 per cent level, a result which is almost too good. For the likelihood

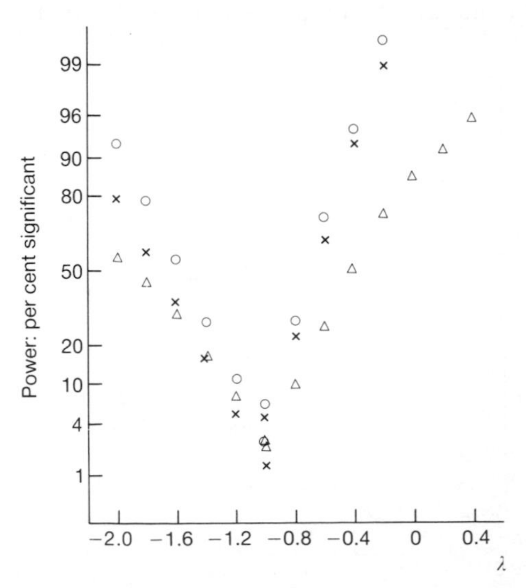


**Fig. 8.12** Example 11: Box and Cox poison data: normal plot of power of three statistics for testing $\lambda = -1$. Proportion of 1000 simulations significant at the 5 per cent level. $\triangle T_A$; $\times T_l$; $\bigcirc T_p$.

statistic and the score statistic the respective numbers significant are 60 and 89. Thus, in the null case when no transformation is required, both tests, and in particular the score test, will suggest the need for a transformation more often than the nominal size of the test indicates.

The results of the analysis of Example 11 in Table 8.2 strongly suggest that the score test is more powerful than the exact tests. This might, in part, be due to the greater size of the score test which is revealed by this null simulation. The simulation was therefore repeated for values of $\lambda$ from $-2$ to 0.4. In all cases the comparison was made for tests of the hypothesis $\lambda = -1$. The results are summarized in Table 8.4 as number of tests significant at the nominal 5 per cent level, divided between the two one-sided tests given by significantly positive and significantly negative values. A normal plot of this summary is shown in Fig. 8.12, where the slope of the plot indicates power, regardless of the size of the test: differences in size lead only to differences in intercept. The main conclusion from this plot is that the likelihood based tests $T_l$ and $T_p$ are similar in power and are both appreciably more powerful than Andrews' exact test $T_A$. Furthermore, $T_A$ is not only the least powerful in the neighbourhood of the null hypothesis, but becomes relatively less powerful as one moves away from $\lambda_0$.

The results of this simulation study complement the earlier results from the analysis of the example. To recapitulate, one conclusion is that the exact tests, that is Andrews' test and Tukey's test, are appreciably less powerful than the likelihood ratio or score tests. The results on the effect of Taylor's series

**Table 8.4** Example 11: Box and Cox poison data. Power of three statistics for testing $\lambda = -1$ when the data are generated from $\lambda = k$. Number out of 1000 simulations significant at the 5 per cent level

| | $T_A(-1)$ | | $T_l(-1)$ | | $T_p(-1)$ | |
|---|---|---|---|---|---|---|
| $k$ | + | − | + | − | + | − |
| −2.0 | 0 | 561 | 0 | 796 | 0 | 925 |
| −1.8 | 0 | 479 | 0 | 588 | 0 | 787 |
| −1.6 | 0 | 316 | 0 | 359 | 0 | 555 |
| −1.4 | 2 | 163 | 0 | 163 | 1 | 281 |
| −1.2 | 3 | 77 | 0 | 51 | 4 | 110 |
| −1.0 | 23 | 26 | 47 | 13 | 64 | 25 |
| −0.8 | 99 | 7 | 211 | 0 | 296 | 0 |
| −0.6 | 264 | 4 | 635 | 0 | 722 | 0 |
| −0.4 | 502 | 0 | 920 | 0 | 943 | 0 |
| −0.2 | 731 | 0 | 989 | 0 | 995 | 0 |
| 0 | 853 | 0 | 997 | 0 | 998 | 0 |
| 0.2 | 925 | 0 | 999 | 0 | 999 | 0 |
| 0.4 | 949 | 0 | 1000 | 0 | 1000 | 0 |

approximation indicate that the loss of power is caused by the substitution of fitted values for observations in the constructed variable. The constructed variables for exact tests also fail to respond to outlying values of the response. □

The advantages of the exact procedures are that the size of the tests is known and that calculation of the constructed variables, especially that for Tukey's test, is simple. The disadvantage, at least for Example 11, is that there is an enormous loss of power. Against this there could be set the fact that the score test, because of its too great size, will sometimes indicate that a transformation is needed when it is not. This effect is unlikely to be misleading as such indications will have to be checked by estimation of the transformation parameter and comparison of the transformed and untransformed models. If the size of a diagnostic test is unavoidably only approximate, it seems preferable that the test should err in the direction of indicating a transformation which is not required, but which can be checked, rather than failing altogether to indicate the need for a transformation.

These conclusions about the tests are based almost solely on the analysis of one example. One reason that was advanced for the observed difference was the effect of replicated observations on the exact tests. For an unreplicated example from Snedecor and Cochran (1967, p. 333), the added variable plots of $w_{\mathrm{p}}(1)$ and $w_{\mathrm{T}}(1)$ shown by Box (1980) reveal much less difference than was found here for Example 11. In the absence of replication there is no evidence about the dependence of within cell variance on the mean that can be suppressed by formation of the exact test. To investigate the relationship between replication and power of the exact tests, this section concludes with a further analysis of Example 8, an unreplicated factorial experiment.

**Example 8 (continued) John's $3^{4-1}$ Experiment 2** In the analysis of these experimental results in Chapter 6 it was shown that evidence for a transformation was generated solely by observation 11. The value of the score statistic $T_{\mathrm{p}}(1)$ was 2.98 and the modified Cook statistic $C_{11}(w_{\mathrm{p}})$ was 6.85. These results are reproduced in Table 8.5. Of the plots in Chapter 6, Fig. 6.11, the added variable plot of $w_{\mathrm{p}}(1)$, shows how observation 11 lies away from the rest of the observations and is the cause of the evidence for a transformation. The highly influential nature of this observation is emphasized by the half-normal plot of $C_i(w_{\mathrm{p}})$ (Fig. 6.12).

Table 8.5 presents in addition analogous quantities for the exact and Taylor series statistics. The results are in line with those from the analysis of the Box and Cox poison data. The Taylor series and score tests give similar evidence of the need for a transformation and both pinpoint observation 11 as most influential. On the other hand, the exact tests, with values near $-0.5$, give no indication of the need for a transformation. In both cases the most

**Table 8.5** Example 8: John's $3^{4-1}$ experiment. Comparison of score statistics from various constructed variables

| Constructed variable | Score statistic | Maximum modified Cook statistic | |
|---|---|---|---|
| | | $i_{max}$ | $C_{i\,max}(w)$ |
| | Original data | | |
| $w_p(1)$ | 2.98 | 11 | 6.85 |
| $w_p(0)$ | 3.47 | 11 | 6.56 |
| $w_A(1)$ | −0.48 | 25 | 3.09 |
| $w_{TS}(1)$ | 2.82 | 11 | 7.24 |
| $w_T(1)$ | −0.49 | 25 | 3.08 |
| | Observation 11 deleted | | |
| $w_p(1)$ | 0.57 | 5 | 2.56 |
| $w_p(0)$ | 0.90 | 5 | 2.29 |
| $w_A(1)$ | 0.14 | 25 | 1.57 |
| $w_{TS}(1)$ | 0.46 | 5 | 2.66 |
| $w_T(1)$ | 0.15 | 25 | 1.58 |
| | Observation 25 deleted | | |
| $w_p(1)$ | 4.07 | 11 | 5.62 |
| $w_p(0)$ | 4.51 | 11 | 5.41 |
| $w_A(1)$ | 0.38 | 11 | 1.58 |
| $w_{TS}(1)$ | 3.95 | 11 | 5.85 |
| $w_T(1)$ | 0.38 | 11 | 1.58 |

influential observation is number 25. The added variable plot for $w_A(1)$, Fig. 8.13, by comparison with Fig. 6.11, again shows observation 11 lying away from the rest of the data, but now in such a position as to have virtually no effect on the slope of the plot.

Table 8.5 also shows the effect of deleting either observation 11 or observation 25, which are the two observations indicated as most influential by the two sets of statistics. Deletion of observation 11 removes the evidence provided by $T_p(1)$ of the need for a transformation and causes a non-significant increase in the values of the exact tests to 0.14 and 0.15. Observation 25 is indicated by the exact tests as most influential, but deletion of this observation has no statistically significant effect on the values of the exact score statistics. This deletion does however increase the values of $T_p(1)$ and $T_{TS}(1)$ from 3 to 4.

The added variable plots of $w_A(1)$, Figs. 8.13 and 8.14, explain these results. Figure 8.13 shows that observation 25 is suggesting a transformation of the data in the opposite direction from observation 11. This differs from the corresponding plot for $w_p(1)$ (Fig. 6.11), where observation 25 is in a position

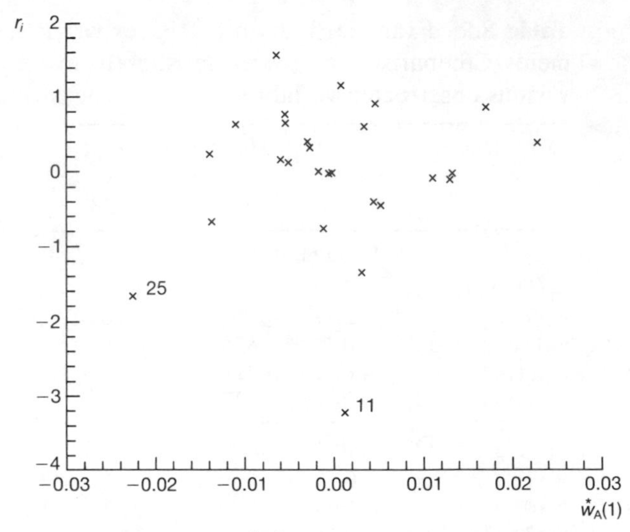


**Fig. 8.13** Example 8: John's $3^{4-1}$ experiment: added variable plot of $w_A(1)$

of low leverage and thus has little effect on the estimated transformation. However, observation 25 does have one of the largest residuals, thereby providing a considerable contribution to the estimate of variance. The result is that observation 25 decreases the value of $T_p(1)$. If the observation is deleted, the plot of $w_A(1)$ (Fig. 8.14) shows the effect of observation 11 against a horizontal cloud of points.

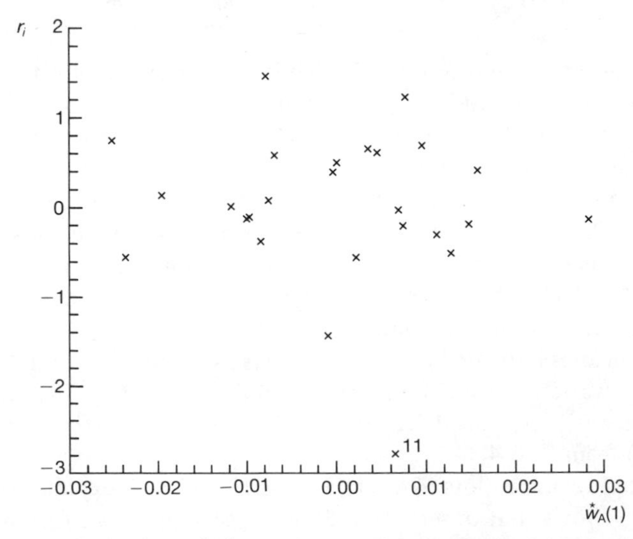


**Fig. 8.14** Example 8: John's $3^{4-1}$ experiment, observation 25 deleted: added variable plot of $w_A(1)$

Comparison of the added variable plots of $w_p(1)$ and $w_A(1)$ in Figs. 6.11 and 8.14 clearly shows the effect of the formation of the exact test on the influence of individual observations. The residuals, the vertical axis in the plots, are the same in both cases, but the effect of using $\hat{y}$ instead of $y$ in the constructed variable is to draw observation 11 towards zero on the horizontal axis, with a consequent reduction in influence. Due to the reduction in leverage, the large residual for observation 11 now merely contributes appreciably to the estimate of variance, but hardly to the numerator of the test statistic. □

These two examples illustrate the lack of power in the exact procedures. The corollary is that the claimed robustness is caused by a lack of response to individual observations. The conclusion is, both for replicated and unreplicated data, that there are considerable disadvantages in use of the exact tests and, it would seem, no compensating advantages. A brief discussion of more conventional procedures for the robust estimation of transformations is presented in Section 12.1.

## 8.2 Quick estimates of the transformation parameter

Calculation of the numerical value of the maximum likelihood estimate of the power transformation parameter $\lambda$ requires iteration. These iterations may take the form of a grid search or they may be relatively sophisticated like the variant of the method of false position described later in this section. In either case some computing is involved which it may not be possible to perform within the confines of a regression package. However, it is possible to obtain an approximation to the value of $\hat{\lambda}$ which requires only the quantities used in calculation of the score statistic $T_p(\lambda)$. This approximation can therefore be calculated in a simple manner. We first derive this quick estimate and then investigate its properties.

In Section 6.4 the score test was found from a Taylor series expansion of the true model about the value $\lambda_0$ which led to the linearized model (6.4.1). For this linearized model the coefficient of regression on the constructed variable $w = w_p(\lambda_0)$ is

$$\hat{\gamma} = w^T Az/w^T Aw.$$

To find a quick estimate we use the linearized model to give an approximation to $\hat{\lambda}$. If we denote the quick estimate by $\tilde{\lambda}$ it follows from the two ways of writing the linearized model in (6.4.1) that we can take as our estimate

$$\tilde{\lambda} - \lambda_0 = -\hat{\gamma}$$

or

$$\tilde{\lambda} = \lambda_0 - \hat{\gamma} = \lambda_0 - w^T Az/w^T Aw. \tag{8.2.1}$$

Comparison of (8.2.1) with the expression for $T_p(\lambda)$ (6.4.3) shows that no additional quantities are required for calculation of the quick estimate $\tilde{\lambda}$.

The most frequent use of $\tilde{\lambda}$ defined by (8.2.1) will be when $\lambda_0 = 1$. For this parameter value the quick estimate is calculated from the residuals of untransformed observations and from the residuals of the constructed variable (6.4.8), both of which should be readily available in a regression package. It is also possible to use (8.2.1) to provide an iterative scheme leading to the calculation of $\hat{\lambda}$. For the sake of clarity we change notation slightly and let $w_0 = w(\lambda_0)$, with the related definition for $z_0$. With this convention (8.2.1) can be rewritten as

$$\tilde{\lambda}_1 = \tilde{\lambda}_0 - w_0^{\mathrm{T}} A z_0 / w_0^{\mathrm{T}} A w_0,$$

when the iteration is

$$\tilde{\lambda}_{k+1} = \tilde{\lambda}_k - w_k^{\mathrm{T}} A z_k / w_k^{\mathrm{T}} A w_k. \tag{8.2.2}$$

A disadvantage of the sequence of estimates given by (8.2.2) is that, after the first iteration, the calculations require the general constructed variable $w(\lambda)$ rather than the simpler $w(1)$. The computational advantage of the quick estimate is then lost. In addition, as we shall see in some examples, the sequence of estimates converges only slowly while in others it diverges. We therefore consider an alternative iteration to the maximum likelihood estimate which involves the same level of calculation as (8.2.2), but which is guaranteed to converge.

The maximum likelihood estimate of $\lambda$ is the value satisfying

$$T_p(\hat{\lambda}) = 0. \tag{8.2.3}$$

The sequence of estimates (8.2.2) corresponds to a Newton method for solving (8.2.3). Such methods are known to have poor properties: they may for example diverge. As an alternative method with better numerical properties we use the method of false position applied to the solution of (8.2.3). Given two values of $\lambda$ at which the values of $T_p(\lambda)$ have opposite signs, a third value of $\lambda$ for which $T_p(\lambda)$ should be zero is found by linear interpolation. If the magnitude of $T_p(\lambda)$ at this new value of $\lambda$ is not sufficiently small, the process is repeated with the two values of $T_p(\lambda)$ which are smallest in magnitude and of opposite sign. The iteration continues until a sufficiently small absolute value of $T_p(\lambda)$ is obtained. The method properly starts with a grid search to locate two values of $\lambda$ bracketing the solution to (8.2.3). However, for this problem, we know that the solution usually lies between 1 and $-1$. Experience shows that it is, in fact, satisfactory to start the method of false position with the values 1 and $\tilde{\lambda}$. With this alternative scheme the estimate after the first iteration is

$$\lambda^+ = \frac{T_p(\tilde{\lambda}) - \tilde{\lambda} T_p(1)}{T_p(\tilde{\lambda}) - T_p(1)}. \tag{8.2.4}$$

Because of the different computational requirements of the two estimates, the estimate given by (8.2.4) is not directly comparable with $\tilde{\lambda}$: although calculation of $\tilde{\lambda}$ only requires quantities calculated at $\lambda_0$, the calculation of $\lambda^+$ requires in addition calculation of the score statistic, and so of the constructed variables, at $\tilde{\lambda}$. One step of the false position iteration is given by

$$\tilde{\lambda}_{k+1} = \frac{\tilde{\lambda}_{k-1} T_p(\tilde{\lambda}_k) - \tilde{\lambda}_k T_p(\tilde{\lambda}_{k-1})}{T_p(\tilde{\lambda}_k) - T_p(\tilde{\lambda}_{k-1})}. \tag{8.2.5}$$

In (8.2.5) successive values of $\tilde{\lambda}_k$ and $\tilde{\lambda}_{k-1}$ are chosen to give values of $T_p(\lambda)$ with opposite signs and the smallest absolute values among the three candidate points.

To compare the convergence of the iterative methods (8.2.2) and (8.2.5) a count was made of the number of evaluations of $T_p(\lambda)$ required to achieve the condition $|T_p(\lambda)| < 10^{-4}$. The numerical results of applying the two algorithms to some of the examples of the previous chapters are collected together in Table 8.6. The picture which emerges from these results is a little blurred. For the Minitab tree data, Example 6, the quick estimates $\tilde{\lambda}$ and $\lambda^+$ are both satisfactory and both iterations converge rapidly. The same general remarks hold for the gasoline data, Example 5. But for the other three examples in the table the two methods do not give similar results. For all examples the method of false position provides a good first estimate and requires no more than five evaluations of $T_p(\lambda)$ for convergence. But the iterative schemes based on the quick estimate either converge so slowly as to be useless or diverge, which is worse.

The conclusion is that if it is required to iterate to the maximum likelihood estimate $\hat{\lambda}$ an effective scheme is provided by the method of false position. These iterations require calculation of the constructed variables $w(\lambda)$ and the transformed response $z(\lambda)$ for general $\lambda$. In all cases except one, Example 8, John's $3^{4-1}$ experiment, the quick estimate $\tilde{\lambda}$ provides a sufficient indication of $\hat{\lambda}$ to suggest a satisfactory simple transformation. It is noteworthy that this one example is the only one in which the value of $\hat{\lambda}$ lies far outside the range $-1$ to 1: the values of both $\hat{\lambda}$ and $\tilde{\lambda}$ are a warning that there is something odd about the data. In some cases the information provided by the value of $\tilde{\lambda}$ about suitable scales for statistical analysis might suffice, particularly if interval estimates for $\lambda$ are not required. Alternatively, if computational simplicity is important and the value of $\hat{\lambda}$ is required, the value of $\tilde{\lambda}$ could be used as the starting point of a grid search for the minimum value of the residual sum of squares $R(\lambda)$.

The discussion and examples of this section have been solely in terms of the power transformation of the response. However, similar quick estimates can be obtained for other parametric families of transformations. In Section 8.4 an example is given of quick estimates for transformations of the explanatory

**Table 8.6** Quick estimates of the transformation parameters for five examples

| Model | Residual sum of squares $\lambda = 1$ | $\lambda = \hat{\lambda}$ | Score statistic $T_p(1)$ | Quick estimates of transformation parameter $\tilde{\lambda}_1$ | $\tilde{\lambda}_2$ | Maximum likelihood estimate $\hat{\lambda}$ | Number of iterations for $\lvert T_p(\lambda)\rvert < 10^{-4}$ | False position estimate $\lambda^+$ | Number of iterations for $\lvert T_p(\lambda)\rvert < 10^{-4}$ |
|---|---|---|---|---|---|---|---|---|---|
| Example 5 Prater's gasoline data | | | | | | | | | |
| Nested | 74.13 | 45.01 | −3.57 | 0.1538 | 0.4414 | 0.4818 | 4 | 0.5837 | 5 |
| Nested − observation 4 | 71.72 | 42.09 | −3.79 | −0.0544 | 0.3396 | 0.4232 | 5 | 0.5729 | 6 |
| Example 6 Minitab tree data | | | | | | | | | |
| $x_1, x_2$ | 421.9 | 135.4 | −7.41 | 0.3943 | 0.3064 | 0.3066 | 3 | 0.3117 | 4 |
| $x_1, x_2$ − observation 31 | 328.8 | 124.8 | −6.84 | 0.2516 | 0.2333 | 0.2342 | 4 | 0.2349 | 3 |
| $x_1, x_2, x_1^2$ | 186.0 | 126.3 | −3.18 | 0.1340 | −0.1534 | −0.0663 | 9 | −0.1830 | 5 |
| $\log x_1, \log x_2$ | 843.0 | 126.7 | −11.40 | 0.1265 | −0.0683 | −0.0672 | 4 | −0.0680 | 3 |
| Example 8 John's $3^{4-1}$ experiment | | | | | | | | | |
| All interactions save $D$ | 24.10 | 21.08 | 2.98 | 22.55 | 5.31 | 8.27 | ∞ | 9.1768 | 5 |
| All interactions save $D$ − observation 11 | 8.53 | 8.45 | 0.57 | 5.84 | −2.52 | 2.75 | ∞ | 2.7432 | 3 |
| Example 9 John's confounded $2^5$ experiment | | | | | | | | | |
| First order | 80.06 | 27.31 | −9.67 | 0.0390 | −0.1711 | −0.0818 | 42 | −0.0538 | 4 |
| First order − observation 3 | 80.04 | 23.24 | −11.51 | −0.0226 | −0.3276 | −0.1884 | > 50 | −0.1428 | 4 |
| Example 10 Brownlee's stack loss data | | | | | | | | | |
| $x_1, x_2, x_3$ | 178.8 | 114.8 | −3.29 | 0.3020 | 0.2960 | 0.2971 | 5 | 0.2976 | 3 |
| $x_1, x_2, x_3$ − observation 21 | 105.6 | 75.8 | −2.53 | 0.4993 | 0.4792 | 0.4810 | 4 | 0.4822 | 3 |
| $x_1, x_2, x_1x_2, x_1^2$ | 120.8 | 61.8 | −5.73 | −0.5488 | −0.0858 | −0.2905 | > 50 | −0.3218 | 4 |

variables. Quick estimates can also be derived from any of the constructed variables of Section 8.1. In calculating the estimates scaling factors such as $2\bar{y}$ in the denominator of (8.1.10) are important, irrelevant though they are in the calculation of the score statistics.

## 8.3 Transformation of the explanatory variables

In all except one of the examples discussed so far, transformations have been confined to the response variable $y$. The exception is the analysis of the Minitab tree data in Section 6.5 where two models were fitted which involved transformations of the explanatory variable. One was the model for the volume of a cone in which the single carrier was of the form $x_1^2 x_2$ and the other was a model with logged carriers. In the remaining two sections of this chapter the discussion centres on new carriers formed by the application of parametric families of transformations to the explanatory variables. The approach, which is close to that of Box and Tidwell (1962), leads to constructed variables and quick estimates of the transformation parameters which are conceptually similar to those already derived for power transformations of the response.

In Section 8.4 we consider simultaneous transformation of both response and explanatory variables. In this section transformation of only the explanatory variables is of interest, procedures for which are simpler than those which also involve the response. Transformation of the response is more complicated because, as was shown in Section 6.2, it is necessary to include the Jacobian of the transformation in the expression for the log-likelihood (6.2.3). Allowance is thus made for the effect of the transformation on the scale of the response variable $y(\lambda)$. But with transformation of explanatory variables, direct comparison can be made between the residual sums of squares of various models. If transformation of only one explanatory variable, $x_k$, is of interest, the parametric family of models is

$$E(Y) = \sum_{j \neq k} \beta_j x_j + \beta_k x_k^{\lambda}, \tag{8.3.1}$$

where, for brevity, the constant term, if there is one, has been included in the summation. The standard model corresponds to setting $\lambda = 1$ in (8.3.1).

For the moment we shall work with the assumption that only one variable is a candidate for a power transformation. The maximum likelihood estimate of $\lambda$ under normal theory assumptions can be found in a straightforward manner by fitting models with various values of $\lambda$ and finding the value for which the residual sum of squares is a minimum. The score test for evidence of this power transformation and a quick estimate of the transformation parameter come from Taylor series expansion of (8.3.1) about $\lambda_0$ which yields the linearized model

$$E(Y) \simeq \sum_{j \neq k} \beta_j x_j + \beta_k x_k^{\lambda_0} + \beta_k(\lambda - \lambda_0) x_k^{\lambda_0} \log x_k. \tag{8.3.2}$$

The test of the hypothesis $\lambda = \lambda_0$ in this linearized model is identical to testing for regression on the constructed variable $x_k^{\lambda_0} \log x_k$ when a term in $x_k^{\lambda_0}$ is already in the model.

Often, interest is in the hypothesis of no transformation, that is $\lambda_0 = 1$, when the linearized model (8.3.2) can be rewritten as

$$E(Y) = \sum_{j=1}^{p} \beta_j x_j + (\lambda - 1)\beta_k x_k \log x_k. \tag{8.3.3}$$

Absence of regression on the constructed variable $x_k \log x_k$ is an indication that a power transformation of $x_k$ is not needed. The influence of individual observations on the evidence for a transformation can be inspected through an added variable plot. A more formal assessment of influence comes from the half-normal plot of the modified Cook statistic derived from the added variable plot.

A quick estimate of the transformation parameter can be obtained from (8.3.3) but the factor of $\beta_k$ must be included in the constructed variable or allowed for in the parameter estimate. This scaling, which is irrelevant for the score statistic, is essential to parameter estimation in just the same way as the factor of $1/2\bar{y}$ was for Tukey's one degree of freedom for non-additivity in Section 8.2. The value of the parameter is, of course, usually unknown. To provide an operational procedure the value is replaced by the estimate $\hat{\beta}_k$ from the model without transformation. This procedure leads to the constructed variable

$$w_{\text{BT}}(1) = \hat{\beta}_k x_k \log x_k \tag{8.3.4}$$

effectively introduced by Box and Tidwell (1962). We shall denote by $T_{\text{BT}}(1)$ the score test for the significance of regression on (8.3.4) or, equivalently, that for regression on $x_k \log x_k$.

Derivation of the quick estimate of the transformation parameter is analogous to that for transformation of the response in Section 8.2, except for a change of sign. Instead of (8.2.1) the general case now yields

$$\hat{\gamma} = (\tilde{\lambda} - \lambda_0)$$

where $\hat{\gamma}$ is the coefficient of regression on $w_{\text{BT}}(\lambda_0)$ found either by fitting (8.3.3) or equivalently from the analogue of (6.4.2) which expresses the regression coefficient in terms of residuals from regression on all other carriers. The quick estimate for the special case $\lambda_0 = 1$ is therefore

$$\tilde{\lambda} = 1 + \hat{\gamma}. \tag{8.3.5}$$

In the derivation of (8.3.5) we use the constructed variable (8.3.4) which includes the estimate $\hat{\beta}_k$. If, instead, the constructed variable is taken as $x_k \log x_k$, the form for $\tilde{\lambda}$ obtained by Box and Tidwell results, namely

$$\tilde{\lambda} = 1 + \hat{\gamma}'/\hat{\beta}_k.$$

Numerical examples of this procedure are deferred to the end of the chapter where we also give examples of the simultaneous transformation of response and explanatory variables. Examples are also given of the simultaneous transformation of several explanatory variables to yield new carriers. In the latter case, if each of $s < p$ explanatory variables is to be transformed to a potentially different power, series expansion of the transformation model analogous to (8.3.2) leads to a model with $s$ added variables, the $k$th corresponding to transformation of $x_k$ to the power $\lambda_k$. Let the indices of these $s$ variables belong to the set $S$. Then, for the hypothesis of no transformation, that is $\lambda_0 = 1$, the model is

$$E(Y) = \sum_{j=1}^{p} \beta_j x_j + \sum_{k \in S} (\lambda_k - 1)\beta_k x_k \log x_k. \tag{8.3.6}$$

The individual constructed variables are therefore the same as those given by (8.3.4). With more than one candidate variable for transformation there is the problem of selection of constructed variables similar to the problems of selection of carriers in a regression equation, introductions to which are given by Weisberg (1980, Chapter 8) and by McCullagh and Nelder (1983, Section 3.9). That is, with a sufficiently large value of $s$ at least some of the variables are very likely to give statistically significant $t$ values even in the absence of any real effects. Before explanatory variables are selected for transformation, it is prudent first to perform an overall $F$ test for the significance of all constructed variables. The results of this test will indicate whether there is significant evidence for the transformation of at least one of the variables.

If a group of explanatory variables are all to be considered for the same transformation, inference is simplified. The simplest case is to suppose that all $s$ explanatory variables in the subset are to receive the same power transformation. Then (8.3.1) is replaced by the model

$$E(Y) = \sum_{j \notin S} \beta_j x_j + \sum_{k \in S} \beta_k x_k^{\lambda}. \tag{8.3.7}$$

The Taylor series expansion of (8.3.7) about $\lambda_0 = 1$ yields the approximate model

$$E(Y) = \sum_{j=1}^{p} \beta_j x_j + (\lambda - 1) \sum_{k \in S} \beta_k x_k \log x_k. \tag{8.3.8}$$

with the obvious generalization to an analogue of (8.3.2) for expansion about more general $\lambda$.

Testing whether all the $s$ explanatory variables should be subject to the same transformation therefore involves the single constructed variable

$$w_{\mathrm{BT}}(1) = \sum_{k \in S} \hat{\beta}_k x_k \log x_k. \tag{8.3.9}$$

An example of the use of this constructed variable is included at the end of the chapter. The extension to a mixture of variables, some of which are transformed in sets and others of which are considered individually, leads to constructed variables consisting in part of individual terms of the form given by (8.3.4), while other constructed variables will include grouped terms similar to those in (8.3.9).

## 8.4 Simultaneous transformation of all variables

If simultaneous transformation of both the response and the explanatory variables is of interest, the methods of the preceding section combine with those of Chapter 6 to produce an enhanced set of constructed variables. For simplicity suppose that the response and all $p - 1$ explanatory variables are to be transformed. The model is that for some vector-valued parameter $\lambda$

$$z(\lambda_{\mathrm{p}}) = \beta_0 + \sum_{j=1}^{p-1} \beta_j x_j^{\lambda_j} + \varepsilon, \tag{8.4.1}$$

where $\beta_0$ is now displayed separately. In (8.4.1) each variable, including the response, could have a separate transformation parameter so that in all there are $p$ such parameters. As with the models of earlier sections, expansion of this model about $\lambda_0$ yields a linearized model. If $\lambda_0$ has the same value for all $p$ variables, the approximation to (8.4.1) is

$$z(\lambda_0) = \beta_0 - (\lambda_{\mathrm{p}} - \lambda_0) w_{\mathrm{p}}(\lambda_0) + \sum_{j=1}^{p-1} \beta_j x_j^{\lambda_0} + \sum_{j=1}^{p-1} (\lambda_j - \lambda_0) \beta_j x_j^{\lambda_0} \log x_j + \varepsilon. \tag{8.4.2}$$

In (8.4.2), $w_{\mathrm{p}}(\lambda_0)$ is the constructed variable (6.4.5) for the power transformation of Section 6.2. A straightforward generalization of (8.4.2) is to expand each item about an individual value $\lambda_{j0}$. But usually a common value will be more appropriate. In particular, the value $\lambda_0 = 1$, that is no transformation, will often be of interest both for the response and for the explanatory variables.

The development of $t$ or $F$ tests for the constructed variables in (8.4.2) is entirely analogous to that of earlier sections. There is however one new case of interest, that of simultaneous transformation of all variables including the response. One example, mentioned in Section 6.1, was the development by Box and Cox (1964) of a model for the wool data, Example 7, in which both response and explanatory variables were logged. A similar model was one of those fitted to the Minitab tree data in Section 6.5. This simultaneous logarithmic transformation is equivalent to a power law for the original response. The development of the present section is to provide a parametric family of transformations which yields these models as special cases. As an example we return, later in the section, to the analysis of the Minitab tree data.

For studying simultaneous transformation of all variables the constructed variable for the hypothesis of no transformation is, from (8.4.2),

$$w_{BT}(1) = \sum_{j=1}^{p-1} \hat{\beta}_j x_j \log x_j - w_p(1), \tag{8.4.3}$$

where the $\hat{\beta}_j$ are the least squares estimates from the first-order model without transformation. The single variable can be used, as other constructed variables have been, to provide both a $t$ test for the hypothesis of no transformation and, if the test is significant, a quick estimate of the overall transformation parameter. The contribution of individual observations can be examined through an added variable plot and, more formally, through calculation of the modified Cook statistic. The procedures are entirely analogous to those of Sections 6.4 and 8.3.

**Example 6 (continued) Minitab tree data 6** At the conclusion of the analysis of the Minitab tree data in Section 6.5 several plausible models had emerged. These included a first-order model with the addition of a quadratic term in $x_1$, the girth of the tree; a model based on the formula for the volume of a cone; a linear model with the response transformed to the $\frac{1}{3}$ power and a model with $\log y$ as response and $\log x_1$ and $\log x_2$ as carriers. We now investigate which of these models are suggested by application of the various constructed variables for transformation of the explanatory variables.

The numerical results are collected in Table 8.7. The first two entries are for

**Table 8.7** Example 6: Minitab tree data. Statistics for transformation of the explanatory variables

| Variable for transformation† | Score statistic $T(1)$ | Quick estimate $\tilde{\lambda}$ | Maximum modified Cook statistic $i_{max}$ | $C_{i\max}(w)$ |
|---|---|---|---|---|
| $x_1$ | 5.88 | 2.56 | 23 | 1.42 |
| $x_2$ | 1.86 | 13.27 | 18 | 5.55 |
| $x_1$ and $x_2$ | 5.89 | 2.52 | 23 | 1.50 |
| $y$ | −7.41 | 0.39 | 31 | 6.30 |
| $x_1$, $x_2$ and $y$ | −7.37 | 0.09 | 31 | 8.30 |

† Where more than one variable is listed, both, or all, are subject to the same power transformation.

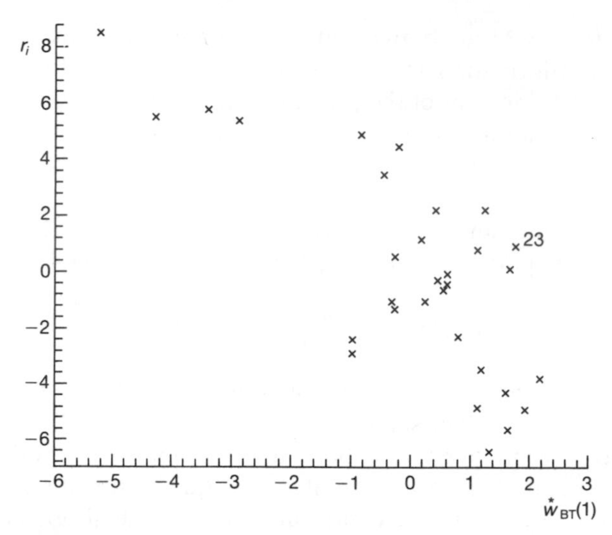


**Fig. 8.15** Example 6: Minitab tree data: added variable plot of $w_{BT}(1)$ for power transformation of $x_1$

transformation of the individual explanatory variables. For $x_1$, girth, and $x_2$, height, the score statistics $T_{BT}(1)$ are respectively 5.88 and 1.86. Thus transformation of $x_1$ is indicated, but not of $x_2$. The quick estimate of the transformation parameter for $x_1$ is 2.56, so that $x_1^2$ and $x_1^3$ might be investigated. The most influential observation is number 23 with $C_{23}(w_{BT}) = 1.42$. Figure 8.15 shows that evidence for a transformation does not depend critically on this one observation. If, instead of individual transformations, $x_1$ and $x_2$ are considered for the same transformation, the constructed variable (8.3.9) is appropriate with both variables included in the set $S$. The results from using the composite variable are virtually identical to those from the variable for $x_1$ alone. The implication is that $x_1$ determines the transformation and that there is little to choose between one transformation of $x_2$ and another.

We have already seen in Section 6.5 that, for the first-order model in $x_1$ and $x_2$, there is appreciable evidence for a power transformation of the response alone. The results of Table 8.7 show that $T_p(1) = -7.41$, a highly significant value. The quick estimate of the transformation parameter is 0.39 which suggests the value of $\frac{1}{3}$ used in Section 6.5. In this case, as was shown in Table 8.6, the quick estimate provides a good indication of the maximum likelihood estimate $\hat{\lambda}$.

The last entry of Table 8.7 is for the constructed variable $w_{BT}(1)$ given by (8.4.3), which leads to a test of the simultaneous transformation of $y$, $x_1$ and $x_2$ all to the same power. There is appreciable evidence for this transformation with the score statistic $T_{BT}(1) = -7.37$. The quick estimate of the trans-

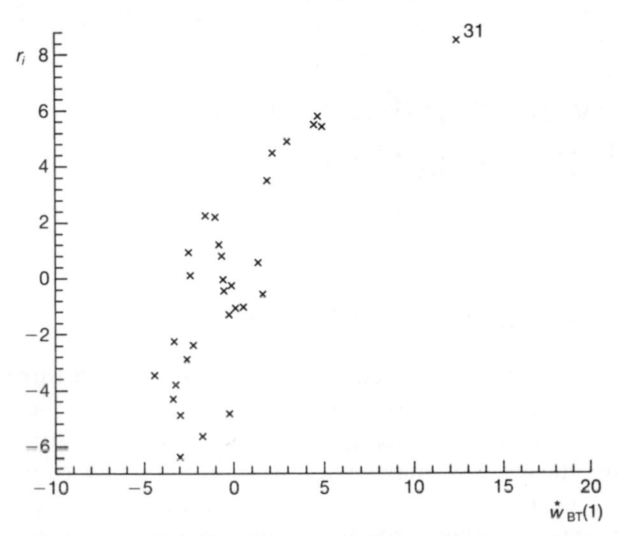


**Fig. 8.16** Example 6: Minitab tree data: added variable plot of $w_{BT}(1)$ for simultaneous power transformation of $x_1$, $x_2$ and $y$.

formation parameter is 0.09, suggestive of the log transformation. Figure 8.16 gives the added variable plot of $w_{BT}(1)$ for the simultaneous transformation. The plot shows that observation 31 is influential, with a value of 8.30 for $C_{31}(w_{BT})$. But evidence of the need for a transformation is also provided by the other observations. As with the power transformation on its own, the largest observation is important, but not of over-riding importance, in suggesting a transformation.

This example demonstrates the usefulness of the constructed variables for transformation of the explanatory variables, singly or together, and with or without the response. In particular, the variety of quick estimates obtained strongly suggests the various models which had earlier emerged by rather less formal methods. □

# 9
# The power transformation with shifted location

## 9.1 Constructed variables for a shift in location

The power transformation studied so extensively in the last three chapters is based on the implicit assumption that the apparent origin of the response is a true lower limit. In the absence of a transformation, the subtraction of a constant from the response may be desirable for numerical convenience. The effect on the fitted model will be to change the value of the estimated constant term while leaving other aspects of the model unchanged. But if a transformation is considered, the effect of subtraction of a constant from the data before a power transformation can be to alter many aspects of the fitted model, including the estimated transformation parameter. The implication is that the units of the response must be physically meaningful. We have already mentioned how for John's $3^{4-1}$ experiment, Example 8, the data of Table 6.5 were reconstructed from the scaled values given by John. In this chapter we consider examples in which the power transformation is perhaps appropriate only after an unknown constant has been added to, or subtracted from, all observed values of the response. For example, in the analysis of survival time experiments, there may be a latent period before a toxic dose has spread throughout the system and starts to act. It is to be expected that, if transformation of the data is appropriate, it will only be so after the latent period has been subtracted from the observed survival times.

In principle, the extension of the power transformation (6.2.1) to allow for a shift parameter $\mu$ is straightforward. The normalized form, equivalent to (6.2.6), is given by

$$z(\lambda, \mu) = \begin{cases} \dfrac{(y+\mu)^{\lambda} - 1}{\lambda\{G(y+\mu)\}^{\lambda-1}} = \dfrac{q^{\lambda} - 1}{\lambda \dot{q}^{\lambda-1}} & (\lambda \neq 0) \\ \log(y+\mu) G(y+\mu) = \dot{q} \log q & (\lambda = 0). \end{cases} \tag{9.1.1}$$

In this expression $G(\cdot)$, as defined in (7.1.2), is the geometric mean function, so that $\dot{q} = G(y+\mu)$ is the geometric mean of the observations after addition of the parameter $\mu$. Use of $q = y + \mu$ in (9.1.1) stresses the relationship with the power transformation without mean shift.

The transformation (9.1.1) is included in the analysis of Box and Cox

(1964) although its properties are not explored in depth. In the present section we proceed, in a straightforward manner, to develop the constructed variable $w_s(\lambda, \mu)$ for the shift parameter $\mu$. The use of this constructed variable is illustrated by two examples.

In the remaining sections of the chapter we explore some of the less straightforward implications of this two-parameter family of transformations. Provided there is a good initial estimate of one of the parameters the behaviour of the transformation is in many ways similar to those of earlier chapters. But if both $\lambda$ and $\mu$ are unknown, compensating changes in the values of the two parameters can give rise to similar models. For example, taking the logarithm of data after the addition of a constant is sometimes virtually indistinguishable from a different transformation without the shift in location. The connection between the two transformations is examined in Section 9.2. The emphasis is on the relationship between the constructed variable for a shift in location and that for a power transformation.

In Section 9.3 the simultaneous estimation of both $\lambda$ and $\mu$ is explored in some detail. A feature of the transformation (9.1.1) is that, since we require $y + \mu > 0$, the range of the observations depends on the value of the parameter $\mu$. This is therefore an example of a non-regular estimation problem of the sort mentioned in Section 6.4. Let $y_{\min}$ be the value of the smallest observation. Then as $\mu$ approaches $-y_{\min}$, there is, as we shall see, a value of $\lambda$ between 0 and 1 for which the residual sum of squares of $z(\lambda, \mu)$ goes to zero. Equivalently, the likelihood is unbounded. Plots of the two-dimensional surface of the residual sum of squares given in Section 9.3 show that two types of surfaces arise: for some examples there is a local minimum whereas in others the sum of squares decreases smoothly to the minimum value of zero.

In the last section of the chapter we return to the use and properties of added variable plots of constructed variables. The transformations studied are those in which groups of observations are subject to the same power transformation but are allowed different shift parameters. But we begin with the simplest case in which there is only one shift parameter, $\mu$.

For the shifted power transformation (9.1.1), differentiation with respect to $\lambda$ yields the constructed variable for the power transformation

$$w_p(\lambda, \mu) = \frac{q^\lambda \log q}{\lambda \dot{q}^{\lambda - 1}} - z(\lambda, \mu)(1/\lambda + \log \dot{q}), \tag{9.1.2}$$

which is $w_p(\lambda)$ (6.4.5) but with $y$ replaced by $q = y + \mu$. The presence of significant regression on $w_p(\lambda, \mu)$ is evidence of the need for a power transformation coupled with addition of a value $\mu$ to all observations.

The constructed variable for the shift parameter $\mu$ is likewise found to be

$$w_s(\lambda, \mu) = \frac{\partial z(\lambda, \mu)}{\partial \mu} = (q/\dot{q})^{\lambda - 1} - (\lambda - 1)z(\lambda, \mu)/hm(q), \tag{9.1.3}$$

where $hm(q) = n/\sum (1/q_i)$ is the harmonic mean of the $q_i$. It is interesting that the constructed variable for the power transformation depends on all observations only through their geometric mean, whereas the variable for a shift in location introduces a second summary statistic, the harmonic mean.

Often interest is in testing hypotheses about both $\lambda$ and $\mu$ when $\mu = 0$. Then $q$ in (9.1.2) and (9.1.3) reduces to $y$ and the constructed variable for the power transformation reduces to $w_p(\lambda)$. We shall similarly use $w_s(\lambda)$ to denote $w_s(\lambda, 0)$. Two special values of $\lambda$ are of interest for the calculation of $w_s(\lambda, \mu)$. The first is $\lambda = 1$, when $w_s(1, \mu) = 1$. If the untransformed model already contains a constant, regression on this constructed variable provides no information about the need for a shift in location. This sensible result arises because it is not possible to estimate $\mu$ in a model already containing a constant. As a consequence the likelihood is constant as a function of $\mu$ and the score statistic, which measures the slope of the log-likelihood, is identically zero. However, the shift in location can be estimated for other values of $\lambda$. In particular, for the log transformation, the constructed variable is

$$w_s(0) = \dot{y}\{1/y + \log y/hm(y)\}. \tag{9.1.4}$$

With two constructed variables we can either use individual $t$ tests for hypotheses about $\lambda$ and $\mu$ or we can use a combined $F$ test for the significance of regression on both variables. The situation is similar to that encountered in Section 8.3 for transformation of the explanatory variables. The examples which follow are used *inter alia* to investigate the relationship between the two sets of tests.

**Example 3 (continued) Simulated factorial experiment 2** To calibrate the score test for a change in location we return to the example of the simulated factorial used in Chapter 4 to introduce diagnostic plots. The design is a $2^4$ factorial with one centre point. We now change the results of Table 4.1 by first exponentiating the response and then adding 30. For this new response $y'$, transformation to $\log(y' - 30)$ leads to recovery of the original data.

Table 9.1 gives the results of applying the tests for transformation and shift in location which were derived in the previous section. The score statistics are the $t$ tests for regression on the individual constructed variables. The analysis of variance tables give the $F$ statistic for regression on both variables which has been broken down into individual degrees of freedom. Two tables are given, corresponding to first fitting $w_p$ followed by $w_s$ and to fitting the variables in reverse order. The total sum of squares explained in the two tables is, of course, the same. In general the $F$ value for fitting the first variable is close to the square of the $t$ value only if fitting the second variable causes no further significant reduction in the residual sum of squares.

From Table 9.1, the score test for the power transformation $T_p(1) =$

$-9.35$, strongly indicating the need for a transformation. For the untransformed data there can be, as was shown above, no information on the shift parameter. But after logs have been taken, the two individual score statistics are $T_p(0) = -7.20$ and $T_s(0) = -11.01$. There is thus appreciable evidence of a need for further transformation, but it is not totally clear whether this should be to some other power than zero, or whether there should be a shift in location before the logarithmic, or other, power transformation is applied. The analysis of variance tables from regression on both constructed variables, also given in Table 9.1, are not especially helpful in this regard. They suggest that a very great reduction in the residual sum of squares will result from either, or both, of these transformations.

We begin by exploring transformations when either $\lambda$ or $\mu$ is constrained to be zero. For the untransformed data the residual sum of squares is 9154.4. The log transformation reduces this to 3552.0 for the residual sum of squares of $z(0)$. If $\mu$ is fixed at zero, the maximum likelihood estimate of $\lambda$ is $-3.50$ with a residual sum of squares of 1335.5. Alternatively, staying with the logarithmic transformation but searching over $\mu$ yields an estimate of $-30.74$ with a residual sum of squares of 369.8. The logarithmic transformation with mean shift thus yields an appreciably smaller residual sum of squares than the power transformation on its own and leads to approximate recovery of the model from which the data were generated, namely $\lambda = 0$ and $\mu = -30$.

The residual sum of squares for the original data of Table 4.1 is 549.0. The value of the residual sum of squares found by minimizing over $\mu$ with $\lambda$ held fixed at zero is appreciably smaller at 369.8. But, as is shown in Section 9.3, smaller values again can be found by simultaneous minimization over both $\lambda$ and $\mu$. In the present example, the procedure of searching over one

**Table 9.1** Example 3: Simulated factorial experiment. Evidence for shift and power transformations

(a) Score statistics

$T_p(1) - 9.35$ $\quad$ $T_p(0) - 7.20$ $\quad$ $T_s(0) - 11.01$

(b) Analyses of variance for regression on constructed variables $w_p(0)$ and $w_s(0)$

| Source | Sum of squares | Degrees of freedom | Mean square | $F$ ratio |
|---|---|---|---|---|
| $w_p(0)$ | 2930.35 | 1 | 2930.25 | 415.86 |
| $w_s(0) \mid w_p(0)$ | 551.32 | 1 | 551.32 | 78.24 |
| Residual | 70.46 | 10 | 7.05 | |
| $w_s(0)$ | 3256.30 | 1 | 3256.30 | 462.13 |
| $w_p(0) \mid w_s(0)$ | 225.28 | 1 | 225.28 | 31.97 |
| Residual | 70.46 | 10 | 7.05 | |

variable with the other held constant leads to an approximate recovery of the model from which the data were generated. But one ambiguity which has been revealed is that, at $\lambda = \mu = 0$, both score statistics point to the need for a further transformation, with a rather muted indication as to whether a shift in location or an alternative power transformation is to be preferred. The two analyses of variance in Table 9.1 likewise indicate that either one or the other of the transformations should have an appreciable effect. These tables also suggest that an extremely small final sum of squares should result from the combined transformation. The general relationship between the statistics for transformation and the sum of squares surface is investigated further in Section 9.3

A fuller analysis of the two transformations for this example would supplement the values of the score statistics with plots of the constructed variables. Rather than expend this attention on simulated data, we now consider an example where the use of plots derived from the two constructed variables leads to insight into data not generated from a known model. □

**Example 12 Brown and Hollander chimpanzee data 1** Brown and Hollander (1977, p. 257) present results on the time, in minutes, for four chimpanzees to learn each of ten words. There are thus forty responses. The data are given in Table 9.2, where it can be seen that the non-negative response ranges from 2 to 476 minutes.

The data are subjected by Brown and Hollander to a standard two-way analysis of variance without interaction. Such an analysis must be suspect since, with non-negative observations and the range given, there is an appreciable chance that the model will result in negative fitted values. The model was criticized by McCullagh (1980) who suggested that either the data should be transformed or a generalized linear model be fitted. In Chapter 11 we consider the choice between these models. For the present we investigate power transformation of the response coupled with a shift in location.

Summary statistics for the analysis of Example 12 are given in Table 9.3. The score test for the power transformation $T_p(1) = -13.2$. The added

**Table 9.2** Example 12: Brown and Hollander chimpanzee data. The response is the time for each of four chimpanzees to learn ten signs

| | | Sign | | | | | | | | | |
|---|---|---|---|---|---|---|---|---|---|---|---|
| | | 1 | 2 | 3 | 4 | 5 | 6 | 7 | 8 | 9 | 10 |
| Chimpanzee | 1 | 178 | 60 | 177 | 36 | 225 | 345 | 40 | 2 | 287 | 14 |
| | 2 | 78 | 14 | 80 | 15 | 10 | 115 | 10 | 12 | 129 | 80 |
| | 3 | 99 | 18 | 20 | 25 | 15 | 54 | 25 | 10 | 476 | 55 |
| | 4 | 297 | 20 | 195 | 18 | 24 | 420 | 40 | 15 | 372 | 190 |

**Table 9.3** Example 12: Brown and Hollander chimpanzee data. Evidence for shift and power transformations

(a) Score statistics and measures of influence

| | All data | Observation 8 deleted |
|---|---|---|
| $T_p(1)$ | −13.21 | −12.65 |
| $T_p(0)$ | 1.18 | −0.50 |
| $C_{i\,max}(w_p)$ | 7.87 | 2.75 |
| $i_{max}$ | 8 | 10 |
| $T_s(0)$ | 1.03 | −2.63 |
| $C_{i\,max}(w_s)$ | 14.76 | 3.47 |
| $i_{max}$ | 8 | 17 |

(b) Analyses of variance for regression on constructed variables. Observation 8 deleted

| Source | Sum of squares | Degrees of freedom | Mean square | $F$ ratio |
|---|---|---|---|---|
| $w_p(0)$ | 411.80 | 1 | 411.80 | 0.87 |
| $w_s(0) \mid w_p(0)$ | 29538.95 | 1 | 29538.95 | 62.75 |
| Residual | 11296.97 | 24 | 470.71 | |
| $w_p(0)$ | 8966.49 | 1 | 8966.49 | 19.05 |
| $w_p(0) \mid w_s(0)$ | 20984.26 | 1 | 20984.26 | 44.58 |
| Residual | 11296.97 | 24 | 470.71 | |

variable plot for the constructed variable $w_p(1)$ (Fig. 9.1) shows that this evidence for the transformation is spread throughout the data and that there is no indication of any unduly influential observation. The maximum likelihood estimate of $\lambda$ is 0.095 and the log transformation is indicated.

The analysis so far parallels many we have seen since the beginning of Chapter 6. As a check on the adequacy of the log transformation, the values of the score statistics for the power and shift transformations at $\lambda = 0$ are $T_p(0) = 1.8$ and $T_s(0) = 1.03$. Both values suggest that the model is satisfactory. However, the maximum values of the modified Cook statistics calculated from these two variables are 7.87 for the power transformation and 14.76 for the shift transformation, both maxima occurring for observation 8. The added variable plot for $w_s(0)$ (Fig. 9.2) shows a cluster of points from which observation 8 is clearly distanced. The half-normal plot of the modified Cook statistic $C_i(w_s)$ derived from this added variable plot, which is not shown here, confirms the influential nature of observation 8.

Observation 8 is the smallest, with a value of the response equal to 2. The next smallest observation is equal to 10. Once the observations have been logged it is to be expected that observation 8 will be highly influential in the

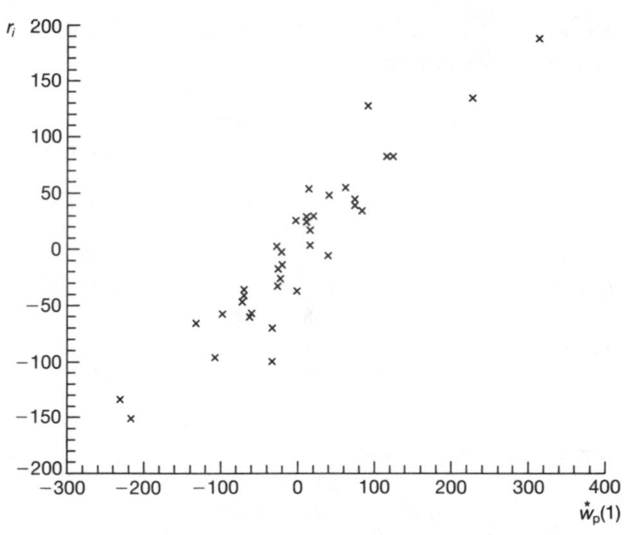


**Fig. 9.1** Example 12: Brown and Hollander chimpanzee data: added variable plot for constructed variable $w_p(1)$

choice of a value of $\mu$. If this observation is deleted, the effect is appreciable. As the results of Table 9.3 show, the evidence for a power transformation remains strong with $T_p(1) = -12.6$. The log transformation is again in order with $T_p(0) = -0.50$. But, with a value of $-2.63$ for $T_s(0)$, there is now evidence of the need for a shift in location if the data are to be logged. The

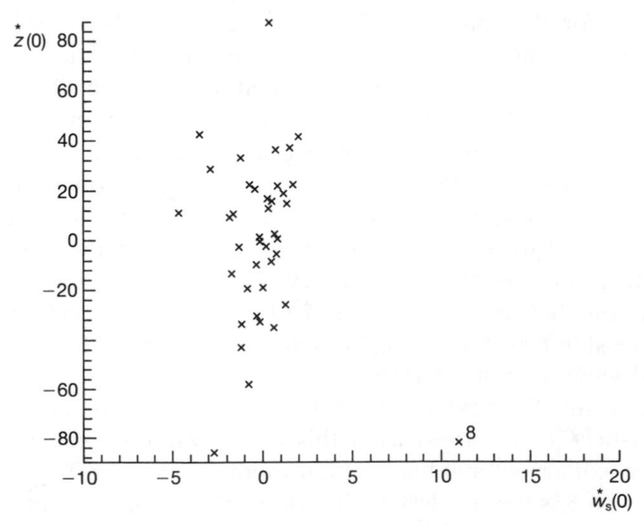


**Fig. 9.2** Example 12: Brown and Hollander chimpanzee data: added variable plot for constructed variable $w_s(0)$

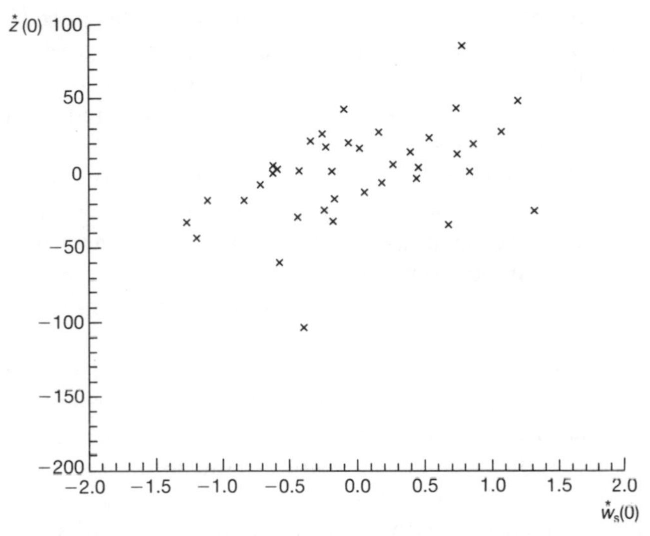


**Fig. 9.3** Example 12. Brown and Hollander chimpanzee data, observation 8 deleted: added variable plot for constructed variable $w_s(0)$

added variable plot for the constructed variable $w_s(0)$ (Fig. 9.3) suggests that evidence for a nonzero value of $\mu$ is not confined to a few observations. This impression is confirmed by the maximum value of the modified Cook statistic $C_i(w_s)$ which has been reduced from 14.76 to 3.47 by the deletion of observation 8. If $\lambda$ is kept fixed at zero, so that only shifted log transformations are considered, the maximum likelihood estimate of $\mu$ is $-8.59$, for which the residual sum of squares is 34 507. When both $\lambda$ and $\mu$ are zero the value is significantly greater at 41 248. The shifted log transformation is therefore justified. These results show that the effect of observation 8 was to nullify the information from the other observations that a shift in location is required.

One further feature of the results of Table 9.3 warrants discussion, namely the structure of the analysis of variance tables for the constructed variables at $\lambda = 0$. If, after observation 8 has been deleted, the power transformation variable $w_p(0)$ is fitted first, the result is not significant, as would be expected from the result for $T_p(0)$. But the subsequent addition of the shift variable $w_s(0)$ yields a highly significant $F$ value of 62.75 on 1 and 24 degrees of freedom. The results for fitting in the reverse order are similar: the shift variable on its own is significant, in line with the value for the univariate score statistic. But addition of the second variable yields a very highly significant regression on both constructed variables. This behaviour is the reverse of that for the simulated data in Table 9.1. There each variable on its own explained much of the residual variation, even though addition of the second variable did yield an appreciable further reduction in the residual sum of squares.

Here the two variables together give a much more significant reduction than either does individually. One interpretation is that simultaneous maximization of the likelihood as a function of $\lambda$ and $\mu$ will lead to an appreciable reduction in the residual sum of squares. This idea is explored further, for this and other examples, in Section 9.3.

The model to which we have been led, in which one observation is deleted and the remaining 39 observations logged after 8.59 has been subtracted from the observed values, seems *a priori* implausible. The bizarre nature of this process may be an indication that the generalized linear model fitted by McCullagh is more appropriate than the transformation model. □

These examples illustrate the use of the two constructed variables in determining a transformation with mean shift. One result from the analysis of the simulated factorial is that when $\lambda = 0$ the two constructed variables can misleadingly give rather similar indications of the need for a transformation. In the example the results of the score tests were not borne out by the fitted models, which yielded markedly different residual sums of squares. In the next section we consider the relationship between the two constructed variables when $\lambda = 0$.

## 9.2 Detailed structure of plots of constructed variables

The added variable plots of residuals against constructed variables shown in this book have tended to look like ordinary regression scatter plots. The appearance of a point separated from the general scatter, supplemented by the half-normal plot of the appropriate modified Cook statistic, has been the chief method of identifying influential observations. But the plots need not look at all random. In this section the structure of the plots is investigated in some detail when $\lambda = 0$, as is the relationship between the constructed variables for the power transformation and for a shift in location.

Figure 9.4 shows an added variable plot of $w_p(0)$ for a simulated log normal sample. The value of the score statistic $T_p(0)$ is $-0.256$ so the log transformation is appropriate, leading to recovery of the original normal sample. Yet the added variable plot, with its clear parabolic shape, is in appearance far from a random scatter of points. The plot is however similar to the added variable plot for the shift parameter, Fig. 9.5, which is again a parabola. With a value of $-1.533$ for $T_s(0)$ there is no significant evidence that the log transformation should be coupled with a shift in location. The two parabolas are similar in the $y$ direction, since the common vertical scale is residual $z(0)$, that is $\overset{*}{z}(0)$. But the plots are also similar in the horizontal direction, suggesting a close relationship between the constructed variables for location and power in the null case.

To investigate this structure we work in terms of $z(0) = \dot{y} \log y$ which we

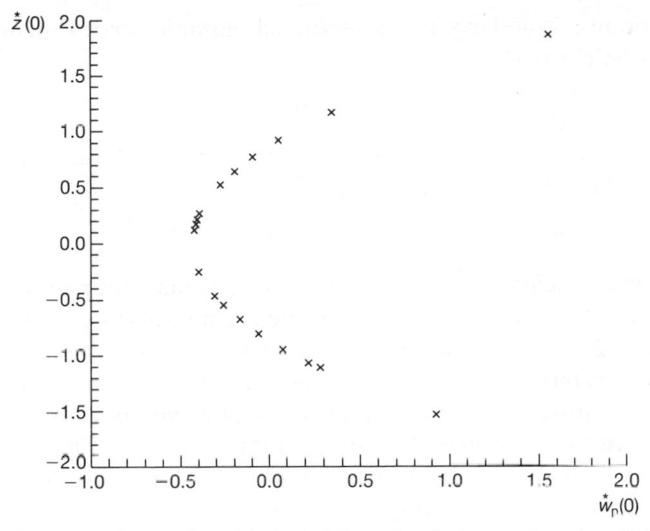


**Fig. 9.4** Lognormal sample: added variable plot for constructed variable $w_p(0)$

shall call $z$. Figure 9.4 is then a plot of residual $z$, that is $\overset{*}{z}$, against residual $w_p(0)$ for a simple random sample. From (6.4.7) the constructed variable can be written in terms of $z$ as

$$w_p(0) = z(z/2\dot{y} - \log \dot{y}), \tag{9.2.1}$$

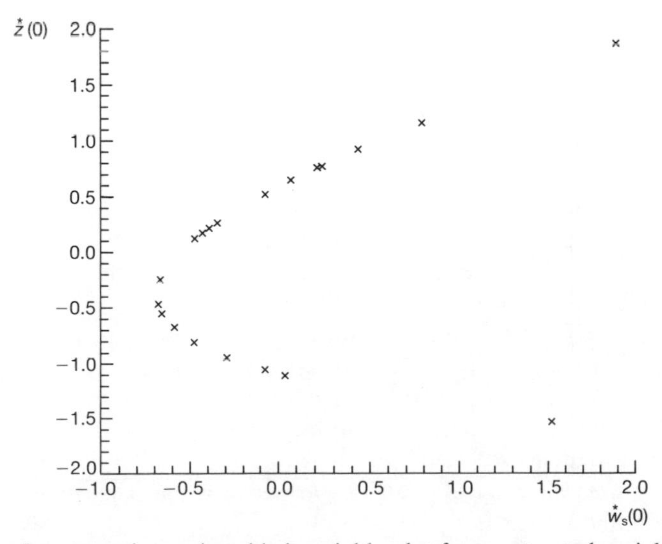


**Fig. 9.5** Lognormal sample: added variable plot for constructed variable $w_s(0)$

a quadratic in $z$. Similarly the constructed variable for a shift in location (9.1.4) can be written as

$$w_s(0) = \dot{y} \exp(-z/\dot{y}) + z/hm(y), \tag{9.2.2}$$

where $hm(y)$ is the harmonic mean of the untransformed observations. Expansion of $\exp(-z/\dot{y})$ in a Taylor series yields

$$w_s(0) = \dot{y} + z\{1/hm(y) - 1\} + z^2/2\dot{y} + \cdots. \tag{9.2.3}$$

For models including an intercept, the residual constructed variable formed from (9.2.3) will not depend on the constant term $\dot{y}$. Comparison of (9.2.1) and (9.2.3) shows that both will give parabolic plots against $z$ with the same quadratic term. In the null case, that is in the absence of any regression of $z$ on the residual constructed variable, the plots will be dominated by the quadratic term. To the extent that the Taylor series expansion yielding (9.2.3) holds, these parabolas will be the same. Figures 9.4 and 9.5 show how similar the parabolas are for one example.

The parabolic structure is obscured when the plots are of residual variables from more complicated models. Figure 9.6 shows an added variable plot for $w_p(0)$ from a two-sample lognormal model. There are therefore two groups of observations with different means and the plot consists of the superposition of two parabolas with differing locations. The plot for $w_s(0)$ is similar in appearance, but not given. For more complicated models, as the examples in this book have shown, the structure is not usually visible, except as a slight curvature in some plots. The implication is that some forms of systematic

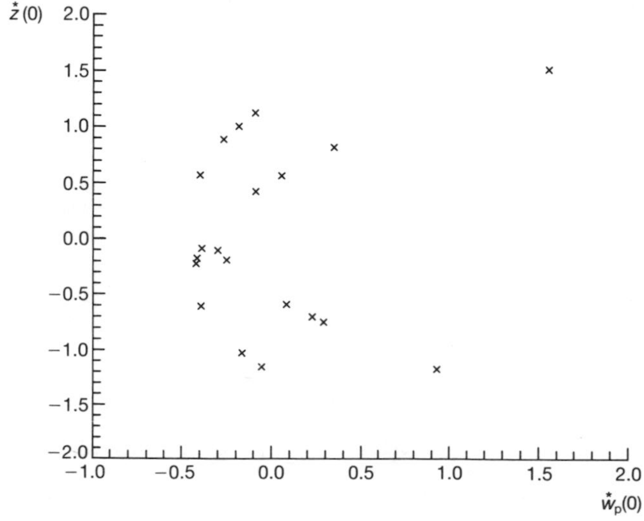


**Fig. 9.6** Two lognormal samples: added variable plot for constructed variable $w_p(0)$

non-randomness in added variable plots may be artefacts of the constructed variables, rather than evidence of departure from the model or transformation.

There is an interesting link between the results of this section and those of Section 8.1. Expansion of the constructed variable $w_s(0)$ in a Taylor series leads to examination of the dependence of residual $z$, that is $\overset{*}{z}$, on $z^2$. The same is true of $w_p(0)$ without expansion. This Taylor series approximation was also used in the investigation of the various constructed variables for the power transformation. In that case the calculations were for $\lambda = 1$, when $z$ reduces to $y$ and $\overset{*}{z}$ to the ordinary residuals $r$. The Taylor series variable $w_{TS}(1)$ then expresses the relationship between the residuals $r$ and $y^2$. The variable derived from Tukey's exact test (8.1.10) is concerned with the relationship between $r$ and $\hat{y}^2$.

The analysis of the constructed variables in this section has been for the null case when there is no significant regression on the constructed variable. But the results do show the relationship between the tests for a further transformation from the log and for a shift in location. They thus provide some explanation of the results of the analysis of Example 3 in Section 9.1 where both tests were highly significant, even though only one form of departure was in fact present.

## 9.3 Simultaneous estimation of shift in location and of the power transformation

The score statistics and related plots of constructed variables in the earlier sections of this chapter describe the individual effect of either the power transformation or of the shift in location coupled with the log transformation. The relationship between these two transformations and the properties of joint estimates of both parameters are most easily understood through contour plots of the log-likelihood surface, or equivalently through plots of the residual sum of squares of $z(\lambda, \mu)$.

For linear least squares problems the contours of the sum of squares in the space of the parameters form a set of concentric ellipsoids centred on the least squares estimate $\hat{\beta}$. For the power transformation of Chapter 6 $R(\lambda)$, the residual sum of squares of $z(\lambda)$ minimized over the parameters of the linear model defined in (6.2.8), behaves asymptotically like the residual sum of squares for a linear problem. The maximum likelihood estimate of $\lambda$ is asymptotically normal and, as Fig. 6.3 shows, the plot of $R(\lambda)$ against $\lambda$ is close to a parabola. But when we move to the two-parameter transformation family we do not obtain elliptical contours. If, analogously to (6.2.8), we define $R(\lambda, \mu)$ to be the partially minimized sum of squares

$$R(\lambda, \mu) = z(\lambda, \mu)^{\mathrm{T}} A z(\lambda, \mu), \tag{9.3.1}$$

plots of $R(\lambda, \mu)$ as a function of $\lambda$ and $\mu$ are far from being a set of concentric ellipses.

Departure from the second-order shape arises because estimation of $\mu$ is not a regular problem. We saw in Section 6.3 that maximum likelihood estimation of $\theta$ for the uniform distribution on $(0, \theta)$ yielded an estimate which asymptotically had an exponential distribution. In the present case we again have a distribution where the range of the observations depends on the value of an unknown parameter. For the power transformation to apply we require $y - \mu > 0$, so that the distribution of the observations $y$ must lie in $(\mu, \infty)$. With the range of the observations dependent on $\mu$, the distribution of the maximum likelihood estimate $\hat{\mu}$ cannot be assumed to be close to normality.

There is an appreciable literature on related problems, some of which are special cases of the power transformation. One example is estimation of the three-parameter log-normal distribution, in which, as in Example 3 in Section 9.1, a mean shift is applied to the data before the logarithm is taken. References include Cheng and Amin (1981) and Voorn (1981). In survival-time experiments the problem of a latent period before a dose becomes effective, mentioned in Section 9.1, is studied by Peto and Lee (1973). The model is an exponential, or other, survival time distribution on $(\mu, \infty)$, where $\mu$ is to be estimated from the data. Cheng and Amin (1983) give references to work on shifted gamma and Weibull distributions. Their paper is concerned with the general problem of estimation for continuous univariate distributions with a shifted origin.

There are two broad approaches to this problem. One is to accept that maximum likelihood estimation will sometimes lead to $-y_{\min}$ as the estimate of $\mu$, where $y_{\min}$ is the smallest observation. For example, Griffiths (1980) uses the likelihood function to obtain interval estimates for the three-parameter log-normal distribution. He also discusses the use of discretization to avoid the problem of an unbounded likelihood. The other approach is to find an alternative method of estimation which avoids estimates on the edge of the parameter region. Cheng and Amin (1983) argue that, after the probability integral transformation has been applied to the shifted observations, the transformed data should appear to be a sample from a uniform distribution. Parameter estimates are accordingly sought which make this distribution as uniform as possible, by effectively maximizing the product of the spacings between the transformed observations on the uniform scale. The properties of the method are not entirely clear, one difficulty being that the estimate is not necessarily a function of the sufficient statistics. In this chapter we consider only use of the maximum likelihood estimator.

Plots of the residual sum of squares of $R(\lambda, \mu)$ fall into two broad classes. In some examples, as we shall see, there is a local minimum of the sum of squares surface, in the region of which the contours are approximately elliptical. For the estimates of the parameters yielding these local minima approximate normality holds. In other examples there is no local minimum but rather the residual sum of squares declines steadily to zero as $\mu$ approaches $-y_{\min}$. Even if there is a local minimum, there is also a region of parameter space in which $R(\lambda, \mu)$ can be made arbitrarily small. It is straightforward to show that there must be a value of $\lambda$ for which $R(\lambda, \mu)$ goes to zero as $\mu$ approaches $-y_{\min}$. For if $0 < \lambda < 1$ both $\lambda$ and $1 - \lambda$ are greater than zero. Then, from (9.1.1), $z(\lambda, \mu)$ can be written as

$$z(\lambda, \mu) = \dot{q}^{1-\lambda}(q^{\lambda} - 1)/\lambda, \tag{9.3.2}$$

where, as before, $q = y + \mu$. As $\mu \to -y_{\min}$ there is at least one value of $q$ which becomes very small, so that $\dot{q}$, the geometric mean of $q$, also becomes small. Since $1 - \lambda$ is assumed to lie between zero and one, $\dot{q}^{1-\lambda}$ in turn becomes small and the residual sum of squares decreases to zero.

Plots of the resulting sums of squares surfaces are not to be found in the statistical literature. Box and Cox (1964) apply the two parameter transformation family to one example for which they give a plot of the posterior density of $\lambda$ and $\mu$. This plot does not indicate any of the problems discussed here, but shows instead a minimum well away from the boundary of the parameter space. Even though the sum of squares surface has a minimum on the boundary, the posterior density does not. Multiplication of the likelihood by the prior distribution of the parameters rules out a minimum on the boundary.

Box and Cox plot the posterior distribution as a function of $\lambda$ and $\mu$. Plots of $R(\lambda, \mu)$ on this scale are not sensitive to the behaviour as $\mu \to -y_{\min}$. To exhibit this behaviour we work instead in the scale defined by

$$\mu = -y_{\min}(1 - 10^{\varepsilon}). \tag{9.3.3}$$

For $\varepsilon = 0$, $10^{\varepsilon} = 1$ and $\mu = 0$. The minimum value of $\varepsilon$ investigated was $-12$ which, on the CDC computer employed, was sufficiently large that $\mu$ could be distinguished from $y_{\min}$. For values of $\varepsilon$ greater than zero, positive values of $\mu$ result. Use of the transformation (9.3.3) leads to plots of $R(\lambda, \mu)$ as a function of $\varepsilon$, which emphasize the behaviour of the sum of squares close to $-y_{\min}$. Although this is an aspect of the transformation which is of interest in the present section, in making inferences about parameter values for a fitted model, values of $\mu$ near zero are often of major importance.

In addition to the contour plots we shall also look at profile plots of $R(\lambda, \mu)$ maximized over $\lambda$ as well as over the parameters of the linear model. For given $\mu$, or equivalently, given $\varepsilon$, the estimate of $\lambda$ is found by the method of false position. The resulting residual sum of squares will be called $R(\mu)$.

Before we begin the study of contour plots for the set of four examples, we return to the relationship between the individual score statistics and the $F$ test for regression on both constructed variables. This relationship is

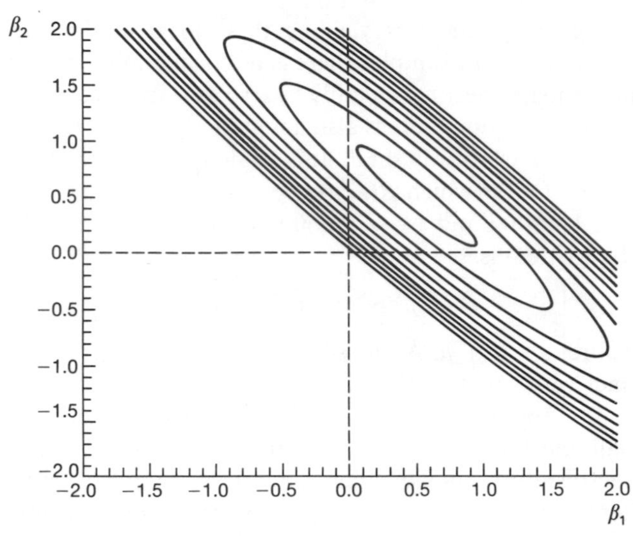


**Fig. 9.7** Contour plot of hypothetical sum of squares surface: either variable significant on its own

illuminated by consideration of the elliptical sum of squares contours which are yielded by a two parameter linear model. As a first example, the results of Table 9.1 for the analysis of Example 3, show a highly significant sum of squares for the first constructed variable to enter the model, whichever it is, Addition of the second variable has an appreciably smaller effect. Figure 9.7 shows how such a situation could arise for regression on two variables in a linear model. The elliptical contours are centred in the north-east quadrant with the major axis of the ellipses running from north-west to south-east. At the origin there is a steep gradient along either axis. Addition of either $x_1$ or $x_2$ to the model causes a reduction of the residual sum of squares to the minimum value along the axis. At these points in parameter space the residual sum of squares is near to the minimum attainable when both variables are in the model. Thus addition of the second variable causes little further reduction in the residual sum of squares.

The reverse situation obtains in Table 9.3 for the analysis of the chimpanzee data where both constructed variables are required before there is a significant reduction. One set of contours giving rise to this behaviour is presented in Fig. 9.8. The centre of the elliptical system of contours is, as in Fig. 9.7, in the north-eastern quadrant, but now the major axis of the system runs south-west towards the origin. For such contours, addition of one of the two explanatory variables has little effect, the minima along the axes being hardly less than the value at the origin. But regression on both variables yields an appreciable reduction.

With these examples in mind we now turn to the study of the more

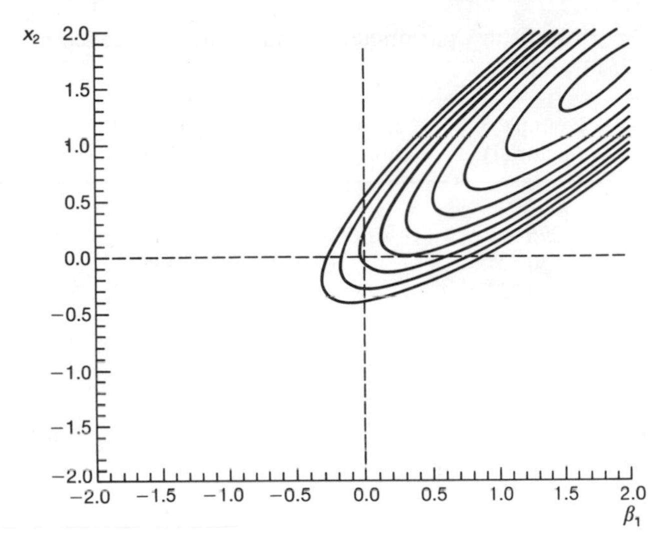


**Fig. 9.8** Contour plot of hypothetical sum of squares surface: both variables required for a significant reduction in the residual sum of squares

complicated systems of contours of the sum of squares surface for the two-parameter transformation family.

**Example 3 (concluded) Simulated factorial experiment 3** We begin by looking again at the simulated factorial experiment used in Section 9.1 to calibrate the behaviour of the two constructed variables. To overcome numerical problems with the particular algorithm used for the estimation of $\lambda$, the data were resimulated. The data, after exponentiation of the normal observations and the addition of 30, are given in Table 9.4 in the standard order of Table 4.1. The centre point comes last.

**Table 9.4** Example 3: Simulated lognormal data for factorial experiment of Table 4.1

| Observation no. | Response | Observation no. | Response |
|---|---|---|---|
| 1 | 32.53 | 10 | 33.13 |
| 2 | 48.19 | 11 | 43.89 |
| 3 | 32.28 | 12 | 134.77 |
| 4 | 56.01 | 13 | 33.45 |
| 5 | 30.32 | 14 | 41.45 |
| 6 | 44.12 | 15 | 88.15 |
| 7 | 40.99 | 16 | 46.03 |
| 8 | 68.68 | 17 | 34.04 |
| 9 | 32.02 | | |

**Table 9.5** Score statistics, parameter estimates and influence measures for Example 3, 7, 11 and 12

| | Example 3 (factorial) | Example 12 (chimpanzees†) | Example 11 (poison) | Example 7 (wool) |
|---|---|---|---|---|
| $\hat{\lambda} \mid \mu = 0$ | −2.09 | −0.057 | −0.750 | −0.059 |
| $R(\hat{\lambda} \mid \mu = 0)$ | 1 296 | 41 050 | 0.32 | 243 413 |
| $y_{min}$ | 30.324 | 10 | 0.18 | 90 |
| $\hat{\lambda} \mid \varepsilon = -12$ | 0.237 | 0.232 | 0.343 | 0.325 |
| $R(\hat{\lambda} \mid \varepsilon = -12)$ | 162 | 2 234 | 0.21 | 115 451 |
| $T_p(1)$ | −8.22 | −12.65 | −13.54 | −18.56 |
| $C_{i\,max}(w_p)$ | 4.48 | 2.12 | 4.16 | 7.66 |
| $i_{max}$ | 12 | 29 | 24 | 19 |
| $R(1, 0)$ | 6 655 | 187 075 | 1.05 | 480 981 |
| $T_p(0)$ | −4.65 | −0.50 | −4.75 | −0.91 |
| $C_{i\,max}(w_p)$ | 2.71 | 2.75 | 3.86 | 2.54 |
| $i_{max}$ | 12 | 10 | 30 | 20 |
| $T_s(0)$ | −5.44 | −2.64 | −4.95 | −0.91 |
| $C_{i\,max}(w_s)$ | 2.76 | 3.47 | 5.61 | 2.62 |
| $i_{max}$ | 12 | 17 | 35 | 9 |

$F$ values from the analysis of variance tables for regression on the constructed variables $w_p(0)$ and $w_s(0)$

| Source | d.f. | $F$ | d.f. | $F$ | d.f. | $F$ | d.f. | $F$ |
|---|---|---|---|---|---|---|---|---|
| $w_p(0)$ | 1 | 43.33 | 1 | 0.87 | 1 | 22.81 | 1 | 0.80 |
| $w_s(0) \mid w_p(0)$ | 1 | 12.04 | 1 | 62.75 | 1 | 1.52 | 1 | 0.02 |
| Residual | 10 | | 24 | | 40 | | 21 | |
| $w_s(0)$ | 1 | 47.63 | 1 | 19.05 | 1 | 24.08 | 1 | 0.79 |
| $w_p(0) \mid w_s(0)$ | 1 | 7.73 | 1 | 44.58 | 1 | 0.25 | 1 | 0.02 |

$R(\lambda, \mu)$ is the residual sum of squares of the transformed response $z(\lambda, \mu)$

† With observation 8 deleted.

Score statistics and sums of squares for these new data are given in Table 9.5. From the value of $-8.22$ for $T_p(1)$ there is clear evidence of the need for a power transformation. At $\lambda = 0$ the fit of the model is not satisfactory and the values of the two score statistics are $-4.65$ for $T_p(0)$ and $-5.44$ for $T_s(0)$. As with the closely related results of Table 9.1, there is evidence that either, or both, of a shift and other power transformation should be used. If $\mu$ is kept equal to zero, the maximum likelihood estimate of $\lambda$ is $-2.088$ with a residual sum of squares of 1296.

The analysis of variance tables for the two constructed variables, given in Table 9.5, show that when both $\mu$ and $\lambda$ are zero, regression on either $w_p(0)$ or $w_s(0)$ has an appreciable effect on the residual sum of squares. This is in line with the $t$ values for the individual score statistics. The difference between the $F$ values and the square of the $t$ values is due to the estimate used for $\sigma^2$. Regression on the second constructed variable is significant at the 5 per cent level for the power transformation and at the 1 per cent level for the shift parameter.

To investigate the effect of simultaneously estimating both $\lambda$ and $\mu$ we look first at the behaviour of the profile residual sum of squares $R(\mu)$, which is minimized over both $\lambda$ and the parameters of the linear model. The limiting value of $\mu$ is $-30.3239 = -y_{\min}$. When $\varepsilon$ (9.3.3) $= -12$, $R(\mu)$ is 161.83, less than one eighth of the value when $\mu = 0$. The plot of $R(\mu)$ (Fig. 9.9) shows that the sum of squares decreases steadily towards zero as $\varepsilon$ approaches $-12$, the numerical limit in single-precision computer arithmetic. The slight point of inflexion near $-2$ corresponds to values of $\mu$ near $-30$, which would lead to recovery of the model from which the data were generated. But this point of inflexion is not sufficiently marked to lead to a local minimum of the sum of squares.

Figure 9.10 is a contour plot of $R(\lambda, \mu)$ for $-1 \leqslant \lambda \leqslant 1$ and for the range of $\varepsilon$ values of Fig. 9.9, namely $-12$ to 0.5, so that the case of special interest $\mu = 0$ is not quite on the boundary. Several features of this plot are common to all contour plots of this section. The first is that the value of $\mu$ is irrelevant for $\lambda = 1$ because the linear model includes a constant. Then $R(1, \mu)$ is a constant and the topmost contour on the plot is a straight line. The second point is that, whatever the structure near $\mu = 0$, as $\mu$ approaches $-y_{\min}$, that is at the

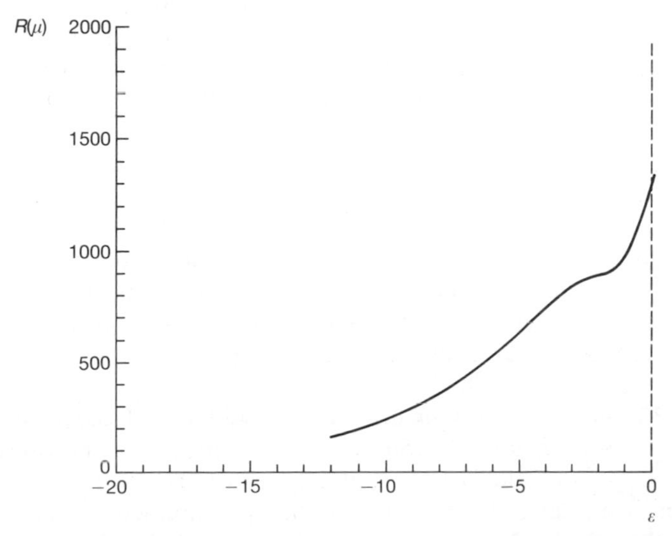


**Fig. 9.9** Factorial with lognormal response, Table 9.4: profile residual sum of squares $R(\mu)$

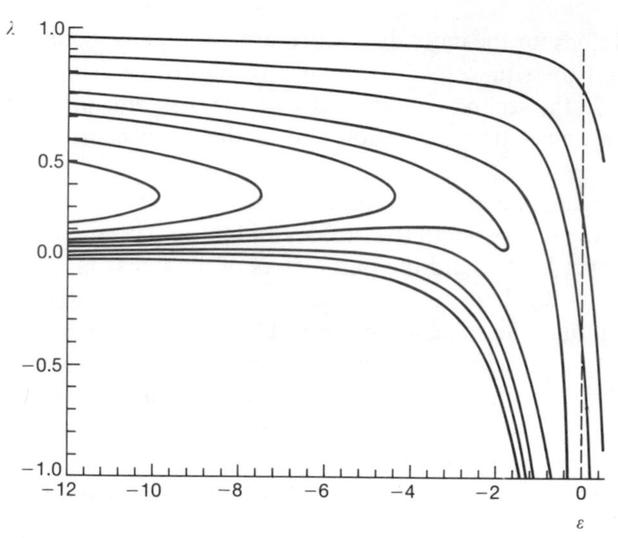


**Fig. 9.10** Factorial with lognormal response, Table 9.4: contour plot of residual sum of squares $R(\lambda, \mu)$

left-hand side of the graph, parabolic contours of decreasing sums of squares are obtained. These are centred on a $\lambda$ value in the approximate range of $\frac{1}{4}$ to $\frac{1}{3}$. For negative $\lambda$ and these values of $\varepsilon$ the sum of squares rises rapidly.

In addition to the general structure, Fig. 9.10 also contains some important specific features. The contours at $\mu = 0$ indicate that the minimizing value of $\lambda$ is less than $-1$ and, as we have seen, $-2.09$ was found. At $\lambda = \mu = 0$ the contours are gently curved away from the origin. The shape is related to the mirror image of Fig. 9.7, which indicates that regression on either constructed variable is appreciable, a conclusion in line with the results of Table 9.5. There is no sign of any local minimum of the surface.

The interpretation of Fig. 9.10 provides a qualitative explanation of Table 9.5 and adds to the insight provided by the statistics. This insight is, of course, gained at the cost of appreciable extra computing effort. In interpreting the contour plots it must be remembered that rather different shapes are seen if, instead of being plotted as a function of $\lambda$ and $\varepsilon$, the plots are presented as a function of $\lambda$ and $\mu$. However, the change from one scale to another cannot cause the appearance or disappearance of local minima. □

**Example 12 (continued) Chimpanzee data 2** In the analysis of the chimpanzee data in Section 9.1 deletion of observation 8 was indicated by the modified Cook statistic for the shift parameter. The results of Table 9.3 subsequently indicated that a change in the shift parameter was desirable. In contrast, the individual score statistic $T_p(0)$ gave no indication of the need for a power transformation other than the logarithm. However, the analysis of variance for regression on both constructed variables suggested that simultaneous estimation of both $\lambda$ and $\mu$ might lead to an appreciable reduction in the residual sum of squares. We now interpret these results with the help of the plots of the residual sum of squares $R(\mu)$ and $R(\lambda, \mu)$.

If, after observation 8 has been deleted, $\mu$ is fixed at zero the estimate of the power transformation is $\hat{\lambda} = -0.057$, which yields a residual sum of squares of 41 050. The results of Table 9.5 show that, for $\varepsilon = -12$, the residual sum of squares is reduced almost twentyfold to 2234. This reduction supports the indication of the analysis of variance for two constructed variables which is repeated in skeleton form in Table 9.5. The plot of the profile sum of squares $R(\mu)$ in Fig. 9.11 shows that the residual sum of squares decreases smoothly as $\mu$ tends to $-y_{\min}$, in this case $-10$. There is not even the hint of a local minimum which was given by the simulated data in the corresponding profile plot of Fig. 9.9. The contour plot for the chimpanzee data, Fig. 9.12, shows a shape which is broadly similar to that of Fig. 9.10 in the hooked contours near $\mu = 0$ and the parabolic contours as $\varepsilon \to -12$. The detailed difference is that, near $\mu = 0$ in Fig. 9.12, the contour is parallel to the $\lambda$ axis so that $T_p(0)$ is approximately zero. There is however a slight decrease in the residual sum of squares in the direction of decreasing $\mu$ so that $T_s(0, 0)$ is negative, with the value $-2.64$. The largest reduction in the residual sum of squares comes from regression on both constructed variables, which corresponds to movement in a north-westerly direction from $\lambda = \mu = 0$, a direction which runs along a valley of the sum of squares surface. In this respect the structure of Fig. 9.12 is similar to that of the elliptical contours of Fig. 9.8. □

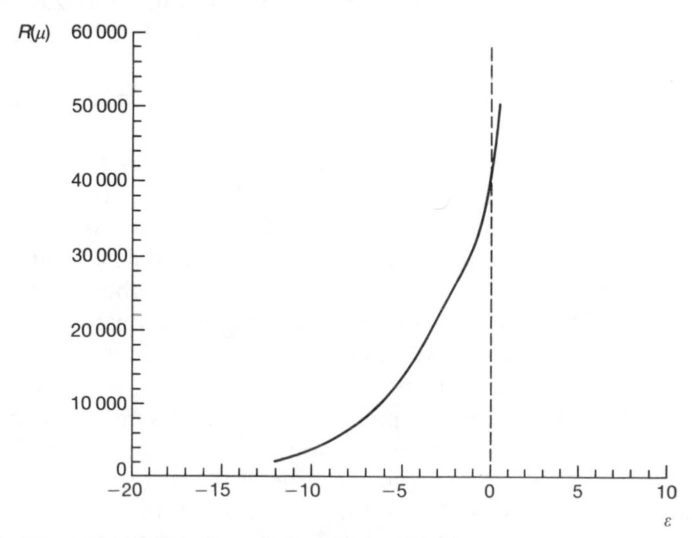


**Fig. 9.11** Example 12: Brown and Hollander chimpanzee data: profile residual sum of squares $R(\mu)$

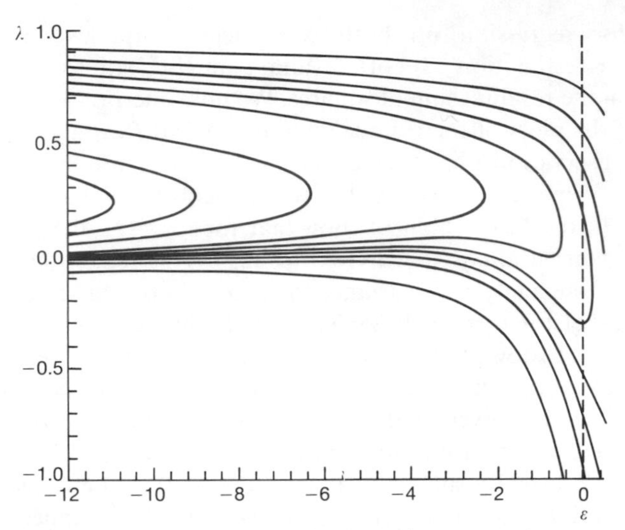


**Fig. 9.12** Example 12: Brown and Hollander chimpanzee data: contour plot of residual sum of squares $R(\lambda, \mu)$

**Example 11 (continued) Box and Cox poison data 2** In both of the preceding examples the full structure of the problem was revealed only by a contour plot of the residual sum of squares, although some of the structure was indicated by regression on the two constructed variables, separately and together. More importantly, the contour plots revealed that there were no local minima to the sum of squares surface. As a result, optimization over $\lambda$ and $\mu$ could lead to as small a value of the residual sum of squares as numerical accuracy would allow.

The third example is rather different. For this we return to the data given by Box and Cox (1964) on the survival times of animals. In Chapter 8 we showed that the reciprocal transformation, $\lambda = -1$, was appropriate. The rejection of $\lambda = 0$ is shown by the value of $-4.75$ for $T_p(0)$ in Table 9.5. But, as with the simulated factorial, Example 3, the value of $T_s(0)$ is similar, in this case $-4.95$. It is accordingly not clear whether a different power transformation should be used or a nonzero value of $\mu$ combined with the log transformation. The analysis of variance table for both constructed variables given in Table 9.5 indicates that one or the other of the further transformations should be considered, but not both.

Figure 9.13 shows the profile plot of the residual sum of squares $R(\mu)$. There is here, in contrast to the two previous profile plots, clear evidence of a local minimum. The contour plot of Fig. 9.14 also exhibits this local minimum which occurs at $\hat{\lambda} = -0.28$ and $\hat{\mu} = -0.10$. At this point the residual sum of squares is 0.3181, compared with 0.3330 for $\lambda = -1$ and $\mu = 0$. Comparison of these two quantities gives an $F$ value close to 2. There is

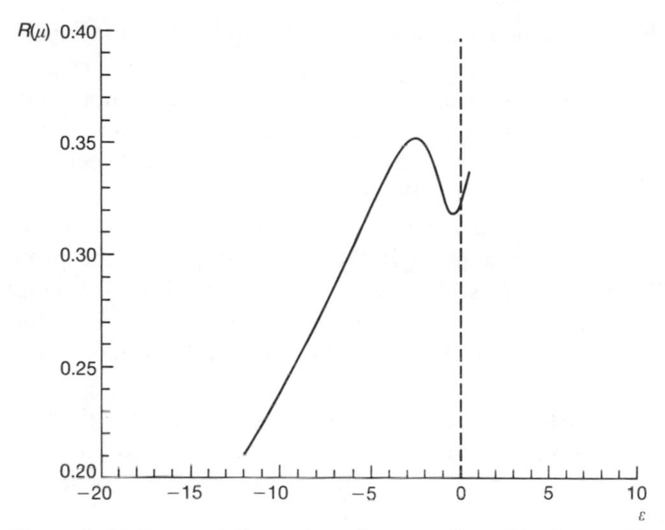


**Fig. 9.13** Example 11: Box and Cox poison data: profile residual sum of squares $R(\mu)$

therefore no reason to consider a shift in location: the inverse transformation is adequate.

The analysis of variance summarized in Table 9.5 suggests, as did the values of the individual score statistics, that one or other of the individual transformations is desirable at $\lambda = \mu = 0$. Comparison of the contours of $R(\lambda, \mu)$ in this region of Fig. 9.14 with those of Fig. 9.10 for the simulated

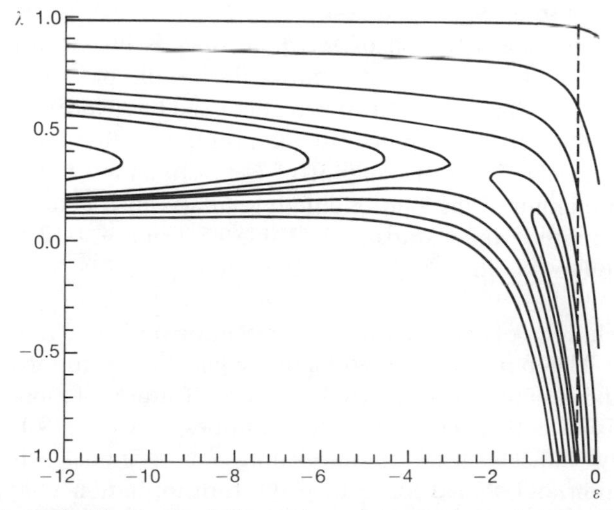


**Fig. 9.14** Example 11: Box and Cox poison data: contour plot of residual sum of squares $R(\lambda, \mu)$

factorial shows that the plots yield a similar explanation for the values of the score statistics.

The plot also shows that, although there is a local minimum, there is also a global minimum as $\mu$ approaches $-y_{\min}$. In this case the minimum value of $y$ is 0.18 and, as $\mu$ approaches $-0.18$, parabolic contours of decreasing sums of squares are again obtained. When $\varepsilon = -12$, the estimate of the power transformation is $\hat{\lambda} = 0.35$, for which the residual sum of squares is 0.21. The strength of the local minimum is shown in the plot of the profile sum of squares. The absence of a scale break in Fig. 9.13 tends to over-emphasize the reduction obtained in the residual sum of squares when $\varepsilon = -12$. But it is clear that, as $\mu \to -y_{\min}$, the residual sum of squares will continue to decrease. □

**Example 7 (concluded) Box and Cox worsted data 5** For the final example of the effect of simultaneous estimation of the shift parameter and of the power transformation we return to the example from Box and Cox (1964) which was used in Chapter 6 to introduce the discussion of transformations of the response

The analysis is concerned only with the first-order model in the three explanatory variables. The results of Table 9.5 confirm those of Chapter 6. There is strong evidence of the need for a power transformation with $T_p(1) = -18.56$. But at $\lambda = \mu = 0$ there is no evidence of need for any further transformation with $T_p(0) = -0.91$ which is also the value of $T_s(0)$. The analysis of variance for the two constructed variables corroborates these results. For either variable the $F$ statistic is about $(0.9)^2$, that is 0.8. Adding the second constructed variable produces virtually no further reduction in the residual sum of squares.

The graphical interpretation of these statistics is in line with the other results of this section. Figure 9.15 shows the profile plot of $R(\mu)$. This is similar in shape to Fig. 9.13 for the poison data and has a local minimum near $\mu = 0$. For values of $\mu$ less than zero the profile plot first increases, before decreasing steadily. The contour plot of Fig. 9.16 shows that $\lambda = \mu = 0$ is close to a minimum of the sum of squares surface and so substantiates the small values for the $t$ and $F$ statistics in Table 9.5. For $\varepsilon = -12$, the value of $\hat{\lambda}$ is 0.325, close to one third. □

The conclusions of this section are straightforward, but not particularly comforting. The purpose of regression diagnostics is to provide simply computed indications of unsuspected or strange features of models and data. One conclusion is that the diagnostic quantities of Section 9.1 may not be adequate to indicate what is happening for a model with a power transformation and shifted location if the transformation is applied in an automatic way. For the textile data there is no indication from the statistics at (0, 0) that any further transformation is necessary, a conclusion which is

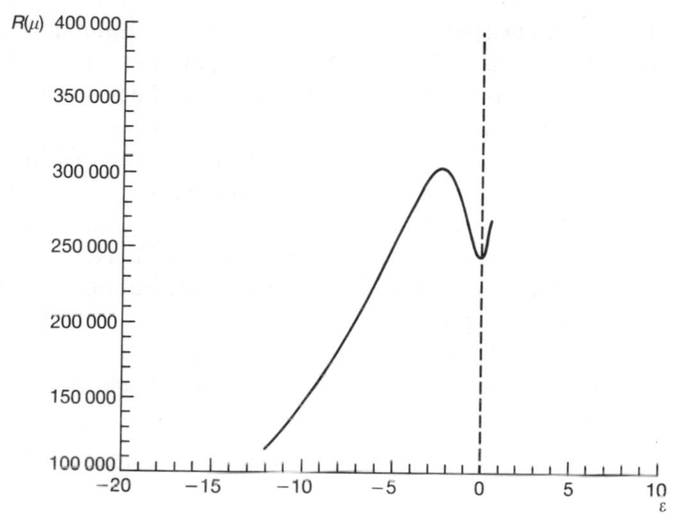


**Fig. 9.15** Example 7: Wool data: profile residual sum of squares $R(\mu)$

supported by the contour plot. But in the other three examples, where further transformation away from (0, 0) is indicated, it is not immediately clear what that further transformation should be. For the two examples with local minima, searches over values of $\lambda$ with $\mu$ held at zero yield sensible models, which the contour plots reveal as close to local minima. Information about the existence of a local minimum seems only to be obtainable from the relatively expensive calculation of a contour plot.

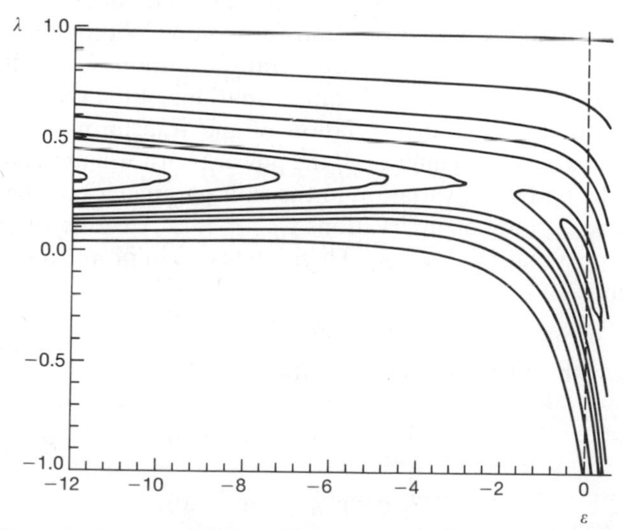


**Fig. 9.16** Example 7: Wool data: contour plot of residual sum of squares $R(\lambda, \mu)$

If a local minimum exists, inference proceeds with the estimated value of $\mu$ in much the same way as if there were no shift parameter or as if the shift parameter were known from previous experience. But the problem remains of how to proceed if there is no local minimum. Minimization of the sum of squares surface in the examples given here leads not only to the shift estimate $\mu = -y_{\min}$ but also to a power transformation close to one third. One solution, typified by Cheng and Amin (1983), is to use an alternative method of estimation. Another suggestion is due to Smith (1985) in a careful analysis of estimation for the three-parameter Weibull distribution. His proposal, transferred to the present problem, is to take $\mu = -y_{\min}$, remove $y_{\min}$ from the data, and to estimate $\lambda$ from the remaining $n - 1$ observations. An advantage of the procedure is that it should yield sensible estimates of the transformation parameter.

## 9.4 Several shift parameters

In the model studied so far in this chapter there has been a single shift parameter, so that it has been assumed that all observations should have the same amount $\mu$ added to them before transformation. But there may be situations where the observations can be divided into groups, each of which requires a separate shift parameter. One natural set of groups consists of the blocks of an experiment, such as litters of animals. Each animal within a litter has the same shift parameter, but there is the possibility of a different parameter in each litter. Similarly, as in the example on which this section is based, data may be presented as differences from an unknown standard, with the standard varying from group to group. Of course, all observations are to be subjected to the same power transformation. Without a relationship across groups, we are left with separate transformations for each group and the problem reduces to a series of transformations of the type introduced in Section 9.1. In that case the variation of the transformation parameters between groups may be of interest. If the power transformation is common to all groups, the pattern of variation in the shift parameter may be important.

Let there be $c$ groups of observations $y_{ij}, j = 1, \ldots, c$ with $n_j$ observations in the $j$th group so that $i = 1, \ldots, n_j$. Then interest is in power transformations of the variable

$$q_{ij} = y_{ij} + \mu_j. \tag{9.4.1}$$

By analogy with (9.1.1) the transformation is

$$z_{ij}(\lambda, \mu) = \begin{cases} \dfrac{q_{ij}^{\lambda} - 1}{\lambda \dot{q}^{\lambda - 1}} & (\lambda \neq 0) \\ \dot{q} \log q_{ij} & (\lambda = 0), \end{cases} \tag{9.4.2}$$

where $\dot{q} = G(y_{ij} + \mu_j)$, the geometric mean of all $n$ observations after the relevant shift in location. Due to the presence of the geometric mean, the transformation within the $j$th group depends on all members of the vector of shift parameters $\mu$.

The constructed variable and test statistic for $\lambda$ are unaltered by the presence of several shift parameters. Written as functions of $q$ they are identical to those of section 9.1. But there are now $c$ constructed variables and test statistics for a shift in location, one for each of the $c$ groups which has its own shift parameter. Each constructed variable has two forms. For observations within the $j$th group the extension of (9.1.3) is

$$w_{sj}(\lambda, \mu) = (q/\dot{q})^{\lambda-1} - (\lambda - 1)(n_j/n)\{hm(q_j)\}^{-1} \tag{9.4.3a}$$

whereas, for observations from other groups,

$$w_{sj}(\lambda, \mu) = -(\lambda - 1)(n_j/n)\{hm(q_j)\}^{-1}. \tag{9.4.3b}$$

In (9.4.3a) and (9.4.3b) $hm(q_j)$, the harmonic mean of the observations in the $j$th group is defined by

$$hm(q_j) = n_j / \sum_{i \in j} (1/q_{ij}).$$

The score statistics and added variable plots derived from the constructed variables are similar to those exemplified from Chapter 6 onwards. They require no further comment. We proceed at once to an example.

**Example 13 John and Draper's difference data** John and Draper (1980) present data on the subjective assessment of the thickness of pipe. In the experiment from which the data of Table 9.6 are taken, five inspectors assessed wall thickness at four different locations on the pipe. The experiment was replicated three times. The sixty responses are a multiple of the difference between the inspector's assessment and the 'true' value determined by an

**Table 9.6** Example 13: John and Draper's difference data

| | | Inspectors | | | | |
|---|---|---|---|---|---|---|
| Replicate | Location | 1 | 2 | 3 | 4 | 5 |
| 1 | 1 | 3.85 | −1.05 | −0.55 | −0.45 | −0.65 |
| | 2 | −1.95 | −0.65 | −2.25 | −0.35 | 0.25 |
| | 3 | −1.10 | −0.10 | 0.10 | 0.40 | −0.30 |
| | 4 | −4.95 | −15.55 | −20.25 | 0.95 | 3.15 |
| 2 | 1 | −2.25 | −0.55 | −0.95 | 0.05 | 9.65 |
| | 2 | −2.05 | −4.55 | −0.25 | −0.15 | −0.85 |
| | 3 | −1.00 | 0.00 | −2.50 | −1.00 | −0.30 |
| | 4 | −23.15 | 2.55 | −19.95 | −0.75 | 5.85 |
| 3 | 1 | −1.75 | 0.05 | 0.55 | −0.45 | 0.65 |
| | 2 | −0.45 | −0.35 | −0.15 | 0.15 | −0.25 |
| | 3 | 0.70 | −0.60 | −1.90 | 1.00 | 1.10 |
| | 4 | 3.45 | −24.15 | −0.35 | 3.15 | −13.75 |

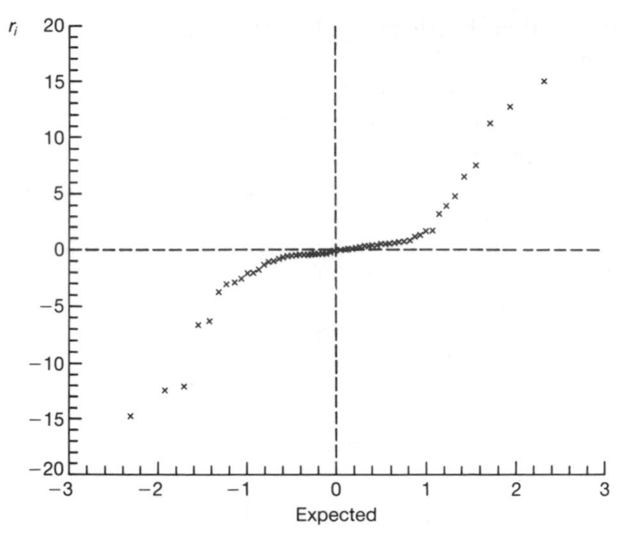


**Fig. 9.17** Example 13: John and Draper's difference data: normal plot of residuals

ultrasonic reader. The ultrasonic readings are no longer known: only the differences are available.

We begin the analysis by fitting an additive model with terms for inspectors, location and replication and include the inspector × location interaction. The assumption is that there is a constant structure over replicates. Figure 9.17 shows a normal plot of the residuals from this model, which clearly indicates that the residuals have a long-tailed distribution. If all of the observations in Table 9.6 had been non-negative, this pattern of residuals might have come from a skew distribution. The power transformation would then have been one way of attempting to obtain symmetry and so a straight residual plot. But the range of the observations is $-24.15$ to 9.65 so that the power transformation cannot be applied. One possibility is the shifted power transformation (9.1.1). John and Draper explore this possibility but reject it in favour of the modulus transformation in which the absolute values of the observations are subjected to a power transformation. The core of the transformation is, in the notation of Section 7.1,

$$u(\lambda) = \{(|y| + 1)^{\lambda} - 1\} \operatorname{sign}(y),$$

that is, a shifted power transformation for the absolute value of the observations. No discussion is given of the choice $\mu = 1$, although some comments are given by Cook and Weisberg (1982, Chapter 2.4.2).

In the present example the scientific meaning of the modulus transformation is not clear. Instead we shall use both (9.1.1) and the power transformation with separate shift parameters (9.4.2). For the justification of (9.1.1) let the unknown estimate of thickness from the $i$th replicate at location $j$ provided by the $k$th inspector be $\xi_{ijk}$ and suppose that the true value of the wall thickness at all locations is $\mu$. Use of the shifted power transformation (9.1.1) is equivalent to a model in which, for some unknown $\lambda$, the difference in the transformed variables $\xi_{ijk}^{\lambda} - \mu^{\lambda}$ satisfies a linear model. We however only have observations

$$y_{ijk} = \xi_{ijk} - \mu, \tag{9.4.4}$$

which differences may not be on the correct scale either for the linear model or for normality. In this formulation $\mu$ can be an unknown constant or can be subject to random variation. In the latter case we argue conditionally on the unknown value. The difference (9.4.4) and the linear model for the differences on the correct scale combine to yield

$$X\theta + \varepsilon = \xi_{ijk}^{\lambda} - \mu^{\lambda} = (y_{ijk} + \mu)^{\lambda} - \mu^{\lambda}.$$

Provided the linear model includes a constant this is equivalent to the model of earlier sections

$$(y_{ijk} + \mu)^{\lambda} = X\beta + \varepsilon. \tag{9.4.5}$$

John and Draper fitted the shifted power transformation (9.4.5) to all 60 observations and obtained parameter estimates $\hat{\lambda} = 8.6$ and $\hat{\mu} = 100$. To begin our analysis we also take $\mu$ as 100 but consider evidence for the power transformation in the additive model with inspector × location interaction. At $\lambda = 1$ the results of Table 9.7 yield the score statistic $T_p(1, 100)$ as 8.08, so there is strong evidence of the need for a power transformation after 100 has been added to all observations. From the untransformed data, as we have already seen, there can be no evidence as to whether 100 is an appropriate value for $\mu$.

The positive value of the score statistic indicates that the maximum likelihood estimate of $\lambda$ will be greater than one. More detailed information about the nature of the evidence for a transformation comes from plots of the constructed variables. Figure 9.18 is the added variable plot for $w_p(1, 100)$. The plot of $w_s(0, 100)$ for the shift parameter after the log transformation has been taken is virtually identical to Fig. 9.18 and so is not given. The plots reveal that the majority of the evidence for a transformation comes from

**Table 9.7** Example 13: John and Draper's difference data. Residual sums of squares and score statistics for the power transformation with one shift parameter

| $\lambda$ | $\mu$ | Residual sum of squares of $z(\lambda, \mu)$ | Score statistics Power $T_p(\lambda, \mu)$ | Shift $T_s(\lambda, \mu)$ |
|---|---|---|---|---|
| 0 | 100 | 1553.4 | 9.394 | 9.243 |
| 1 | 100 | 1345.6 | 8.083 | — |
| 9 | 105 | 828.5 | 0.003 | −0.003 |

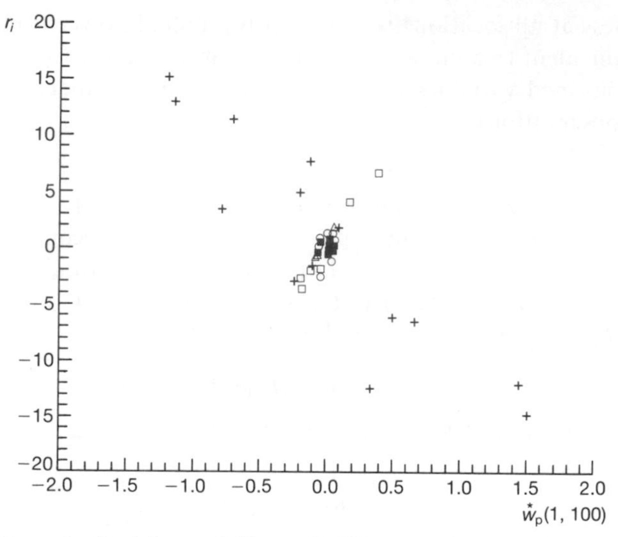


**Fig. 9.18** Example 13: John and Draper's difference data: added variable plot for constructed variable $w_p(1, 100)$: location 1, □; location 2, ○; location 3, △; location 4, +

location 4, with the other locations, especially location 1, indicating a power or shift transformation in the opposite direction. The similarity of evidence for the two transformations echoes the results of Section 9.2 where agreement was between two plots for $\lambda = 0$. In the present example the added variable plot for $w_p(1, 100)$ is similar to that for $w_p(0, 100)$ because the value of $\hat{\lambda}$ is far from 1.

Maximization of the likelihood over values of $\lambda$ and $\mu$ leads to the parameter estimates $\hat{\lambda} = 9$ and $\hat{\mu} = 105$. The residual sum of squares for $z(\lambda)$, 828.5, is close to the value at the estimates 8.6 and 100 reported by John and Draper. The likelihood is therefore flat in this region. These estimates suggest that all is not well. As a general rule, observations should not be raised to anything like the ninth power. Amongst other causes of this result is the possibility of an outlier: in the analysis of John's $3^{4-1}$ experiment, Example 8, the value of 8.27 for $\hat{\lambda}$ was caused by the presence of a single outlier. Another possibility is that the model may be systematically inadequate in some way. Inspection of the added variable plot for $w_p(9, 105)$ (Fig. 9.19) fails to reveal any one outlying observation. But the plot, which is similar to that for $w_s(9, 105)$, shows that the effect of the transformation has been to find a set of parameter values for which the evidence for either a mean shift or a power transformation is the same at location 1 as at location 4, but of opposite sign. The two effects cancel, with the other two locations contributing little to the values of the statistics.

It is clear from Fig. 9.19 that the implausible value for $\hat{\lambda}$ is due to systematic differences between the locations. To allow for these differences we employ

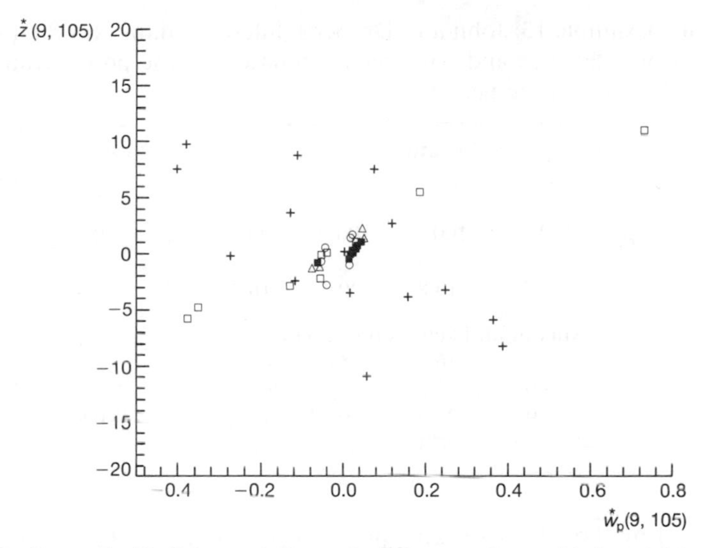


**Fig. 9.19** Example 13: John and Draper's difference data: added variable plot for constructed variable $w_p(9, 105)$: location 1, □; location 2, ○; location 3, △; location 4, +

the transformation (9.4.2) which, in the present case, is equivalent to letting the thickness of the pipe at location $j$ be $\mu_j$. The model (9.4.4) for the observed $y_{ijk}$ is replaced by

$$y_{ijk} = \xi_{ijk} - \mu_j \qquad (j = 1, \ldots, 4). \tag{9.4.6}$$

The hope is still that, after an unknown common power transformation, the difference $\xi_{ijk}^{\lambda} - \mu_j^{\lambda}$ will, at least approximately, satisfy a normal theory linear model.

With four shift parameters, one per location, there are four score statistics $T_{sj}(\lambda, \mu_1, \ldots, \mu_4)$. To begin analysis of this extended model we start with the statistics for $\lambda = 0$ and all $\mu_j = 100$ given in Table 9.8. These suggest values of the shift parameters appreciably less than 100 for the first three locations. For location 4 a value in excess of 100 is indicated. Table 9.8 also gives values of the maximum likelihood estimates of the shift parameters for a series of $\lambda$ values in the range 0.5 to $-1$. Over these values of $\lambda$, which are *a priori* plausible, the value of the residual sum of squares varies only slightly with $\lambda$. A value of 0, corresponding to the log transformation, would not be inappropriate. Such a value is much to be preferred to the value of 9 obtained with a single shift parameter.

Elaboration of the model to include four shift parameters reduces the residual sum of squares from a minimum of 828.5, Table 9.7, to 242.8 in Table 9.8 and produces a reasonable estimate for $\lambda$. But the values of $\mu$, which range from 16.5 to 204, seem surprising and disturbing as the $\mu_j$ are the thicknesses of the walls of a pipe at four locations. Further disquiet about this model is

**Table 9.8** Example 13: John and Draper's difference data. Residual sums of squares, score statistics and parameter estimates for the power transformation with four shift parameters

| | | Location | | | | Power | | Residual sum of squares |
|---|---|---|---|---|---|---|---|---|
| | | 1 | 2 | 3 | 4 | | | |
| Parameter | $\mu_j$ | 100 | 100 | 100 | 100 | $\lambda$ | 0 | |
| Score statistic | $T_{sj}$ | −5.77 | −14.8 | −20.3 | 16.7 | $T_p$ | 9.39 | 1553.4 |
| | | Maximum likelihood estimates $\hat{\mu}_j$ | | | | | $\lambda$ | |
| | | 39.0 | 10.3 | 5.0 | 600 | | 0.5 | 248.7 |
| | | 48.0 | 23.6 | 16.5 | 204 | | 0 | 243.3 |
| | | 37.9 | 25.2 | 19.4 | 110 | | −0.5 | 242.8 |
| | | 38.7 | 29.3 | 23.6 | 92.3 | | −1 | 243.5 |

engendered by the added variable plot of $w_p(0, \hat{\boldsymbol{\mu}})$ in Fig. 9.20, in which the behaviour of the four groups is still distinct. The added variable plots for the shift parameters have a structure which is more difficult to interpret. Because of the bipartite nature of the variables given by (9.4.3a) and (9.4.3b) some structure might be expected. But the plots for $w_{s2}(0, \hat{\boldsymbol{\mu}})$ and $w_{s4}(0, \hat{\boldsymbol{\mu}})$ given in Figs. 9.21 and 9.22 show a degree of difference among the locations which is sufficient to suggest that a satisfactory model has not been found by these transformation methods.

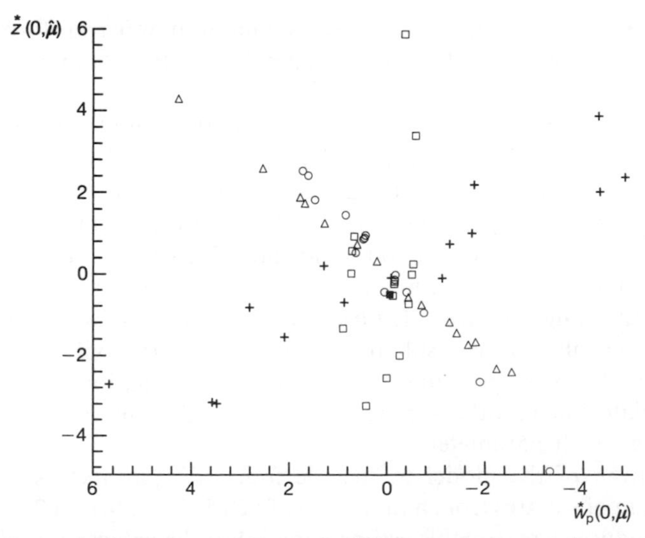


**Fig. 9.20** Example 13: John and Draper's difference data: added variable plot for constructed variable $w_p(0, \hat{\boldsymbol{\mu}})$: location 1, □; location 2, ○; location 3, △; location 4, +

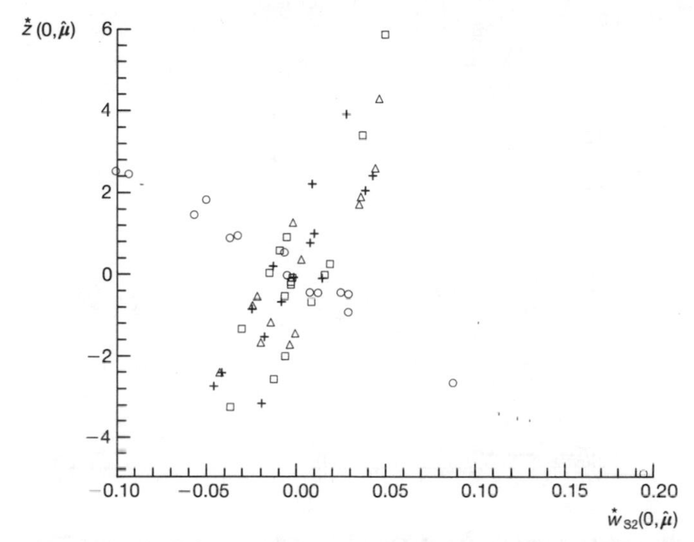


**Fig. 9.21** Example 13: John and Draper's difference data: added variable plot for constructed variable $w_{s2}(0, \hat{\boldsymbol{\mu}})$: location 1, □; location 2, ○; location 3, △; location 4, +

The suggestion that a transformation might be appropriate came from the normal pot of the residuals, Fig. 9.17. A shortcoming of this plot is that it does not indicate the group from which each residual comes. Since the four locations have turned out to be of extreme importance in the analysis of the

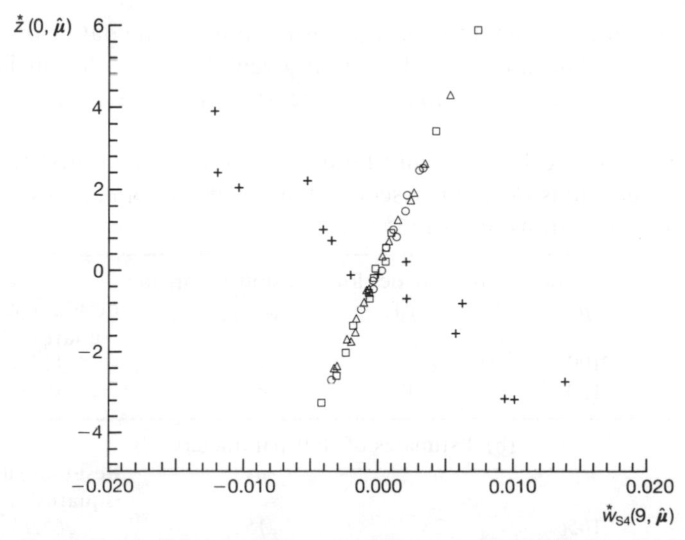


**Fig. 9.22** Example 13: John and Draper's difference data: added variable plot for constructed variable $w_{s4}(0, \hat{\boldsymbol{\mu}})$: location 1, □; location 2, ○; location 3, △; location 4, +

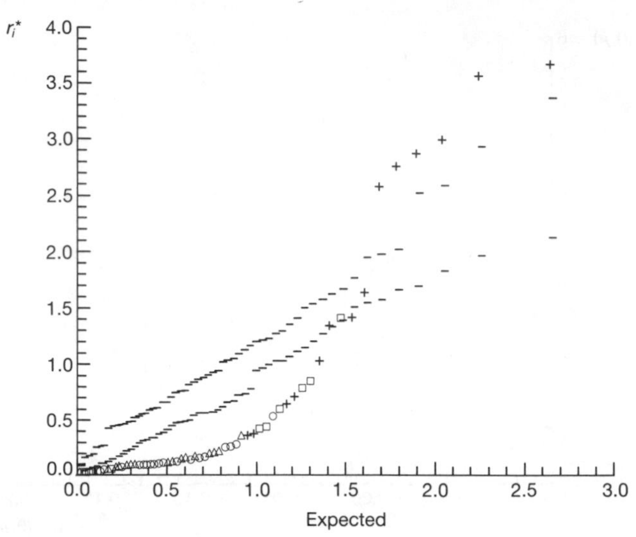


**Fig. 9.23** Example 13: John and Draper's difference data: half-normal plot of deletion residuals $r_i^*$: location 1, □; location 2, ○; location 3, △; location 4, +

data, we again plot the residuals, but this time include an indication of location. Figure 9.23, the resulting half-normal plot of the deletion residuals $r_i^*$, reveals a clear pattern: the eight largest values of $r_i^*$, and ten out of the largest 11, come from location 4. The plot suggests that the data from location 4 should be analysed separately from that from the other three locations.

We now repeat the analysis of transformations for the data from the first three locations. This analysis echoes that given above for all four locations. The resulting parameter estimates, score statistics and residual sums of

**Table 9.9** Example 13: John and Draper's difference data, first three locations. Residual sums of squares, score statistics and parameter estimates for the shifted power transformation

| | | (a) Score statistics for one shift parameter | | |
|---|---|---|---|---|
| $\lambda$ | $\mu$ | $T_s(\lambda, \mu)$ | $T_p(\lambda, \mu)$ | Residual sum of squares of $z(\lambda)$ |
| 1 | 100 | — | $-6.26$ | 112.5 |
| 0 | 100 | $-5.80$ | $-5.83$ | 106.5 |
| | | **(b) Estimates of shift parameters** | | |
| $\lambda$ | $\mu_1$ | $\mu_2$ | $\mu_3$ | Residual sum of squares of $z(\lambda)$ |
| 0.5 | 165 | 28 | 15 | 65.94 |
| 0 | 101 | 43 | 32 | 65.22 |
| $-0.5$ | 79 | 46 | 37 | 64.53 |

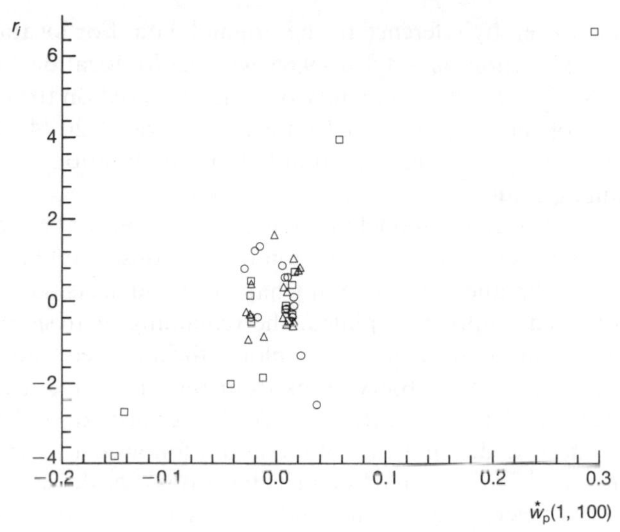


**Fig. 9.24** Example 13: John and Draper's difference data, locations 1–3: added variable plot for constructed variable $w_p(1, 100)$: location 1, □; location 2, ○; location 3, △

squares are given in Table 9.9. For the model with a single shift parameter there is evidence in the value of $-6.26$ for $T_p(1, 100)$ in Table 9.9(a) that a power transformation is needed if $\mu = 100$. The related added variable plot of $w_p(1, 100)$ (Fig. 9.24) has something of the structure of Fig. 9.18 for the four groups. This time the three groups behave differently, with most of the evidence for a transformation coming from location 1. The results in Table 9.9(b) for three shift parameters, one per location, show, as did the results of Table 9.8, that there is little to choose between the log and neighbouring transformations. In this case the log transformation with separate shift parameters for each location requires the estimation of four extra parameters: as a result of adding these parameters the residual sum of squares is reduced, from 112.5 for the untransformed data, to 64.5. The data from location 4, on the other hand, provide no evidence of the need for a transformation. For $\lambda = 1$ the residual sum of squares of the observations is 1232.0 and $T_p(1, 100) = 1.55$. The added variable plot for $w_p(1, 100)$, not shown here, does not reveal any particularly informative observations for this transformation. The need for a transformation has therefore not been masked by an outlying observation.

These results are quite unsatisfactory for providing a physically plausible model of the data. The evidence from the first three locations is that a power transformation with separate shift parameters should be applied, but the log transformation is just rejected for location 4 with a value of 2.10 for the score statistic. It is implausible that the two sets of measurements of the same quantity should be treated on different scales. That the two sets are very different can be seen by reference to the original data. For locations 1–3 the range of the observations is −4.55 to 9.65, whereas for location 4 the range is −24.15 to 5.85. The $F$ test for equality of variance based on the two residual mean squares of the untransformed data has the value 20.44 on 8 and 28 degrees of freedom. The 0.2 per cent point of this distribution is 4.69, so there is overwhelming evidence of variance heterogeneity.

To try to find a common model for all locations we now abandon transformations and consider instead the deletion of a few observations. We employ the diagnostic techniques described in Chapters 3 and 4. As these techniques have already been amply exemplified, the remaining analysis is somewhat brief. Figure 9.25 shows the half-normal plot of the modified Cook statistic $C_i$ for the 45 untransformed observations from the first three locations. The largest value belongs to observation 20 which is clearly an outlier from this model. The effect of deleting this observation, followed by the deletion of observations 1 and 22 is summarized in Table 9.10. After these three observations have been deleted the half-normal plot for the remaining 42 observations, which is not shown, contains no obvious outliers. By deletion of the three observations, which can be regarded as the addition of three extra parameters to the model, the residual sum of squares is reduced from 112.50 to 12.89. This is to be compared with 65.2 for the transformation model which required four extra parameters. A similar analysis for location 4 shows that deletion of observation 7 reduces the residual sum of squares from 1232.0 to 804.1. If the purpose of the analysis is to find a model with low residual sum of squares in

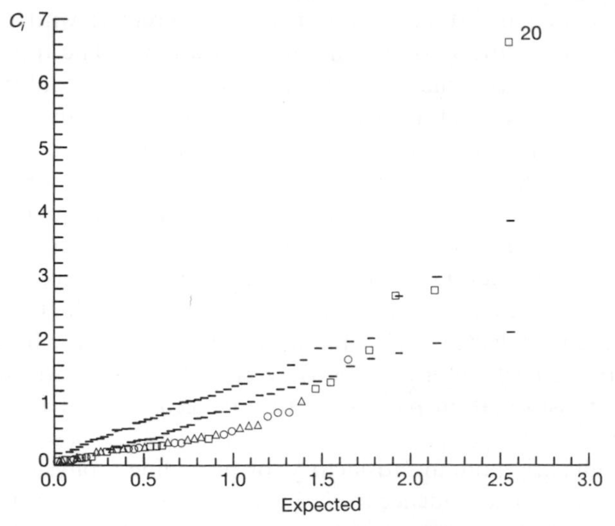


**Fig. 9.25** Example 13: John and Draper's difference data, locations 1–3: half-normal plot of modified Cook statistic $C_i$ for untransformed data: location 1, □; location 2, ○; location 3, △

**Table 9.10** Example 13: John and Draper's difference data. Effect of deletion of observations on the residual sum of squares of the untransformed data

(a) Locations 1, 2 and 3

| Number of observations | Observations deleted | Maximum modified Cook statistic $C_i$ | Residual sum of squares |
|---|---|---|---|
| 45 | — | 6.63 | 112.50 |
| 44 | 20 | 5.14 | 42.77 |
| 43 | 1, 20 | 3.82 | 20.84 |
| 42 | 1, 20, 22 | 3.49 | 12.89 |
| Log transformation with three shift parameters | | | 65.22 |

(b) Location 4

| Number of observations | Observations deleted | Maximum modified Cook statistic $C_i$ | Residual sum of squares |
|---|---|---|---|
| 15 | — | 1.93 | 1232.0 |
| 14 | 7 | 2.37 | 804.1 |

which all locations are assessed on the same scale, deletion of a few observations is more effective than power transformations.

In the absence of further information about the data it is pointless to speculate about the differences between the two groups. For locations 1, 2 and 3 the deletion procedure found successively the observations with values 9.65, 3.85 and $-4.55$. The range of the remaining 42 observations is $-2.50$ to 1.10, with a residual mean square of 0.42. For location 4 the residual mean square after deletion of observation 7 is 114.9. □

There are several negative conclusions to be drawn from this analysis, some of which have been stated above. Another is that our analysis has failed to establish the usefulness of the power transformation with several shift parameters. The desirability of the modulus transformation came from a misinterpretation of the normal plot of residuals, Fig. 9.17. The more detailed plot of Fig. 9.23, combined with the subsequent analysis, showed that the long-tailed distribution of residuals is caused by variance heterogeneity. But it should be evident from inspection of the data of Table 9.6 that the variances are not homogeneous and, in particular, that there is something different about location 4. Ignoring this point led to unnecessary complications in the analysis. Despite the appearance of Fig. 9.17, not all departures in the distribution of residuals can be transformed away.

# 10
# Further diagnostic quantities

## 10.1 Multiple deletion diagnostics

The discussion in Section 9.4 of transformation models with several shift parameters concludes the second major part of this book. In the remaining three chapters some more specialized and advanced topics are explored.

In the present chapter we return to the diagnostic measures for the linear regression model which were the subject of the first five chapters. This section is concerned with the effect of multiple deletion on diagnostic calculations. In the next section consideration is given to measures of influence other than Cook's distance for all the parameters in the linear model. The last section of the chapter outlines extensions of the methods to nonlinear least squares and to generalized linear models. In all three sections the emphasis is on the development of statistical methods, rather than on their application to examples.

The two remaining chapters contain examples of the use of some other diagnostic techniques. Chapter 11 describes one extension of the technique of constructed variables to generalized linear models. The chimpanzee data, Example 12, are used as an illustration. The last chapter returns to the linear regression model with further results on transformations, including robust transformations, and on non-constancy of variance. But we begin this section with the multiple deletion analogues of the deletion residuals $r_i^*$ and of the modified Cook statistic $C_i$.

In Section 3.1 the deletion residual was derived for the effect of deletion of a single observation $i$. We now look for the effect of deletion of a group of $m$ observations indexed by $I$. The $t$ test (3.1.1) defining $r_i^*$ is for agreement between the observed value $y_i$ and the prediction $x_i^T\hat{\beta}_{(i)}$. An alternative derivation comes from consideration of the reduction in the residual sum of squares when an extra parameter is added to the model to explain the $i$th observation. The $F$ test for the effect of this parameter, on 1 and $n-p-1$ degrees of freedom is the square of (3.1.1) or equivalently of (3.2.2). The carrier for this extra term thus consists of zeros everywhere except for the $i$th element which is unity. The resulting residual sum of squares is therefore identical to that which results from deletion of the $i$th observation, since the residual for the $i$th observation is zero. We now give this derivation for a subset $I$ of $m$ observations.

For all $n$ observations the residual sum of squares is $R(\hat{\beta})$. In the augmented model with $m$ extra parameters, or equivalently when the set $I$ of observations is deleted, the residual sum of squares is, in a somewhat cumbersome notation, $R_{(I)}(\hat{\beta}_{(I)})$. The decrease in the residual sum of squares due to fitting the augmented model is

$$R(\hat{\beta}) - R_{(I)}(\hat{\beta}_{(I)}).$$

From (2.2.3) the estimate of $\sigma^2$ from the augmented model is defined by

$$(n - p - m)s_{(I)}^2 = R_{(I)}(\hat{\beta}_{(I)}).$$

The $F$ test for the extra parameters, that is for deletion of the $m$ observations, is therefore given by

$$F_I = \frac{\{R(\hat{\beta}) - R_{(I)}(\hat{\beta}_{(I)})\}/m}{s_{(I)}^2},$$

which, from (2.2.7) can be written as

$$mF_I = r_I^{\mathrm{T}}(I - H_I)^{-1} r_I / s_{(I)}^2. \tag{10.1.1}$$

For $m = 1$ this $F$ test, as was asserted above, reduces to the square of the deletion residual (3.1.2).

Instead of (10.1.1) Cook and Weisberg (1980) give the analogue of the standardized residual (2.1.14) in which $s_{(I)}^2$ in (10.1.1) is replaced by $s^2$, the estimate from the model before deletion. The relationship between the two, since we are dealing with an $F$ distributed quantity, is given by the square of (3.1.3), with the degrees of freedom adjusted to $n - p - m$ (Cook and Weisberg 1982, p. 30).

The analogue of the Cook statistic $D_i$ is, in a similarly straightforward manner, defined to be

$$D_I = (\hat{\beta}_{(I)} - \hat{\beta})^{\mathrm{T}} X^{\mathrm{T}} X (\hat{\beta}_{(I)} - \hat{\beta})/ps^2.$$

From the expression for $\hat{\beta}_{(I)} - \hat{\beta}$ given by (2.2.5) the analogue of (3.2.4) for multiple deletion is

$$D_I = r_I^{\mathrm{T}}(I - H_I)^{-1} H_I (I - H_I)^{-1} r_I / ps^2. \tag{10.1.2}$$

Use of $s_{(I)}^2$ rather than $s^2$ to estimate $\sigma^2$ in (10.1.2) would have a similar effect to that for deletion of a single observation and would lead to a modified Cook statistic.

Both of the statistics (10.1.1) and (10.1.2) consist of quadratic forms in the residuals. By use of the diagonalization of $H_I$ it is possible to write both expressions as sums of $m$ terms, each of which is like the square of a statistic for the deletion of a single observation. But the linear combinations of residuals which result do not have an easy interpretation. The details are given by Cook and Weisberg (1982, pp. 30, 136).

In the numerical examples of the use of single deletion diagnostics summarized in Tables 3.1–3.3, the values of $r_i^*$ and $C_i$ were supplemented by lists of the leverage measures $h_i$ and $h_i/(1 - h_i)$. Comparison of the single deletion formulae with their multiple deletion versions (10.1.1) and (10.1.2) suggests that the obvious analogue of $h_i/(1 - h_i)$ is the $m \times m$ matrix $H_I(I - H_I)^{-1}$. To collapse this matrix to a scalar measure of leverage Cook and Weisberg (1980; 1982, pp. 139–142) suggest use of the trace of the matrix, that is $\operatorname{tr}\{H_I(I - H_I)^{-1}\}$. An alternative would be the determinant $\det\{H_I(I - H_I)^{-1}\}$. This ambiguity in the definition of a single quantity for multi-observational leverage suggests that leverage may be a concept which is most useful for observations considered singly rather than in groups.

There are two disadvantages to the routine use of $F_I$ and $D_I$, or related diagnostic measures. The minor one is that plotting methods for these quantities have not yet been developed, so that the analyst of data will have to rely on, perhaps lengthy, lists of diagnostic quantities. A start in the provision of plots has been made by Cook and Weisberg (1982, p. 145) who provide a semi-graphical display for the deletion of pairs. The major disadvantage, mentioned at the beginning of this chapter, is the number of possibilities to be considered. Methods for overcoming this problem have something in common with methods for the selection of carriers in multiple regression. In multiple regression there are $p$ possible carriers, a subset of which is to be chosen. This can be found by backwards elimination, which corresponds to dropping one variable at a time from the equation, or by consideration of the effect of dropping all possible subsets of size $m$. A method for doing this systematically is described by Furnival and Wilson (1974).

The combinatorial problems are even worse in trying to detect anomalous observations because there are $n$ candidate points as opposed to $p$ carriers. The number of pairs, triplets etc, is therefore correspondingly much greater. A disincentive to undertaking this computing is that the value of the multiple deletion diagnostics is not clearly established. In theory there are examples where observations are jointly, but not individually, influential. A plot of such a set of data is given by Cook and Weisberg (1980, Fig. 1; 1982, p. 135). Another problem which can occur is that the importance of a particular observation may not be apparent until some other observation has been deleted. In the presence of such a masking effect, finding subsets which contain observations which are individually influential may not reveal all of the structure in the data. However, the hope is that the sequential use of single deletion statistics will usually serve to identify strange observations. A guide to the statistical properties of such sequential methods is given by Dempster and Gasko-Green (1981). Experience confirming the hope that sequential methods are often adequate is reported by Cook and Weisberg (1980) and by Belsley *et al.* (1980, p. 31). If it is thought necessary to calculate multiple deletion diagnostics, the problem can be reduced to some extent by first deleting observations which are individually influential or outlying. Cook and Weisberg (1980; 1982, p. 146) give bounds on the value of $D_I$ which can significantly reduce the number of $m$-tuples to be investigated.

An alternative method of combating the combinatorial problem in the detection of multiple outliers is suggested by Hawkins *et al.* (1984). The method samples subsets of $p$ observations, to each of which a model is fitted. The fitted models are then used to predict the remaining $n - p$ observations. Repeated sampling means that a distribution of residuals is built up from which multiple outliers can be detected.

Although in this book only one at a time diagnostics have been considered, the potential importance of multiple outliers and diagnostics remains, as is stressed by Andrews and Pregibon (1978). Synthetic examples for two-way layouts are given by Gentleman and Wilk (1975a, 1975b) where the outliers are caused by disturbances in the responses. There is thus no problem of leverage, so that only statistics related to $F_I$ (10.1.1) are considered. Simulation is used by Gentleman and Wilk to establish the null distribution of their statistics and to provide plots to aid interpretation. An example in which multiple deletion diagnostics are important, but without the combinatorial complexity of choosing $m$ out of $n$ observations, is in case control studies. Each observation on an experimental unit, usually a patient, is matched with $m - 1$ controls. Deletion of the single patient involves simultaneous deletion of all controls in the matching set.

## 10.2 Other measures of influence

### *10.2.1 A subset of parameters*

The derivation of the Cook statistic in Section 3.2 started from the formula for the confidence region for all the parameters in the model, given by (3.2.1). This is appropriate if all parameters are of equal interest. But it may be that only a subset of the parameters is important. One example is when the parameters can be divided into two groups, one set describing the response to a treatment, the other allowing for the effects of blocking and of concomitant variables. The influence of groups of observations on conclusions about the treatment parameters would then be of direct interest. A second example is in multiple regression where, often, the coefficients of the explanatory variables are of greater interest than the value of the constant term.

A general formulation is to suppose that interest is in $t$ contrasts which are the elements of the vector $\theta = A^{\mathrm{T}}\beta$, where $A^{\mathrm{T}}$ is a $t \times p$ matrix of known constants, with rank $t \leqslant p$. The covariance matrix of the least squares estimate $\hat{\theta} = A^{\mathrm{T}}\hat{\beta}$ is $\sigma^2 A^{\mathrm{T}}(X^{\mathrm{T}}X)^{-1}A$. The $100(1-\alpha)$ per cent confidence region for $\theta$ is, corresponding to (3.2.1), given by those values satisfying

$$(\theta - \hat{\theta})^{\mathrm{T}}\{A^{\mathrm{T}}(X^{\mathrm{T}}X)^{-1}A\}^{-1}(\theta - \hat{\theta}) \leqslant ts^2F_{t,\nu,\alpha}. \qquad (10.2.1)$$

The measure of influence analogous to the Cook statistic for this set of contrasts is therefore

$$D_I(\theta) = (\hat{\theta}_{(I)} - \hat{\theta})^{\mathrm{T}}\{A^{\mathrm{T}}(X^{\mathrm{T}}X)^{-1}A\}^{-1}(\hat{\theta}_{(I)} - \hat{\theta})/ts^2, \qquad (10.2.2)$$

where $\hat{\theta}_{(I)} = A^{\mathrm{T}}\hat{\beta}_{(I)}$. Substitution for $\hat{\beta}_{(I)} - \hat{\beta}$ from (2.2.5) leads to the somewhat opaque general expression

$$ts^2 D_I(\theta) = r_I^{\mathrm{T}}(I - H_I)^{-1}X_I(X^{\mathrm{T}}X)^{-1}A\{A^{\mathrm{T}}(X^{\mathrm{T}}X)^{-1}A\}^{-1}A^{\mathrm{T}}(X^{\mathrm{T}}X)^{-1} \times X_I^{\mathrm{T}}(I - H_I)^{-1}r_I. \qquad (10.2.3)$$

Simplification occurs when only one observation is deleted. Let

$$M = (X^{\mathrm{T}}X)^{-1}A\{A^{\mathrm{T}}(X^{\mathrm{T}}X)^{-1}A\}^{-1}A^{\mathrm{T}}(X^{\mathrm{T}}X)^{-1}, \qquad (10.2.4)$$

when, for the deletion of a single observation, (10.2.3) can be rewritten as

$$D_i(\theta) = \frac{r_i^2}{ts^2(1 - h_i)^2}x_i^{\mathrm{T}}Mx_i = \frac{r_i'^2}{t}\frac{x_i^{\mathrm{T}}Mx_i}{(1 - h_i)}. \qquad (10.2.5)$$

In this form $D_i(\theta)$ is easily computed because $M$, the kernel of the quadratic form in $x_i$, is a $p \times p$ matrix which does not depend on $i$. When all parameters are of interest, that is when $A = I$, (10.2.5) reduces to $D_i$ given by (3.2.4).

One special case of (10.2.5) which corresponds to the first example above, with some parameters describing response to a treatment, is when $A$ defines a subset of parameters which are of interest. In this case

$$A^{\mathrm{T}} = (I_t \quad 0) \qquad (10.2.6)$$

and the set of contrasts can be written explicitly as

$$A^{\mathrm{T}}\beta = (I_t \quad 0)\beta = (\beta_1 \quad \beta_2 \quad \dots \quad \beta_t)^{\mathrm{T}}.$$

For this set of contrasts it follows, from the expression for the inverse of a partitioned matrix given in the Appendix, that $M$ in (10.2.4) reduces to

$$M = \begin{bmatrix} P & Q \\ Q^{\mathrm{T}} & Q^{\mathrm{T}}P^{-1}Q \end{bmatrix} = \begin{bmatrix} P & Q \\ Q^{\mathrm{T}} & U - (U - Q^{\mathrm{T}}P^{-1}Q) \end{bmatrix}.$$

But from (A2) applied to

$$\begin{bmatrix} P & Q \\ Q^{\mathrm{T}} & U \end{bmatrix}^{-1} = X^{\mathrm{T}}X,$$

$$U - Q^{\mathrm{T}}P^{-1}Q = (X_2^{\mathrm{T}}X_2)^{-1}$$

so that

$$M = (X^TX)^{-1} - \begin{bmatrix} 0 & 0 \\ 0 & (X_2^TX_2)^{-1} \end{bmatrix}. \qquad (10.2.7)$$

This matrix result can now be substituted in (10.2.3) to yield the influence measure

$$ts^2D_I(\theta) = ps^2D_I - r_I^T(I - H_I)^{-1}G_I(I - H_I)^{-1}r_I, \qquad (10.2.8)$$

where $G = X_2(X_2^TX_2)^{-1}X_2^T$ is the $n \times n$ hat matrix for regression on $X_2$ alone and $D_I$ is the Cook statistic for all parameters as defined in (10.1.2).

When only one observation is deleted (10.2.8) can be written in the more revealing form

$$D_i(\theta) = \frac{r_i'^2(h_i - g_i)}{t(1 - h_i)}, \qquad (10.2.9)$$

where $r_i'$ is the standardized residual (2.1.14) and $g_i$ is the $i$th diagonal element of the hat matrix $G$. The comparison with the definition of $D_i$ (3.2.4) is striking. To investigate the influence of a single observation on a subset of parameters quantities are required from two regressions: on the full model and on the complement of the submodel of interest in the full model.

This interpretation of (10.2.9) is particularly helpful in deriving the statistic for all the parameters in a regression model excluding the constant. The complement of the parameters of interest in the full model is therefore the constant, so that regression on $X_2$ is equivalent to averaging the observations. The resulting hat matrix $G$ has diagonal element $1/n$ and

$$D_i(\theta) = \frac{r_i'^2(h_i - 1/n)}{(p - 1)(1 - h_i)}. \qquad (10.2.10)$$

This new measure of influence is therefore the same as the Cook statistic $D_i$, but with $1/n$ subtracted from the value of $h_i$ in the numerator. The modification is not particularly interesting and has not been much employed. For example, Cook and Weisberg (1982, p. 126) derive (10.2.10) but continue to use $D_i$ in the analysis of data.

There is a set of results analogous to the interpretation of (10.2.9) in the theory of the optimum design of experiments described by Silvey (1980). The $D$-optimum design is appropriate if all parameters are of interest. From the general equivalence theorem these designs minimize the maximum of the variance of the predicted response, that is $h_i$, over the design region. $D$-optimum designs for the set of contrasts $A^T\beta$ are called $D_A$-optimum (Sibson 1974). For the special case in which the set of contrasts reduces to a subset of the parameters, the criterion is called $D_s$-optimality. The analogue of the variance of the predicted response in the general equivalence theorem for this criterion is the difference between the variance for the full model and that for the complement of the subset of interest, that is $h_i - g_i$. The resemblance to (10.2.9) is striking.

The results leading to (10.2.9) apply when interest is in a subset of parameters in a fitted model. The simplest case is when interest is in the influence of the observations on the estimate of only one of the parameters. The Cook statistic (10.2.2) then reduces to looking at the change in the estimate of the parameter on deletion of the $i$th observation, divided by the estimated standard error of the parameter estimate. To see this suppose that the parameter of interest is $\beta_j$, and let the $j$th diagonal element of $V = (X^TX)^{-1}$ be $v_j$. Then $A$ is a column vector of zeros with unity in the $j$th position and $A^T(X^TX)^{-1}A = v_j$. The Cook statistic reduces from (10.2.2) to

$$D_i(\beta_j) = \frac{(\hat{\beta}_{j(i)} - \hat{\beta}_j)^2}{s^2 v_j}, \tag{10.2.11}$$

which is related to the square of the $t$ statistic for testing hypotheses about the value of $\beta_j$.

As with the other generalizations of the Cook statistic derived in this section, (10.2.11) could be modified, and improved, by use of the variance estimate $s^2_{(i)}$. From (2.2.8), this suggests that the analogue of the modified Cook statistic $C_i$ (3.2.5), when it is required to assess influence for a single parameter, is given by the absolute value of

$$\begin{aligned} C_i(\beta_j) &= \frac{\sqrt{(n-p)}\{(X^TX)^{-1}x_i\}_j r_i}{s_{(i)}(1-h_i)\sqrt{v_j}} \\ &= \frac{\sqrt{(n-p)}\{(X^TX)^{-1}x_i\}_j |r_i^*|}{\sqrt{\{v_j(1-h_i)\}}}, \end{aligned} \tag{10.2.12}$$

where $\{(X^TX)^{-1}x_i\}_j$ is the $j$th element of the vector $(X^TX)^{-1}x_i$. Apart from the factor of $\sqrt{(n-p)}$, (10.2.12) is the absolute value of the quantity which Belsley *et al.* (1980, p. 13) call $DFBETAS_{ij}$. Either $D_i(\beta_j)$ or the modified form (10.2.12) could usefully be displayed as an index plot. In Chapter 6, Figs. 6.38 and 6.39 show index plots of the change in the parameter estimate $\Delta\hat{\beta}_j = \hat{\beta}_{j(i)} - \hat{\beta}_j$. An advantage of the forms derived in this section is that scaling by the standard error of the parameter estimate provides some guidance to the statistical significance of the observed changes on deletion.

### *10.2.2 Confidence regions*

Influence has so far been judged by how much a vector of parameter estimates, or a linear combination of such estimates, is changed by deletion. In the present section references are given to work on the effect of the deletion of observations on the volume of a confidence region.

The $100(1-\alpha)$ per cent normal theory confidence region for $\beta$ was given in (3.2.1). The volume of this ellipsoid, dependent on the determinant of $X^TX$ and on the estimate of $\sigma^2$, is given by

$$E \propto \{s^{2p}/|X^TX|\}^{1/2}. \tag{10.2.13}$$

When the $i$th observation is deleted the volume is

$$E_{(i)} \propto \{s_{(i)}^{2p}/|X_{(i)}^TX_{(i)}|\}^{1/2}.$$

Belsley *et al.* (1980, p. 22) give the name *COVRATIO* to $\{E_{(i)}/E_i\}^2$. By taking the logarithm of the ratio a statistic is obtained which has the value zero if the volume is unchanged by deletion. Cook and Weisberg (1980; 1982, p. 159) use the log ratio but adjust by the ratio of $F$ values in the definition of the confidence regions.

Although this new influence measure is derived from a starting point quite distinct from those considered earlier in this book, it can be shown to be another combination of the basic building blocks $r_i$ and $h_i$ which repeatedly occur in the derivation of diagnostic quantities.

The interpretation of $\log(E_{(i)}/E)$ is that an appreciable negative value means that deletion of observation $i$ has caused an appreciable decrease in the volume of the confidence ellipsoid. Deletion of an outlier for which $r_i^*$ is large but which is at a point of low leverage would cause such a decrease. An appreciable positive value would result if $h_i$ were near 1 and $r_i^*$ small. This would mean that a leverage point had been deleted leading to an increase in the volume of the confidence region. For most observations the value would be expected to be near zero. This discussion suggests that deletion of an observation with appreciable leverage and a large value of $r_i^*$, which is certainly an observation one would like to know about, might leave the value virtually unchanged. Figure 4.4.1(d) of Cook and Weisberg (1982) shows that this is indeed the case. The indication is therefore that the statistic is not reliable as an omnibus measure of influence.

Extensions of these ideas to measures for deletion of a subset $I$ of observations, and to interest in the vector of contrasts $A^T\beta$, are discussed in Chapter 4 of Cook and Weisberg (1982), which is devoted to alternative approaches to influence. As well as the confidence region approach of this section, these alternatives include a determinantal criterion proposed by Andrews and Pregibon (1978) and a predictive approach due to Johnson and Geisser (1983). All can be expressed as functions of the residuals $r_I$ and of the elements $H_I$ of the hat matrix. The conclusion drawn by Cook and Weisberg from the comparison of these various measures of influence is that the Cook statistic $D_i$, which is easiest to compute and interpret, is to be preferred. In the earlier chapters of this book reasons have been given for preferring the modified Cook statistic $C_i$.

## 10.3 Extensions and generalizations

### *10.3.1 Nonlinear least squares*

So far it has been assumed that the parametric model relating the response $y$ to the explanatory variables $x$ is of the linear form given by (2.1.1). A more general formulation is to suppose that the model is instead

$$E(Y) = f(x, \theta) \tag{10.3.1}$$

where the model is now not linear in the vector of unknown parameters $\theta$. A simple example is the first-order growth model in which

$$E(Y) = 1 - e^{-\theta x} \qquad (x \geqslant 0).$$

If the error assumptions of additivity, independence and constancy of variance given in (2.1.2) hold, then it is appropriate to estimate $\theta$ by $\hat{\theta}$, the least squares estimate which minimizes

$$R(\theta) = \sum_{i=1}^{n} \{y_i - f(x_i, \theta)\}^2. \tag{10.3.2}$$

Because the problem is nonlinear, the minimum is found by an iterative method. Numerical methods for the estimation of parameters by nonlinear least squares are discussed in Kennedy and Gentle (1980, Chapter 10).

One consequence of the nonlinearity of the problem is that the parameter estimates and residuals are no longer linear functions of the observations. The deletion relationships of Chapter 2 therefore no longer hold and the diagnostic quantities described so far in this chapter are not readily calculable. However, interesting analogues of these earlier results can be obtained by linearization of the model.

Let $\theta_k$ be an estimate of $\theta$, perhaps at the $k$th stage of an iterative procedure for minimizing (10.3.2). Then the nonlinear model in (10.3.1) can be expanded in a Taylor series around $\theta_k$ as

$$f(x, \theta) = f(x, \theta_k) + z^{\mathrm{T}}(k)(\theta - \theta_k) + \cdots, \tag{10.3.3}$$

where

$$z^{\mathrm{T}}(k) = \left.\frac{\partial f(x, \theta)}{\partial \theta_j}\right|_{\theta = \theta_k} \qquad (j = 1, \ldots, p).$$

In the linearized model (10.3.3) the place of the carriers is taken by the vector of partial derivatives $z(k)$ which are functions of $\theta_k$. The $i$th residual, also dependent on $\theta_k$, is

$$r_i(k) = y_i - f(x_i, \theta_k)$$

and, from (10.3.3), the linearized form of the model analogous to (2.1.1) is

$$r_i(k) = z_i^{\mathrm{T}}(k)(\theta - \theta_k) + \varepsilon_i,$$

or, for all $n$ observations in matrix form,

$$r(k) = Z(k)(\theta - \theta_k) + \varepsilon. \tag{10.3.4}$$

The least squares estimates of the parameters in this linearized model are

$$\tilde{\theta} - \theta_k = \{Z^{\mathrm{T}}(k)Z(k)\}^{-1}Z^{\mathrm{T}}(k)r(k)$$

which can be rewritten in terms of $\theta_{k+1}$ as

$$\theta_{k+1} = \theta_k + \{Z^{\mathrm{T}}(k)Z(k)\}^{-1}Z^{\mathrm{T}}(k)r(k). \tag{10.3.5}$$

The iteration defined by (10.3.5) is the Gauss–Newton method for the estimation of parameters by nonlinear least squares.

Although the Gauss–Newton algorithm is not, in general, the computationally most efficient method of finding the least-squares estimate $\hat{\theta}$, procedures based on (10.3.5) provide a useful method of calculating deletion diagnostics. Let $\hat{\theta}_{(i)}$ be the least squares estimate of $\theta$ when the $i$th observation is deleted. That is, $\hat{\theta}_{(i)}$ minimizes $R_{(i)}(\theta)$, which is (10.3.2) summed over all observations except the $i$th. The first stage in the Gauss–Newton estimation of $\hat{\theta}_{(i)}$, with starting value $\hat{\theta}$, yields

$$\hat{\theta}^1_{(i)} = \hat{\theta} + (\hat{Z}^{\mathrm{T}}_{(i)}\hat{Z}_{(i)})^{-1}\hat{Z}_{(i)}\hat{r}_{(i)}, \tag{10.3.6}$$

where $\hat{Z} = Z(\hat{\theta})$ and $\hat{r} = r(\hat{\theta})$. The parameter estimate $\hat{\theta}^1_{(i)}$ which results from using (10.3.6) to move one Gauss–Newton step from $\hat{\theta}$ is called a one-step estimate.

It is possible to use (10.3.6) to define an iteration scheme, analogous to (10.3.5), which leads to the fully iterated deletion estimate $\hat{\theta}_{(i)}$. Pregibon (1981) and Cook and Weisberg (1982, p. 189) report that, for the examples they investigate, convergence usually requires only three iterations. A more important use of (10.3.6) is the analogy with deletion diagnostics for the linear model. From (2.2.8) it follows that (10.3.6) can be rewritten to yield the one-step estimate as

$$\hat{\theta}^1_{(i)} = \hat{\theta} - (\hat{Z}^{\mathrm{T}}\hat{Z})^{-1}\hat{z}_i\hat{r}_i/(1 - \hat{h}_i), \tag{10.3.7}$$

where $\hat{h}_i$ is the $i$th diagonal element of the hat matrix $\hat{Z}(\hat{Z}^{\mathrm{T}}\hat{Z})^{-1}\hat{Z}^{\mathrm{T}}$. Linearization of the model at the least squares estimate $\hat{\theta}$ can thus provide one-step approximations to the deletion residuals and to Cook's statistic, as well as to the other measures of influence mentioned in this chapter. For example, the one-step deletion residual would be given by

$$r_i^{*1} = \frac{\hat{r}_i}{\hat{s}_{(i)}\sqrt{(1 - \hat{h}_i)}},$$

with

$$(n - p - 1)\hat{s}^2_{(i)} = R(\hat{\theta}) - \hat{r}_i^2/(1 - \hat{h}_i).$$

Thus $r_i^{*1}$ is defined solely in terms of quantities calculated at the least squares estimate $\hat{\theta}$.

If the model were linear, the contours of the sum of squares surface would be ellipsoidal and the one-step approximations would be exact. To the extent that the contours are non-ellipsoidal, $\hat{\theta}^1_{(i)}$ will provide a poor estimate of the effect on parameter estimation of deletion of the $i$th observation. Likewise, unless the contours are ellipsoidal, the Cook statistic $D_i$, or any of its modified forms, may provide a poor estimate of the effect on the residual sum of squares of moving from $\hat{\theta}$ to $\hat{\theta}_{(i)}$. The effect of nonlinearity on parameter estimation and inference, without regard to deletion, is discused by Beale (1960) and by Bates and Watts (1980).

Some possibilities for deletion diagnostics in the presence of appreciable nonlinearity are discussed by Cook and Weisberg (1982, Section 5.3). These alternatives to use of $\hat{\theta}^1_{(i)}$ require calculation of the fully iterated estimate $\hat{\theta}_{(i)}$ for the deletion of each of the $n$ observations and, in some cases, other quantities such as the vector of predictions $\hat{Y}_{(i)}$. The computational simplifications resulting from the deletion formulae of Chapter 2 are thus lost and the diagnostics become that much less attractive.

Once $\hat{\theta}_{(i)}$ is available the deletion residuals can be calculated using (3.1.1) but with the predicted value given by

$$\hat{y}_{(i)} = f(x_i, \hat{\theta}_{(i)}).$$

The variance of this prediction is calculated from the linearized model with $X^{\mathrm{T}}_{(i)}X_{(i)}$ in (3.1.1) replaced by $\hat{Z}^{\mathrm{T}}_{(i)}\hat{Z}_{(i)}$ calculated, not as in (10.3.6) at $\hat{\theta}$, but at $\hat{\theta}_{(i)}$.

There are several choices for the analogue of the Cook statistic. One is, by analogy with (3.2.2), to use a quadratic form in $\hat{\theta}_{(i)} - \hat{\theta}$ with kernel $\hat{Z}^{\mathrm{T}}\hat{Z}$. This is a good approximation to the metric for the confidence region only if the contours of the sum of squares surface are approximately ellipsoidal. Since the reason for using the fully iterated estimate $\hat{\theta}_{(i)}$, rather than the one-step estimate $\hat{\theta}^1_{(i)}$, is that the contours are sufficiently non-ellipsoidal to render the one-step estimate a poor approximation, it seems unwise subsequently to base a diagnostic measure on the ellipsoidal approximation.

The other two choices for Cook's distance measure do not depend on the ellipsoidal approximation. One is to use the alternative form of $D_i$ as a measure of changes in prediction (3.2.3) to suggest looking at the sum of squared changes in prediction

$$\sum_{i'=1}^{n} \{f(x_{i'}, \hat{\theta}_{(i)}) - f(x_{i'}, \hat{\theta})\}^2.$$

The second choice is to return to the derivation of $D_i$ from the confidence region for $\theta$. For the nonlinear model, the approximate $100(1-\alpha)$ per cent normal theory confidence region for $\theta$ is, in place of (3.2.1), given by those values for which the difference in residual sum of squares satisfies

$$R(\theta) - R(\hat{\theta}) \leqslant ps^2 F_{p,\nu,\alpha}. \tag{10.3.8}$$

Although the shape of this region is exactly that of a likelihood contour, the content is only approximately $100(1-\alpha)$ per cent. For a linear model (10.3.8) reduces to (3.2.1). For a nonlinear model the analogue of the Cook statistic defined in (3.2.2) is

$$D_i(\theta) = \{R(\hat{\theta}_{(i)}) - R(\hat{\theta})\}/ps^2,$$

where $R(\hat{\theta}_{(i)})$ is the residual sum of squares for all $n$ observations calculated at the parameter value $\hat{\theta}_{(i)}$. The modified Cook statistic analogous to (3.2.4) is therefore

$$C_i(\theta) = \left\{\frac{n-p}{p}\frac{R(\hat{\theta}_{(i)}) - R(\hat{\theta})}{\hat{s}^2_{(i)}}\right\}^{1/2}. \tag{10.3.9}$$

The diagnostics generated by these methods, although available, have not been widely applied. An example is given by Cook and Weisberg (1982, Section 5.3), who report very good agreement between $\hat{\theta}_{(i)}$ and $\hat{\theta}^1_{(i)}$. The plot of the sum of squares surface which they give in their Fig. 5.3.1 shows the contours to be only slightly non-elliptical (there are only two parameters). It is therefore to be expected that this would be an example in which one-step diagnostics using (10.3.7) would be almost as good as fully iterated ones such as (10.3.9). What is needed, of course, is a characterization of the problems in which fully iterated diagnostics, with their additional computational burden, can be avoided. In the example of the generalized linear model given by Pregibon (1981) where one-step and fully iterated diagnostics also arise, there is some evidence that there is a loss of power, in the statistical sense, in using the one-step estimates. But the reduction in responsiveness to outlying and influential observations does not seem great enough to render the one-step diagnostics ineffective.

### *10.3.2 Generalized linear models*

Generalized linear models provide a powerful method for the unified handling of data from exponential family distributions such as the Poisson, binomial, gamma and inverse Gaussian, as well as the normal, in the presence of explanatory variables. For the $n$ observations $Y$, let $E(Y) = \mu$. The vector of means $\mu$ is related to the linear model $\eta = X\beta$ through the link function $\eta = g(\mu)$. For the linear regression model central to this book, the error distribution is the normal and the link is the identity, which leads to the model (2.1.1). For other error distributions, other links are appropriate.

For example, for Poisson counts with structure in the means $\mu$, the identity link is not appropriate as particular values of $x_i^T\beta$ could be negative. It is therefore natural to take $\mu = \exp(\eta) = \exp(X\beta)$, which guarantees non-negative values of $\mu$. In terms of the link function $g(\mu)$ the relationship is

$$X\beta = \eta = \log \mu, \tag{10.3.10}$$

the log link. A second example is the binomial distribution with probability of success $\theta$, that is with expected value $E(Y) = n\theta = \mu$. The logit link

$$\log \{\theta/(1-\theta)\} = X\beta \tag{10.3.11}$$

provides the powerful logistic model for the analysis of binary data which forms the intellectual centre of Cox (1970).

Generalized linear models were formulated by Nelder and Wedderburn (1972). A brief introduction is given by Wetherill (1981, Chapter 17) and a more extended introduction by Dobson (1983). The treatment by McCullagh and Nelder (1983) is more advanced. For the exponential family see Cox and Hinkley (1974, Section 2.2).

The linear structure of the models combined with error distributions within the exponential family results in a simple form for the likelihood. Maximum likelihood estimation of $\beta$ can then be achieved by iterative weighted least squares, the method used in the GLIM system of computer programs (Baker and Nelder, 1978). Because of the relationship with least squares, analogues of linear regression diagnostics are available. Pregibon (1981) gives a development of diagnostic quantities for the generalized linear model, analogous to those derived in this chapter, with examples of single deletion diagnostics for the logistic model for binary data (10.3.11). McCullagh and Nelder (1983, Chapter 11) describe several diagnostic techniques for checking generalized linear models. Details of the diagnostics for some generalized linear models are also given by Cook and Weisberg (1982, Section 5.4) who give a logistic example and an example with exponential survival times.

Although the general structure of these diagnostic procedures for generalized linear models is similar to that for the linear regression model, there are some complications. In particular, because the estimation method is iterative, there is the choice between one-step and fully iterated procedures, just as there was for nonlinear least squares in the previous section. Another problem is that residuals are not uniquely defined. A further development is presented by Jørgensen (1983) who considers generalized nonlinear models in which, by analogy with (10.3.1), the linear structure is replaced by $g(\mu) = \eta = f(x, \theta)$. Jørgensen mentions the possibility of diagnostic methods for this more general class of models, but the details are not developed.

The diagnostic techniques presented by Pregibon and by Cook and Weisberg are concerned with residuals and with influence on inferences about the parameters of the linear model. A very different question is inference about the correct form of the link function. Pregibon (1980) derived a test for the link function which is related to procedures for the choice of a power transformation. This topic is discussed in Chapter 11.

# 11
# Goodness of link tests

## 11.1 Constructed variables for tests of the link in a generalized linear model

The main material of this chapter describes the extension of the technique of constructed variables to tests of the form of the link function in generalized linear models. As an example we return to the chimpanzee data, Example 12. In Chapter 9 this was analysed using a normal model after the log transformation of the response. In this section we present an alternative analysis based on the assumption of a gamma model with log link. As part of the background for this example, a brief description is given of relevant aspects of the theory of generalized linear models not included in Section 10.3.2.

In the Section 11.2 we consider the choice between the gamma model and the previous lognormal model. This not only introduces a Monte-Carlo test of separate, or non-nested, models but serves as a reminder that the transformation models of Chapters 6–9 are not the only class of models with linear structure in the predictor. In this section we first describe the gamma model in the context of generalized linear models with some reference to the relationship with normal theory linear models. We then develop a diagnostic procedure for the family of link tests proposed by Pregibon (1980).

The $n$ observations $Y$ in a generalized linear model have expectation $E(Y) = \mu$. These means are connected to the linear predictor $\eta = X\beta$ by the link function $\eta = g(\mu)$. We shall be particularly interested in the log link for which $\eta = \log \mu$, with inverse $\mu = \exp(\eta)$. The random variables $Y$ have a distribution which is a member of the exponential family with density

$$f(y; \theta, \phi) = \exp[\{y\theta - b(\theta)\}/a(\phi) + c(y, \phi)]. \qquad (11.1.1)$$

The mean and variance of $Y$ in (11.1.1) are

$$E(Y) = \mu = b'(\theta)$$

and

$$\text{Var}(Y) = b''(\theta)a(\phi) = V(\mu)a(\phi)$$

with $V(\mu)$ the variance function. For the gamma and normal distributions, amongst others, $a(\phi) = \phi$, when $\phi$ is called the dispersion parameter. From now on we shall assume that $a(\phi)$ does have this simple form. For the normal distribution $\phi = \sigma^2$. The normal distribution combined with the identity link $\mu = \eta = X\beta$ yields the linear regression model of Chapters 1–5.

The gamma distribution with index $\alpha$ has density

$$f(y;\alpha,\mu)=\frac{1}{\Gamma(\alpha)}\left(\frac{\alpha}{\mu}\right)^{\alpha}y^{\alpha-1}\exp\left(-\frac{\alpha y}{\mu}\right),\tag{11.1.2}$$

which is of the exponential family form. The log-likelihood for a single observation is

$$l(\alpha,\mu;y)=\alpha(-y/\mu-\log\mu)+(\alpha-1)\log y+\alpha\log\alpha-\log\Gamma(\alpha).\tag{11.1.3}$$

For the general model (11.1.1) the log-likelihood of a single observation is

$$l(\theta,\phi;y)=\{y\theta-b(\theta)\}/\phi+c(y,\phi).\tag{11.1.4}$$

Comparison of this expression with (11.1.3) shows that for the gamma distribution $\theta=-1/\mu$ and $b(\theta)=-\log(-\theta)$. The parameterization is therefore such that $E(Y)=\mu$ and $\text{Var}(Y)=\mu^2/\alpha$. The variance function is $\mu^2$ and the dispersion parameter $1/\alpha$.

The log-likelihood for the whole sample is found by summing (11.1.4) over all $n$ observations. The maximum likelihood estimates of the parameters $\beta$ of the linear predictor are those values for which the derivative of this sum with respect to $\beta$ is zero. This set of $p$ equations can be written

$$D^{\mathrm{T}}V^{-1}\{y-\mu(\hat{\beta})\}=D^{\mathrm{T}}V^{-1}r=0,\tag{11.1.5}$$

where $D$ is the $N\times p$ matrix of derivatives $\mathrm{d}\mu/\mathrm{d}\beta$. For the identity link $D$ reduces to the matrix of carriers $X$. If the observations are independent, $V$ is a diagonal matrix with elements $V(\mu)=b''(\theta)$. This set of equations for $\hat{\beta}$ is the analogue of the normal equations in linear regression given by (2.1.3), namely $X^{\mathrm{T}}(y-X\hat{\beta})=X^{\mathrm{T}}r=0$. In this special case, the parameter estimates do not depend on $\sigma^2$ and it is clear from (11.1.5) that in general $\hat{\beta}$ does not depend on the dispersion parameter $\phi$.

Although (11.1.5) is a weighted least squares equation for the parameter estimates $\hat{\beta}$, it is a nonlinear equation as the values of $D$ and $V$ will, in general, depend upon the current estimates of the parameters. To solve (11.1.5) the Newton–Raphson method can be used to provide an iteration analogous to the Gauss–Newton iteration for nonlinear least squares (10.3.5). This iteration yields the sequence of estimates

$$\beta_{k+1}=\beta_k+(D^{\mathrm{T}}V^{-1}D)^{-1}D^{\mathrm{T}}V^{-1}\{y-\mu(k)\},$$

where all quantities on the right-hand side are calculated using the parameter estimate $\beta_k$. An equivalent way of writing the iteration is

$$\beta_{k+1}=(D^{\mathrm{T}}V^{-1}D)^{-1}D^{\mathrm{T}}V^{-1}\{D\beta_k+y-\mu(k)\}.\tag{11.1.6}$$

This estimation method is often called iteratively re-weighted least squares, although McCullagh and Nelder (1983) prefer the name iterative weighted least squares. Green (1984) describes uses of the method in solving a wide range of estimation problems.

For a generalized linear model the canonical link (McCullagh and Nelder 1983, p. 24) is that for which $\theta$ in (11.1.4) equals $\eta$. For the gamma model this is the inverse link for which $\theta = 1/\eta$. To provide a satisfactory model for the chimpanzee data we shall instead use the log link $\mu = \exp(X\beta)$. For the gamma model with this link (11.1.6) simplifies appreciably since the $i$th row of $D$ is $d\mu_i/d\beta = \mu_i x_i^T$. The $i$th diagonal element of $V$ is $\mu_i^2$, $D^T V^{-1} D = X^T X$ and (11.1.6) reduces to

$$\hat{\beta} = (X^T X)^{-1} X^T t, \tag{11.1.7}$$

where $t = (t_1, \ldots, t_n)^T$, $t_i = \hat{\eta}_i + (y_i - \hat{\mu}_i)/\hat{\mu}_i$ and $\hat{\eta} = X\hat{\beta}$ with $\hat{\mu}_i = \exp(\hat{\eta}_i)$. Thus for the gamma model with log link, iterative least squares is required, but without weights.

The estimates $\hat{\beta}$ were defined as those maximizing the log-likelihood. For a normal theory linear model an identical definition of $\hat{\beta}$ is as the value which minimizes the residual sum of squares $(y - X\beta)^T(y - X\beta)$. The analogue of the residual sum of squares in the generalized linear model is the deviance which is proportional to the likelihood ratio statistic for testing the goodness of fit of the model.

The maximum value of the log-likelihood (11.1.4) for all $n$ observations is obtained when $\theta$ is a vector of $n$ parameters, so that the model fits exactly. Call this vector of parameters $\theta(y)$ and let $\hat{\theta}$ be the vector corresponding to the maximum likelihood estimate $\hat{\beta}$. The likelihood ratio statistic for testing the goodness of fit of the model is, from (11.1.4),

$$2\sum [y\{\theta(y) - \hat{\theta}\} - b\{\theta(y)\} + b(\hat{\theta})]/\phi = D(y, \hat{\mu})/\phi,$$

where $D(y, \hat{\mu})$ is the deviance of the model. For the gamma model the deviance is

$$2\sum \{-\log(y_i/\hat{\mu}_i) + (y_i - \hat{\mu}_i)/\hat{\mu}_i\}. \tag{11.1.8}$$

The final term in this expression is identically zero at the maximum likelihood estimate and so can be ignored. It will not however necessarily be zero for some other estimate, for example a one-step estimate used in the calculation of diagnostic quantities.

Assessment of the significance of the deviance requires an estimate of the dispersion parameter $\phi$ as does calculation of confidence intervals for the parameters $\beta$. The dispersion parameter can be estimated either by maximum likelihood or, as in the GLIM computer package (Baker and Nelder, 1978), by the method of moments. McCullagh and Nelder (1983, pp. 157–8) give reasons for preferring the latter method.

The discussion of generalized linear models in this chapter has been in terms of error distributions which are members of the exponential family. In least squares theory an alternative to the distributional assumption of normality is the second-order assumption of the constancy of variance, independently of the value of the mean. Assumptions of other relationships between mean and variance, coupled with the structure of a linear model, lead to the idea of quasi-likelihood and to estimation methods which are those for the related generalized linear models. For example, the assumption that the variance is proportional to the square of the mean, that is that the coefficient of variation is constant, leads to the estimation procedure derived in this chapter. The theory of quasi-likelihood is discussed by McCullagh and Nelder (1983, Chapter 8) and by McCullagh (1983).

So far we have assumed that the correct form of the link function is known. In Pregibon (1980) goodness of link tests are developed in which the assumed link is embedded in a more general parametric family. Taylor series expansion of the link at the null hypothesis yields a constructed variable, regression on which provides a test for the goodness of the link.

Let the link function $g_0(\mu)$ be embedded in a parametric family $g(\mu, \lambda)$ indexed by the scalar parameter $\lambda$, where $g_0 = g(\mu, \lambda_0)$. Taylor expansion of $g(\mu, \lambda)$ at $\lambda_0$ yields the approximation

$$\begin{aligned} g(\mu, \lambda) &\simeq g(\mu, \lambda_0) + (\lambda - \lambda_0)w_1(\lambda_0) \\ &= X\beta + \gamma w_1(\lambda_0). \end{aligned} \tag{11.1.9}$$

The constructed variable for the link function $w_1(\lambda_0)$ is given by

$$w_1(\lambda_0) = \left.\frac{\partial g(\mu, \lambda)}{\partial \lambda}\right|_{\lambda=\lambda_0}.$$

Since $w_1$ depends upon the unknown vector $\mu$, estimates have to be substituted for calculation of the constructed variable. Use of $\hat{\mu}$, the fitted value under the link $g_0(\mu)$, leads to a constructed variable which will be called $\hat{w}_1$. For a vector parameter $\lambda$ there will be a set of constructed variables, one for each element of $\lambda$.

To test the adequacy of the proposed link, Pregibon suggests fitting the model with the approximate extended link (11.1.9) and looking at the change in deviance. Although the fit of this model may be iterated to achieve convergence, the constructed variable is calculated only once. The proposed test is thus a one-step approximation to the likelihood ratio test for $\lambda$ in the extended link $g(\mu, \lambda)$. Athough the likelihood ratio test could be found by iteratively updating the constructed variable $w_1(\lambda)$, the advantage over the one-step approximation is liable to be negligible.

The calculations and discussion in Pregibon (1980) are in terms of the GLIM package which lends itself to calculation of likelihood ratio tests through the comparison of deviances. In earlier chapters we used approximate score tests, rather than likelihood ratio tests, for testing hypotheses about transformations. One advantage of score tests is that the direction of departure from the model is indicated by the sign of the score statistic. The calculation of score tests in GLIM is described by Pregibon (1982) who uses special features of the calculation and storage of intermediate results to reduce computation.

Before these methods are applied to testing the log link in the gamma model, we briefly describe the two applications of the theory given by Pregibon. In one example the identity link $g_0(\mu) = \mu$ for the normal linear model is embedded in the shifted power transformation family

$$g(\mu; \alpha, \lambda) = \{(\mu + \alpha)^\lambda - 1\}/\lambda,$$

which is the unnormalized version of (9.1.1). This generalized link function contains two parameters, $\alpha$ and $\lambda$, so there are two constructed variables. The sum of squares for testing the identity link $g_0(\mu) = g(\mu; 1, 1)$ yields an exact $F$ test with two degrees of freedom in the numerator.

The second example is for binomial data with $g_0(\mu)$ the logit link $\eta = \log\{\mu/(n - \mu)\}$. This link is again embedded in a two-parameter family, although not of the power transformation type. In this case the change in deviance on fitting the extended model yields a test for which the $\chi^2$ distribution holds only asymptotically. This test is therefore not exact.

In both examples only aggregate statistics are calculated. Information is thus not obtained on the contribution of the individual observations to the value of the statistic. In the further analysis of the chimpanzee data we obtain this information from an added variable plot of the constructed variable $w_1(\lambda)$. The method is analogous to the use of plots of the constructed variable $w_p(\lambda)$ to augment calculation of the score statistic for transformations $T_p(\lambda)$.

**Example 12 (continued) Brown and Hollander chimpanzee data 3** In Chapter 9 the chimpanzee data were subject to power transformations with and without mean shift. The alternative of the gamma distribution with log link was proposed by McCullagh (1980) who made some comparisons with the power transformation model. Both models are plausible in that they avoid the possibility of predicting negative values for the response, which is learning time.

To test the log link let the parametric family of links be the one-parameter power transformation family

$$g(\mu, \lambda) = (\mu^\lambda - 1)/\lambda,$$

when the null link can be written $g_0(\mu) = g(\mu, 0) = \log \mu$. The constructed variable for the hypothesis $\lambda = 0$ is

$$\hat{w}_1(0) = \left.\frac{\partial g(\mu, \lambda)}{\partial \lambda}\right|_{\lambda = 0, \mu = \hat{\mu}} = -\log^2 \hat{\mu}/2 = -\hat{\eta}^2/2. \qquad (11.1.10)$$

The last form for $\hat{w}_1$ stresses the formal relationship with Tukey's one degree of freedom.

When the model with log link is fitted to all 40 observations of the chimpanzee data, the deviance is 14.97. The value of the dispersion parameter, 0.5545, suggests the physically plausible model of a gamma distribution with index 2, corresponding to two independent learning stages, each of which is exponential. Pregibon's test for the link consists of adding the variable (11.1.10) to the model and calculating the resulting decrease in deviance. For this augmented model the deviance is 14.75. To determine the significance of the decrease we take the index of the distribution as 2, so that the dispersion parameter equals 0.5. Twice the decrease in deviance is 0.4406. Comparison with the $\chi^2$ distribution on one degree of freedom shows that there is no evidence of departure from the log link.

In the original analysis of these data in Section 9.1, evidence of the need for a shift in location following the log transformation was suppressed by observation 8. The added variable plot of Fig. 9.2 served to reveal the effect of this one observation on the test statistic. To determine whether the evidence for a different link is similarly being suppressed by one or a few observations we supplement the one-step test with the appropriate added variable plot.

In general the expression for $\hat{\beta}$ (11.1.6) yields weighted least squares so that the added variable plot would have to be adjusted for the weights of the individual observations. However, as we have seen, (11.1.7) shows that, for the present example, the estimation method is unweighted least squares. Thus an alternative to the one-step likelihood ratio test is to calculate the score statistic in a manner directly analogous to that used in Chapter 6. From (11.1.7) the response variable is $t$, with the constructed variable given by (11.1.10). As well as the score test, the related quick estimate of the parameter in the generalized link function can be calculated in the manner of (8.2.1). The value of this quick estimate is $\tilde{\lambda} = -0.166$, close to the log link. The $t$ value for the score statistic is $-0.6814$, the square of which, 0.4643, is in close agreement with the $\chi^2$ value from Pregibon's one-step test.

The added variable plot has $-\hat{\eta}^2/2$ as the constructed variable and $t = \hat{\eta} + (y - \hat{\mu})/\hat{\mu}$ as the response. Since the model contains a constant, calculation of residuals from regression on $X$ causes the response to simplify to $\overset{*}{t}$, which is the residual of $(y - \hat{\mu})/\hat{\mu}$. The resulting plot (Fig. 11.1) appears devoid of structure. This impression is confirmed by the corresponding half-normal plot of the modified Cook statistic, not shown here, which fails to reveal any influential observations. The results of the tests and plots are therefore in agreement that there is no evidence of any departure from the log link.

The main feature of Fig. 11.1 is the large value of 1.90 for $\overset{*}{t}_5$. This value is produced by an observed value of 225, which is by far the largest value in Table 9.2 for sign 5. The other values are 24, 15 and 10. Observation 8, which was so important for the lognormal model, gives a value of $-0.90$

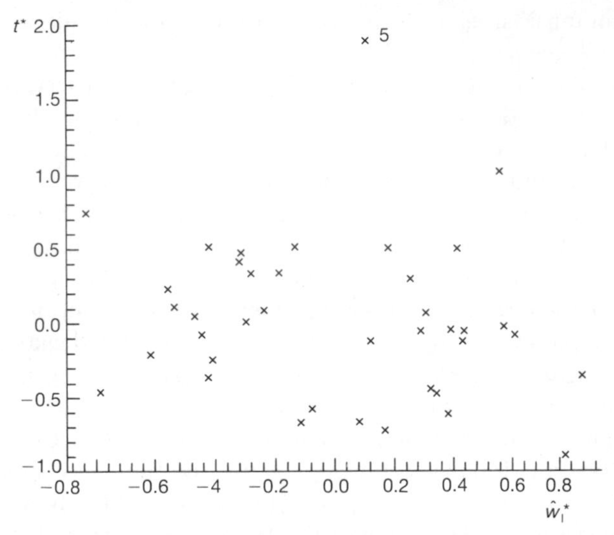


**Fig. 11.1** Example 12: Brown and Hollander chimpanzee data: added variable plot of the constructed variable $\hat{w}_1(0)$ for the log link

for $t^*$, which, although small, does not lie clear of the other values. One interpretation is that large values of the response are important for the gamma model, whereas small values are important for the lognormal model. However, this interpretation of different behaviour in the two tails of the distribution is slightly complicated by the effect of deleting observation 8. For the model with log link the deviance is reduced by 3.0, corresponding to a $\chi^2$ value of 6.0. This large value suggests that, if the gamma model is to be used for these data, a diagnostic analysis of the kind outlined in Section 10.3.3 should be undertaken.

## 11.2 A Monte-Carlo test for separate families of hypotheses

The analyses of the chimpanzee data given above and in the previous chapter do not indicate which of the two models, the gamma and the lognormal, is to be preferred. In this section we discuss the choice between the two models and exemplify a graphical procedure for presenting the results of a Monte-Carlo test.

Many of the procedures in this book, such as plots of residuals, are aids towards the informal choice of a model. At the more formal level, such as the choice of a suitable transformation, we have used significance tests based on, or derived from, the likelihood ratio. The basis of these tests is to consider a parametric family of models which includes special cases of interest and then to test whether the restriction on the parameters implied by a specific special case is acceptable. The asymptotic distributions of the resulting tests have either been $\chi^2$, or normal in the case of score tests. But the problem in choosing between the lognormal and gamma models is that neither is a special case of the other. The models are said to be separate in the sense that it is not possible, for all parameter values, to approximate one distribution arbitrarily closely by the other. A wide range of problems of this type is described by Cox (1961, 1962).

One way of developing a test for separate models is to embed the two models in a more general family of which they are special cases. A simple example is the parametric family of power transformations introduced in Chapter 6, which makes it possible to use standard statistical procedures to choose a suitable transformation. For the particular example of the gamma and lognormal distributions Prentice (1974) provides a general distribution of which these two, as well as many others, are special cases. In the obvious way of parameterizing the distribution, the lognormal occurs on a boundary of the parameter space. But an inverse parameterization brings this special case to the centre of the region. Within this family tests of hypotheses about parameter values which correspond to individual models of interest can be conducted in the usual manner. Some examples are given by Farewell and Prentice (1977) and by Kalbfleisch and Prentice (1980, Section 3.9).

The general model proposed by Prentice includes a wide range of distributions. Tests about the parameters in the distribution may therefore have low power if interest is in testing one model against a specific alternative. In this case a more general method of embedding the two models (Atkinson, 1970) is appropriate. Let the densities of the two distributions be $f(y, \alpha)$ and $g(y, \beta)$, with no restrictions on the dimensionality of the two parameter vectors $\alpha$ and $\beta$. The combined distribution is

$$f_\lambda(y) = \frac{\{f(y, \alpha)\}^\lambda \{g(y, \beta)\}^{1-\lambda}}{\int \{f(z, \alpha)\}^\lambda \{g(z, \beta)\}^{1-\lambda} \, dz}. \tag{11.2.1}$$

This power combination is more suitable for distributions belonging to the exponential family than, for example, the linear combination $\lambda f(y, \alpha) + (1 - \lambda) g(y, \beta)$. Testing the hypothesis that $\lambda$ in (11.2.1) equals one is equivalent to testing the hypothesis that the observations come from the distribution $f(y, \alpha)$ with power against the alternative distribution $g(y, \beta)$. It would be possible, in principle, to perform a likelihood ratio test for hypotheses about the value of $\lambda$. In practice such a procedure, which would require the calculation of the estimate $\hat{\lambda}$, would need a prohibitive amount of computing. In particular, the integral in the denominator of (11.2.1) usually has to be calculated by quadrature. An alternative, which appreciably reduces computation, is to develop score tests for the hypotheses $\lambda = 0$ and $\lambda = 1$. For the null hypothesis $\lambda = 0$ the score test for independent and identically distributed observations is

$$T_{\mathrm{f}} = \frac{\sum [F_i(\hat{\alpha}) - G_i(\hat{\beta}) - E_{\hat{\alpha}}\{F(\hat{\alpha}) - G(\beta_{\hat{\alpha}})\}]}{[n\{V_{\alpha}(F - G) - C_{\alpha}^2(F - G, F_{\alpha})/V_{\alpha}(F_{\alpha})\}]^{1/2}}. \tag{11.2.2}$$

In this expression

$$F_i = \log f(y_i, \alpha), \qquad G_i = \log g(y_i, \beta), \qquad F = \log f(z, \alpha)$$

$$F_{\alpha} = \partial/\partial\alpha\{\log f(z, \alpha)\} \qquad \text{and} \qquad E_{\alpha}(H) = \int H(z) f(z, \alpha)\, \mathrm{d}z.$$

The definitions of the variances and covariances $V_{\alpha}$ and $C_{\alpha}$ are similar to that for the expectation $E_{\alpha}$. The parameter estimates $\hat{\alpha}$ and $\hat{\beta}$ are the usual maximum likelihood estimates under the two separate models. If the true model is $g(y, \beta)$, the estimate $\hat{\beta}$ converges to $\beta$. On the other hand, if the true model is $f(y, \alpha)$, $\hat{\beta}$ will converge to some other value, which we call $\beta_{\alpha}$. Since $\alpha$ is not known, but estimated by $\hat{\alpha}$, $\beta_{\alpha}$ is replaced by $\beta_{\hat{\alpha}}$ in order to provide an operational test.

The form (11.2.2) was developed by Cox (1961, 1962). The derivation from the score statistic does not specify what estimate of the parameter of the alternative model should be used. Atkinson (1970) suggested use of $\beta_{\hat{\alpha}}$ rather than $\hat{\beta}$ in a statistic which was otherwise identical to (11.2.2). A discussion of the surprising difference in the behaviour of the two tests is given by Pereira (1977).

Although (11.2.2) is a complicated expression, the basic structure is clear. The observations enter directly through $\sum \{F_i(\hat{\alpha}) - G_i(\hat{\beta})\}$, which is the difference of the log-likelihood of the two models. If the models were nested, twice this difference would have a $\chi^2$ distribution. But, because the families are separate, the observed value of the difference can be either positive or negative and, in general, the distribution of the difference is not known. The numerator of the asymptotically standard normal test statistic (11.2.2) consists of this difference, from which is subtracted its expectation under the null hypothesis. The denominator of the statistic is the estimated standard error of the difference.

There are two disadvantages to (11.2.2). One is that evaluation of the statistic can sometimes require fairly complicated calculation. Shen (1982) mentions a simple example in which calculation of the variance requires more than 40 terms, each of which is a function of the di-, tri-, tetra- and penta-gamma functions of the parameter estimates. Another disadvantage is that, even for moderate sample sizes, the distribution of the test statistic can be far from the asymptotic standard normal. Simulation may therefore have to be used to find the null distribution of the test statistic, so that the significance of an observed value can be determined.

Two simpler alternatives to (11.2.2) have been suggested. Shen (1982) is concerned to test the Dirichlet model for compositional data against the much richer logistic-normal class. Her procedure consists of approximating the Dirichlet model by the logistic normal and then testing hypotheses within that class. Provided the Dirichlet can be adequately approximated, a test of the distribution results which avoids the problems of the separate families test. A second possibility is to use the separate families test but to use simulation to estimate the required mean and variance. An example of such a test is given by Shen (1982). The disadvantage of such a procedure is that it does not escape the assumption that the statistic has a normal distribution. The alternative which we follow here (Atkinson, 1981) is to use the simulated values of the differences as an estimate of the distribution of the test statistic, rather than attempting to summarize this distribution by its first two moments. The Monte-Carlo procedure is to fit one model to the data and to use the estimated model to generate 100 samples, for each of which the maximized log-likelihood is calculated for both of the models. A plot is made of these 100 simulated points and the observed value. If the model holds, the observed value will lie within the simulated set. The procedure is then repeated for testing the second model by using the second fitted model to generate the data.

**Example 12 (concluded) Chimpanzee data 4** We now apply this general method to testing the suitability of the lognormal and gamma models for the chimpanzee data. For the lognormal model the maximized log-likelihood is the residual sum of squares of the logged responses whereas for the gamma model it is the deviance given by (11.1.8). For an arbitrary generalized linear model the Monte-Carlo procedure could be costly, involving fitting the model by iterative weighted least squares for each simulated sample. But for the present example it is straightforward as, from (11.1.7), $\hat{\beta}$ is a linear function of $t$. Therefore only one matrix inversion is required for the whole study. Further, the iterative solution of (11.1.6) starts from $\mu_0 = \log y$, so that the first stage provides the fit to the lognormal model. Another simplification, which follows if the dispersion parameter is taken as 2, is that the required random variables are readily generated. For then, if $U_1$ and $U_2$ are pseudo-random numbers, $-0.5 \log (U_1 U_2)$ has a gamma distribution with mean 1 and index 2.

The plots of 100 simulated values of the two log-likelihoods together with the two observed values are shown in Figs. 11.2 and 11.3. Both figures show the values of the log-likelihoods varying in an almost linear manner, yielding simulation envelopes which are long narrow ellipses. The observed value falls near the centre of both systems and the plots provide no evidence of departure from either model. If the two figures are superposed, preferably after one has been copied, it will be seen that there is a large overlap between the two ellipses with the observed values of the statistics towards the centre of

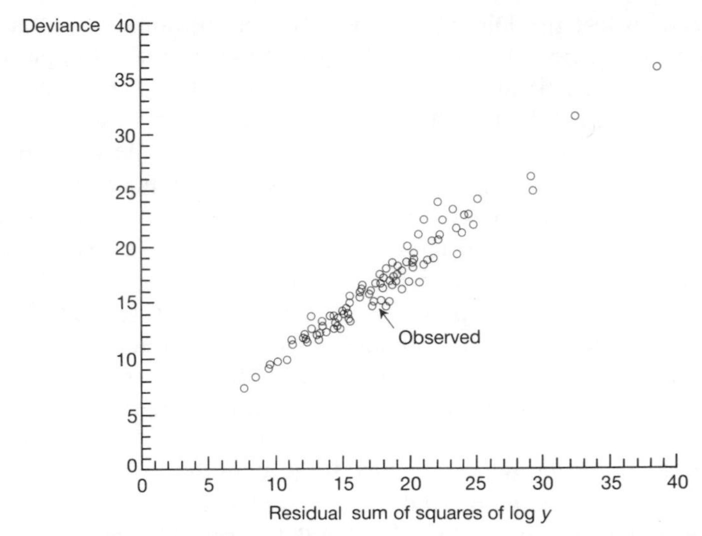


**Fig. 11.2** Example 12: Brown and Hollander chimpanzee data: Monte-Carlo test for separate families with data simulated from lognormal distribution

the overlapping region. The conclusion of this test is that not only is there nothing to choose between the models for the description of these data, but that for this sample size and these parameter values, discrimination between the two models is virtually impossible. □

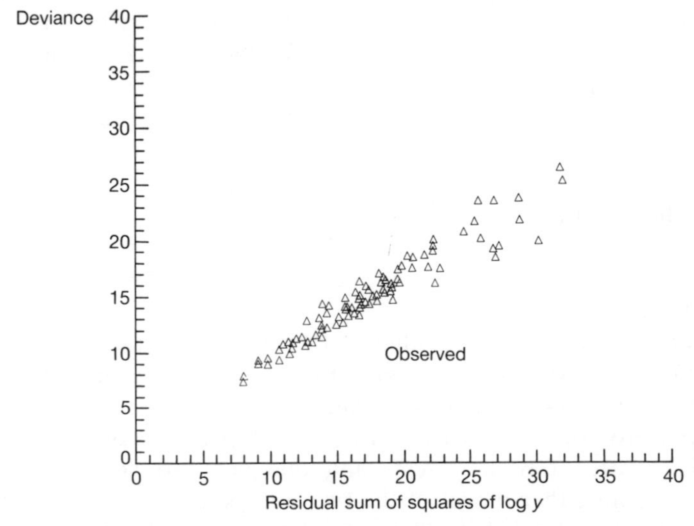


**Fig. 11.3** Example 12: Brown and Hollander chimpanzee data: Monte-Carlo test for separate families with data simulated from gamma distribution

The relationship of the Monte-Carlo test to the score test (11.2.2) suggests that the two procedures should have approximately the same power. The advantages of the Monte-Carlo test are that it is easier to calculate and that simulation provides an empirical distribution in which the observed statistic can be placed. Unlike the score test, the asymptotic distribution is not of interest. In some ways interpretation of the plots would be easier if a one-dimensional ordering were achieved. One possibility would be normal or other plots of the difference in the log-likelihoods.

# 12
# Further problems, other analyses

## 12.1 Robust transformations

In this last chapter we consider briefly some further developments of the methods of the earlier chapters for the linear regression model. The material in this section outlines the combination of the parametric family of power transformations of Chapter 6 with the methods of robust regression. The next section describes the development of constructed variables for non-constancy of variance. In Section 12.3 we study the effect of leverage on the added variable techniques for transformations which form a large part of the second half of this book. Comments are given in Section 12.4 on other analyses of some of the book's 14 examples. The book ends with some speculations on the development of diagnostic techniques and their integration into a systematic analysis of data. But we begin with robust transformations.

In Chapter 6 the transformation parameter $\lambda$ was chosen to minimize the residual sum of squares of the normalized variables $z(\lambda)$. The main intention of robust regression is to provide a method of estimation and inference which is almost as efficient as least squares when the errors are normally distributed, but much more efficient than least squares under alternative error distributions, particularly those that are long tailed. The derivation and properties of robust estimators are described by Huber (1981). For an introduction to the robust analysis of data see Hoaglin *et al.* (1983).

To begin, we give a brief description of robust regression. For the extension to transformations we argue conditionally on the particular value of $\lambda$, as we did in the calculation of the least squares parameter estimates $\hat{\beta}(\lambda)$.

The least squares estimates $\hat{\beta}$ minimize the sum of squares

$$(y - X\beta)^{\mathrm{T}}(y - X\beta)$$

which can also be written as the sum of squared residuals

$$R(\beta) = \sum r_i^2, \tag{12.1.1}$$

where $r_i = y_i - x_i^{\mathrm{T}}\beta$. Minimization of (12.1.1) leads to the estimating equation $X^{\mathrm{T}}r = 0$. For robust regression let us define $\rho(z)$ by

$$(\partial/\partial z)\rho(z) = \rho'(z) = \psi(z). \tag{12.1.2}$$

One example of $\psi(z)$ is Huber's proposal 2 for which

$$\psi(z) = \begin{cases} -c & (z < -c) \\ z & (-c \leqslant z < c) \\ c & (z \geqslant c). \end{cases}$$

so that

$$\rho(z) = \begin{cases} z^2/2 & (|z| \leqslant c) \\ c|z| - c^2/2 & (|z| > c). \end{cases} \tag{12.1.3}$$

The analogue of the sum of squared residuals is given by

$$D(\beta) = \sum \rho(r_i/\sigma). \tag{12.1.4}$$

Often $c$ is taken in the range 1.2 to 1.6, depending on the anticipated degree of departure from normality (Huber 1981, pp. 144–5). As $c$ in (12.1.3) goes to infinity, (12.1.4) approaches the sum of squares $R(\beta)$.

Differentiation of (12.1.4) with respect to $\beta$ yields the estimating equation

$$X^{\mathrm{T}}\psi(\tilde{r}) = 0. \tag{12.1.5}$$

For least squares and, as we noticed in the previous chapter, for generalized linear models, the estimates $\hat{\beta}$ which satisfy the equation $X^{\mathrm{T}}r = 0$ do not depend on the value of $\sigma$. In robust regression the estimates $\tilde{\beta}$ do depend on $\sigma$, which has therefore to be estimated along with $\beta$. Although the dependence of the estimates $\tilde{\beta}$ on $\sigma$ is not strong, inferences about the parameters based on tests and confidence intervals do depend appreciably on the method of estimation of $\sigma$. A discussion of the estimation of $\sigma$ is given by Schrader and Hettmansperger (1980, Section 3.2).

For inference about models, tests are used which are of the maximum likelihood type. That is, for any $\psi$, or equivalently any $\rho$, there is a distribution for which the estimating equation (12.1.5) yields maximum likelihood estimates. Thus for Huber's proposal 2 (12.1.3), the distribution has a normal centre and exponential tails. Likelihood ratio tests of hypotheses about the value of $\beta$ are given by comparing values of $D(\beta)$.

The application of these results to transformations does not present any new problems. For given $\lambda$ let $y(\lambda)$, as in Chapter 6, be $(y^{\lambda} - 1)/\lambda$. The estimating equation for $\beta(\lambda)$ is

$$X^{\mathrm{T}}\psi[\{y(\lambda) - X\beta(\lambda)\}/\sigma] = 0,$$

with a similar extension to include $y(\lambda)$ in the estimation of $\sigma$. Solution of these two equations for fixed $\lambda$ yields the estimate $\tilde{\beta}(\lambda)$. Because $\sigma$ is estimated separately, yielding the estimate $\tilde{\sigma}(\lambda)$, the values of $D\{\tilde{\beta}(\lambda)\}$ can be compared directly for different values of $\lambda$. The difficulties of scaling which led to use of the normalized variables $z(\lambda)$ do not occur. The values of $D\{\tilde{\beta}(\lambda)\}$ can then be used to test for evidence of a transformation.

The likelihood ratio test used by Carroll (1980) is of this form. Schrader and Hettmansperger (1980) describe tests and analogues to the analysis of variance for robust analysis of linear models. The likelihood ratio tests are corrected by a factor to allow for robust estimation under the normal distribution. A similar factor is used by Carroll (1982).

In Carroll (1980) these methods are applied to the Box and Cox poison data, Example 11, which was extensively analysed in Section 8.1. Amongst other aspects, simulations again reveal the low power of Andrews' method. The second example studied by Carroll is the textile data, also from Box and Cox (1964), which was introduced in Chapter 6. Interpretation of the results of Carroll's paper is facilitated when it is realized that his Figures 1 and 2 have been interchanged. In a second paper on robust transformations, Carroll (1982) compares the behaviour of tests based on two robust estimates with the likelihood ratio test resulting from least squares. The examples are the two factorial experiments from John (1978) which were analysed in Chapter 6.

Carroll's results show that the robust methods work in the sense of performing nearly as well as least squares for normal data in the absence of outliers, and better than least squares when outliers are present. This improvement in performance over least squares is obtained at the cost of extra computation. It is not clear, for the small sets of data used in the examples, whether diagnostic methods leading to the identification of outliers would not perform equally well. For larger data sets inspection of individual diagnostic measures may become increasingly difficult and the desirability of robust methods will increase. But, from the standpoint adopted in this book, it is quite clear that an adequate robust analysis requires more than a list of robust point estimates of the parameters. For example, the robust fit could be accompanied by a diagnostic analysis of the weights attached to the observations. One possibility is an index plot of $\psi(\tilde{r}_i)/\tilde{r}_i$, or its reciprocal, which has the value 1 for observations making a full contribution to the fit. Cook and Weisberg (1982, Section 5.5) give an example of a diagnostic analysis of robust regression.

The four examples used by Carroll are all factorial experiments, so that the extreme effects of leverage are absent. The effect of leverage on the form of robust regression outlined here can however be as great as it can be for least squares. Particularly if the iterative estimation of $\tilde{\beta}$ leading to the solution of (12.1.5) starts from the least squares fit, observations with high leverage will have small residuals. As a result, these observations will lie on the central part of $\psi$ and will not be reduced. For Example 2, the synthetic data due to Huber introduced in Chapter 1, Fig. 1.7 shows that the effect of small values of $c$ would be to reduce the effect of observations 1, 4 and 5, but to leave the leverage point 6 unchanged.

The effect of leverage on robust estimation in regression is discussed by Huber (1981, Section 7.9) who suggests what is, in effect, replacement of the residuals in (12.1.5) by the robust form of the standardized residuals (2.1.14). Although use of this method reduces the effect of leverage, the expression for the Cook statistic $D_i$ (3.2.4) suggests that observations with high leverage will still have appreciable influence on the fitted equation. One way of bounding the influence that any observation can have is the limited influence regression described by Krasker and Welsch (1982). Huber (1983) describes some properties and problems of this approach, which are further explored by the discussants of Huber's paper. A diagnostic use of limited influence regression is to provide an omnibus method of screening for outliers and influence. If comparison of the robust fit with least squares reveals important differences, the more detailed diagnostic methods of this book can be used to investigate the cause.

For robust transformations for regression data it may also be desirable to use an estimating equation in which allowance is made for leverage. The properties of such methods are explored by Carroll and Ruppert (1985) who include a further analysis of the salinity data, Example 4.

## 12.2 Constructed variables for constancy of variance

The diagnostic procedures described so far in this book are intended to reveal differences between the data and the systematic part of the fitted model. The main sources of such differences have been outlying observations, systematically inadequate linear models and the need for a transformation. But the belief has been that, perhaps after transformation and outlier deletion, a model will be obtained for which the errors are independent with constant variance. In this section we consider methods for detecting failure of the assumption of constant variance.

In Section 1.2 several plots were mentioned which, in an informal way, exhibit the presence of non-constant variance. One possibility is to look at plots of residuals against individual explanatory variables, another is to plot residuals against predicted values. These plots, like the others given in the first chapter, are well-known and widely used. However they are not necessarily the graphical equivalent of a test for a specific departure. In this they differ from many of the plots in later chapters. One example is the plot of residuals against a carrier not in the model. The score test for inclusion of the carrier leads instead to the added variable plot. The score tests and related plots of this section are for rather general departures from constancy of variance which we assume cannot be achieved, for example, by transformation of the response. In some simple cases the general method yields plots closely related to well-known procedures. The development follows that of Cook and Weisberg (1983).

Suppose that we have the usual linear regression model but write it, for convenience, as

$$Y = \mathbf{1}\beta_0 + X\beta + \varepsilon, \tag{12.2.1}$$

where $X$ is the $n \times p$ matrix of carriers and $\mathbf{1}$ is an $n \times 1$ vector of ones. The full model including the constant therefore has rank $p + 1$.

Previously we have assumed that the errors $\varepsilon$ follow a multivariate normal distribution with mean 0 and covariance matrix $\sigma^2 I$. We now assume instead that the covariance matrix of $\varepsilon$ is $\sigma^2 W$, where the elements of the diagonal matrix $W$ are such that all $w_i > 0$. To establish a score test requires that the $w_i$ have a parametric form. Let this be that

$$w_i = w(z_i, \lambda), \tag{12.2.2}$$

where the parameter $\lambda$ can be a vector and $z_i$ is a vector of known constants. If (12.2.2) is such that there is a value $\lambda_0$ of $\lambda$ for which $w(z, \lambda) = 1$ for all $z$, a test of constancy of variance is equivalent to a test of the hypothesis $\lambda = \lambda_0$. Cook and Weisberg discuss several possible forms for (12.2.2). One is to take

$$w(z_i, \lambda) = \exp\left\{\sum_j \lambda_j z_{ij}\right\}, \tag{12.2.3}$$

which does not impose any restrictions on the values of the $z_{ij}$.

In applications, interest is often in the dependence of the variance on the columns of $X$ which then furnish the values of the $z_{ij}$ in (12.2.3). Dependence of the variance on the expected response is modelled by taking

$$w_i = w(\lambda x_i^{\mathrm{T}}\beta)$$

with $\lambda_0 = 0$. One example is the special case

$$w_i = \exp(\lambda x_i^{\mathrm{T}}\beta). \tag{12.2.4}$$

Because the linear model (12.2.1) is written in partitioned form, the intercept $\beta_0$ is not included in the weight function (12.2.4). Inclusion of the intercept leads to an overparameterized model which yields the same score test as (12.2.4).

The score test for constancy of variance is found in the same way as the score test for transformations. That is, by evaluation of the derivative of the log-likelihood with respect to $\lambda$ at the null hypothesis $\lambda = \lambda_0$. To calculate this derivative let $U$ be the vector of squared residuals $r_i^2/\hat{\sigma}^2$, with $r_i = y_i - \hat{\beta}_0 - \hat{\beta}^{\mathrm{T}}x_i$ the least squares residual under the null hypothesis $W = I$. The estimate of variance is $\hat{\sigma}^2 = \sum r_i^2/n$ and so is the maximum likelihood, rather than the least squares, estimator. If the dimension of $\lambda$ is $q$, the vector of derivatives of the weights (12.2.2) is given by

$$w'(z_i, \lambda_0) = \left\{\frac{\partial w(z_i, \lambda)}{\partial \lambda_j}\bigg|_{\lambda = \lambda_0}\right\}.$$

Further, define $D$ to be the $n \times q$ matrix with $i$th row $\{w'(z_i, \lambda_0)\}^{\mathrm{T}}$ and let $\bar{D} = D - \mathbf{1}\mathbf{1}^{\mathrm{T}}D/n$ be the $n \times q$ matrix obtained from $D$ by subtraction of column averages. Then Cook and Weisberg (1983) show that the score test for the hypothesis $\lambda = \lambda_0$ is

$$S = U^{\mathrm{T}}\bar{D}(\bar{D}^{\mathrm{T}}\bar{D})^{-1}\bar{D}^{\mathrm{T}}U/2. \tag{12.2.5}$$

Comparison of (12.2.5) with earlier expressions such as (2.1.6) for the residual sum of squares from a regression shows that the score test is one half of the sum of squares for regression of $U$ on $D$ in the constructed model

$$U = \mathbf{1}\gamma_0 + D\gamma + \varepsilon_u. \tag{12.2.6}$$

For the general weight function (12.2.3), $w_i'(z_i, \lambda) = z_i$ when $\lambda_0 = 0$. For the weight function (12.2.4) in which dependence is on the expected response, the constructed variable is proportional to $x_i^{\mathrm{T}}\hat{\beta}$, that is to $\hat{y}_i$, the fitted value when the variance is assumed constant. In this and other examples, the variables $z_i$ are columns, or linear combinations of columns, of $X$. The matrix of variables $D$ is then readily calculated and, from (12.2.6), the score statistic is found by the standard regression methods which underlie so many of the diagnostic procedures described in this book.

The null asymptotic distribution of $S$ is $\chi^2$ on $q$ degrees of freedom. When $q = 1$ the distribution of the square root of $S$ is therefore asymptotically standard normal. For this important special case we simplify notation by writing $w'(z_i, \lambda_0)$ as $w_i^0$ with $\bar{w}^0 = \sum w_i^0/n$. In this informative notation the score statistic becomes

$$T(\lambda_0) = S^{1/2} = \frac{\sum (w_i^0 - \bar{w}^0)(r_i^2/\hat{\sigma}^2 - 1)}{\sqrt{\{2\sum (w_i^0 - \bar{w}^0)^2\}}}. \tag{12.2.7}$$

As would be expected from the constructed model (12.2.6), the score statistic (12.2.7) is proportional to the normal test for the regression of the response $r_i^2/\hat{\sigma}^2$ on the constructed variable $w'(z_i, \lambda_0)$. The difference in form between (12.2.7) and earlier score statistics such as $T_{\mathrm{p}}(\lambda)$ (6.4.3) arises because the regression equation (12.2.6) includes an intercept.

A graphical version of the score test for scalar $\lambda$ follows from the interpretation of (12.2.7) as a test for regression. In previous chapters this interpretation led to an added variable plot. But here (12.2.7) is not a regression coefficient adjusted for variables already in the model. So, instead of an added variable plot, the graphical equivalent is a plot of $r_i^2/\hat{\sigma}^2$ against $w'(z_i, \lambda_0)$.

In some special cases the new procedure therefore reduces to something close to the familiar plots of $r_i$ against $x_i$ or $\hat{y}_i$. But an important difference is that the procedure plots $r_i^2/\hat{\sigma}^2$ against the variable of interest. Plotting the values of the squared residuals has the advantage of doubling the optical density of the plots, which may be important for small data sets. In this way there is an analogy with the preference for half-normal plots of $r_i^*$ and $C_i$ in Chapter 4. In the case of the score test (12.2.7) there is no loss of information about non-constant variance from plotting squared values. However, there may be loss of information about other forms of departure, such as the presence of curvature in an explanatory variable. But one of the messages of this book is that specific plots should be designed to detect particular departures of interest. Portmanteau or general procedures, designed for simultaneously detecting many departures, may have unacceptably low power against many specific departures.

Examples and extensions of the general method are described by Cook and Weisberg (1983). One extension is to the use of regression to find a variable $z_i$ which is a linear combination of the columns of $X$. With $r_i^2/\hat{\sigma}^2$ as the response, regression on $X$ yields the coefficients $\hat{\delta}_j$. The resulting variable $z_i = \sum \hat{\delta}_j x_{ij}$ is then used to form a score statistic and related plot. The significance of this plot should, however, be assessed from the score test for all the carriers which make up $z_i$. Testing for the significance of $z_i$ alone is analogous to testing the most significant contrast in an analysis of variance without allowing for the effect of selection. An illustration of the method given by Cook and Weisberg yields results which are significant in a direction which is a linear combination of two of the carriers. This example serves as a reminder that there is no *a priori* reason why non-constancy of variance, when considered in the $p$-dimensional space of $X$, should necessarily lie along one or more of the axes of the explanatory variables. It is however to be hoped that the direction of non-constancy is physically meaningful.

Cook and Weisberg also develop improved plots which allow for the effect of leverage. The recommendation is to scale the values of $r_i^2/\hat{\sigma}^2$ by a factor $(1 - h_i)^{-1}$ and to estimate $\sigma^2$ not by the maximum likelihood estimate $\hat{\sigma}^2$, but by the least squares estimate $s^2$. Since the model (12.2.1) contains $p + 1$ parameters $(n - p - 1)s^2 = n\hat{\sigma}^2$. An argument is also presented for scaling the constructed variable, but this time by $(1 - h_i)$ rather than its inverse. The resulting plot is of the squared standardized residuals $r_i'^2$ (2.1.4) against $(1 - h_i)w'(z_i, \lambda_0)$. Another possibility is to replace the standardized residuals with the deletion residuals $r_i^*$. The consequences of this suggestion have not been investigated.

## 12.3 Transformations and leverage

Added variable plots of constructed variables were introduced in Chapter 6 to assess the contribution of individual observations to the score statistic. Examples in that and succeeding chapters show how useful the plots can be. However the plots can fail to identify observations influential for a transformation if the observation is also a leverage point for the linear model. In this section we first describe why this happens and then derive and compare quick estimates of the effect of individual observations on the estimated transformation.

In the majority of examples in which the added variable plot has been used to identify influential observations, the data came from a designed experiment. Depending on the fitted model all observations had similar values or the same low value of the leverage measure $h_i$. The score statistic and the related added variable plot then depend hardly at all on the values of the $h_i$. But for a leverage point, that is one with a value of $h_i$ near one, both the residuals $r_i$ and the residual constructed variable $\overset{*}{w}_i$ tend to be small. They accordingly make only a slight contribution to the score statistic and added variable plot. This is so even if the observation is important in determining a transformation. Cook and Wang (1983) give an example of data containing such an observation. An analysis of their data is given later in this section.

The importance of a set of observations to the estimated transformation can be assessed by the change in the score statistic or in the estimate of the transformation parameter when the observations are deleted. Although the change in these quantities could be calculated for each subset $I$, something simpler is required for a diagnostic measure. Ideally the measure should be derived from the fit of all the data to the untransformed model. In what follows we derive approximations of this kind to $\hat{\lambda}_{(I)}$. Expressions for the change in the score statistic can be derived in a similar manner, although they are slightly more complicated to describe.

To begin, we gather together several results which have been widely used in this book. For the linear model

$$E(Y) = X\beta \tag{12.3.1}$$

the least squares estimates of the parameters are given by $\hat{\beta} = (X^{\mathrm{T}}X)^{-1}X^{\mathrm{T}}y$. Addition of an extra carrier $w$ to (12.3.1) yields the partitioned model

$$E(Y) = X\beta + w\gamma. \tag{12.3.2}$$

From (5.2.5) and (5.2.7) the least squares estimate of the coefficient of $w$ is

$$\hat{\gamma} = \frac{w^{\mathrm{T}}(I - H)y}{w^{\mathrm{T}}(I - H)w} = \frac{w^{\mathrm{T}}Ay}{w^{\mathrm{T}}Aw} = \frac{\overset{*}{w}^{\mathrm{T}}r}{\overset{*}{w}^{\mathrm{T}}\overset{*}{w}} \tag{12.3.3}$$

where the residual carrier $\overset{*}{w} = (I - H)w$.

In Section 6.4 these relationships were applied to the constructed variable for the power transformation. Expansion of the transformation in a Taylor series about $\lambda_0$ yields the approximation

$$z(\lambda) \simeq z(\lambda_0) + (\lambda - \lambda_0)w(\lambda_0).$$

The approximate linear model is then

$$z(\lambda_0) = X\beta - (\lambda - \lambda_0)w(\lambda_0) + \varepsilon. \tag{12.3.4}$$

Application of (12.3.3) to this linear model gives the quick estimate of the transformation parameter

$$\tilde{\lambda} = \lambda_0 - w^{\mathrm{T}}Az/w^{\mathrm{T}}Aw = \lambda_0 - \overset{*}{w}{}^{\mathrm{T}}\overset{*}{z}/\overset{*}{w}{}^{\mathrm{T}}\overset{*}{w}, \tag{12.3.5}$$

which provides a one-step approximation to $\hat{\lambda}$. In (12.3.5) both $w$ and $z$ are calculated at $\lambda_0$.

These results have been repeatedly applied in this book. To find the effect of individual observations on the transformation we now develop the analogous one-step approximation to $\hat{\lambda}_{(I)}$, the maximum likelihood estimate of $\lambda$ when $m$ observations in the set indexed by $I$ are deleted.

The deletion results of Section 2.2 applied to the full model (12.3.1) give the changes in all important quantities. From (2.2.6) the residual sum of squares from fitting (12.3.1) is

$$\begin{aligned}(n-p)s^2 &= y^{\mathrm{T}}y - \hat{\beta}^{\mathrm{T}}X^{\mathrm{T}}y \\ &= y^{\mathrm{T}}(I-H)y \\ &= y^{\mathrm{T}}Ay = r^{\mathrm{T}}r. \end{aligned} \tag{12.3.6}$$

After deletion of $m$ observations (2.2.7) shows that the residual sum of squares is given by

$$(n-p-m)s_{(I)}^2 = y^{\mathrm{T}}Ay - r_I^{\mathrm{T}}(I-H_I)^{-1}r_I. \tag{12.3.7}$$

Here $H_I$ is the $m \times m$ submatrix of the hat matrix given by $H_I = X_I(X^{\mathrm{T}}X)^{-1}X_I^{\mathrm{T}}$ and $r_I$ is the vector of residuals of the deleted observations.

We now consider the partitioned model (12.3.2) in order to obtain the effect of deletion on the parameter estimate $\hat{\gamma}$. From (12.3.3) it is clear that $\hat{\gamma}_{(I)}$ is the ratio of the residual sum of cross products of $w$ and $y$, when the subset $I$ is deleted, to the sum of squares of $w$, similarly after deletion. These sums of squares and products can be found by direct calculations similar to those of Section 2.2. However we know from the analysis of covariance that adjustment for the covariate $w$ is made by performing identical operations on sums of squares of $y$, which yield the usual analysis of variance, and on the sums of squares and products of $y$ and $w$. By analogy with (12.3.7) which gives the effect of deletion on the residual sum of squares, it follows that

$$\hat{\gamma}_{(I)} = \frac{y^{\mathrm{T}}Aw - r_I^{\mathrm{T}}(I-H_I)^{-1}\overset{*}{w}_I}{w^{\mathrm{T}}Aw - \overset{*}{w}{}_I^{\mathrm{T}}(I-H_I)^{-1}\overset{*}{w}_I}, \tag{12.3.8}$$

with $\overset{*}{w}_I$ the vector of residual constructed variables for the deleted observations.

Application of (12.3.8) to the approximate linear model (12.3.4) yields the required approximation to $\hat{\lambda}_{(I)}$. There is however the difficulty that the standardized response variables $z(\lambda_0)$ depend on the Jacobian for all $n$

observations. That is

$$z(\lambda) = (y^{\lambda} - 1)/(\lambda \dot{y}^{\lambda-1}) \qquad (\lambda \neq 0), \tag{12.3.9}$$

where $\dot{y}$ is the geometric mean of all $n$ observations. If the $m$ observations indexed by $I$ are deleted, the Jacobian will change. In general the change in $\dot{y}$ can be expected to be negligible if $m$ is small relative to $n$. An exception is if one or more of the deleted observations are much closer to zero than are the rest. As our purpose is to provide readily calculable diagnostic quantities we ignore this problem in our theoretical development. The effect of this approximation is examined for a series of examples in which we compare the quick estimate $\tilde{\lambda}_{(i)}$ for deletion of a single observation with the maximum likelihood estimate $\hat{\lambda}_{(i)}$.

With this caveat in mind, application of the expression for $\hat{\gamma}_{(I)}$ (12.3.8) to the approximate model (12.3.4) yields the one-step estimate

$$\tilde{\lambda}_{(I)} = \lambda_0 - \frac{z^{\mathrm{T}}Aw - \overset{*}{z}{}_I^{\mathrm{T}}(I - H_I)^{-1}\overset{*}{w}_I}{w^{\mathrm{T}}Aw - \overset{*}{w}{}_I^{\mathrm{T}}(I - H_I)^{-1}\overset{*}{w}_I}, \tag{12.3.10}$$

where the residuals $\overset{*}{z} = (I - H)z$ are, like all other quantities, evaluated at $\lambda_0$.

In the usual diagnostic applications of (12.3.10) $\lambda_0 = 1$ so that $\overset{*}{z}_I = r_I$, the residuals of the untransformed observations. Cook and Wang (1983) develop this expression for $\lambda_0 = \hat{\lambda}$. At this value of $\lambda$ the score statistic for the transformation is zero so that $z^{\mathrm{T}}Aw = 0$. The one-step estimate is then given by

$$\hat{\lambda}^1_{(I)} = \hat{\lambda} + \frac{\overset{*}{z}{}_I^{\mathrm{T}}(I - H_I)^{-1}\overset{*}{w}_I}{w^{\mathrm{T}}Aw - \overset{*}{w}{}_I^{\mathrm{T}}(I - H_I)^{-1}\overset{*}{w}_I}. \tag{12.3.11}$$

A disadvantage of basing the one-step estimate on $\hat{\lambda}$ is that iterative calculation of the maximum likelihood estimate is not avoided. In this way the diagnostic procedure is made less appealing.

For the deletion of a single observation (12.3.10) becomes

$$\tilde{\lambda}_{(i)} = \lambda_0 - \frac{\overset{*}{z}{}^{\mathrm{T}}\overset{*}{w} - \overset{*}{z}_i\overset{*}{w}_i/(1 - h_i)}{\overset{*}{w}{}^{\mathrm{T}}\overset{*}{w} - \overset{*}{w}{}_i^2/(1 - h_i)}. \tag{12.3.12}$$

This expression suggests how the added variable plot may be adjusted to take account of leverage. The added variable plot for all observations is of $\overset{*}{z}$ against $\overset{*}{w}$, which has slope $\tilde{\lambda}$. For a point with high leverage (12.3.12) shows that $\overset{*}{w}_i$ and $\overset{*}{z}_i$ may both be small, but that deletion of them, when weighted by $(1 - h_i)^{-1}$, can cause an appreciable change in the estimated transformation. This suggests that a plot of $z_i/\sqrt{(1 - h_i)}$ against $w_i/\sqrt{(1 - h_i)}$ might be a useful diagnostic tool. Because these quantities are proportional to standardized residuals, we call the plot a standardized added variable plot. The idea is close to that of Cook and Weisberg (1983), described in Section 12.2, in which the plot for non-constant variance was adapted to take account of leverage. For balanced designs, the standardized added variable plot reduces to the added variable plot of previous chapters.

**Example 14 Cook and Wang's 'data'** To begin the illustration of these ideas we look at an example due to Cook and Wang (1983). The synthetic data, given in Table 12.1, consist of 11 observations with a single explanatory variable. For all 11 observations the score statistic for the power transformation $T_p(1) = -0.92$, so there is no significant evidence of the need for a power transformation. Yet if observation 11 is deleted, the value of the score statistic increases in magnitude to $-3.40$. The estimate of the transformation parameter at the same time changes from 0.71 for $\hat{\lambda}$ to $\hat{\lambda}_{(11)} = -0.182$. Deletion of observation 11 therefore changes the analysis from one in which there is no evidence for a transformation to one in which the response should be subjected to a log transformation.

In the analysis of John's $3^{4-1}$ experiment, Example 8, the added variable plot for $w_p(1)$ clearly showed the effect that one observation, in that case also number 11, was having on the evidence for the transformation. But, for this synthetic data set, the added variable plot does not reveal the importance of observation 11. From the added variable plot for $w_p(1)$, Fig. 12.1, it might seem as if observation 10 were the most important observation, but there is no clear pattern. Figure 12.2 is the standardized added variable plot suggested earlier in this section, that is the plot for the standardized variables $r_i/\sqrt{(1-h_i)}$ and $\mathring{w}_p/\sqrt{(1-h_i)}$. This standardization slightly accentuates the

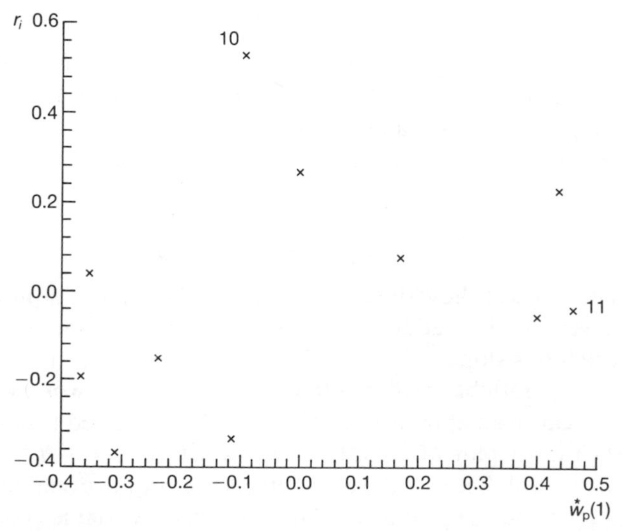


**Fig. 12.1** Example 14: Cook and Wang's 'data': added variable plot of $w_p(1)$

**Table 12.1** Example 14: Cook and Wang's 'data'

| Observation | $x$ | $y$ |
|---|---|---|
| 1 | 0.3 | 1.49 |
| 2 | 0.4 | 1.42 |
| 3 | 0.5 | 1.77 |
| 4 | 0.6 | 2.18 |
| 5 | 0.8 | 2.01 |
| 6 | 0.9 | 2.41 |
| 7 | 1.0 | 2.41 |
| 8 | 1.1 | 2.80 |
| 9 | 1.2 | 3.25 |
| 10 | 1.5 | 4.39 |
| 11 | 2.6 | 6.2 |

importance of observation 11 relative to Fig. 12.1, but gives nothing like a full indication of the importance of this one point.

Some indication of the importance of observation 11 can be gathered from standard diagnostic techniques. Figure 12.3 is a scatter plot of $y$ against $x$. This shows that observation 11 is clearly distanced, in $x$ space, from the remaining 10 observations. Observation 11 is similar, in this respect, to observation 6 in Huber's 'data', Example 2. There is also some indication, from a careful study of the scatter plot, that observations 1 to 10 fall on a

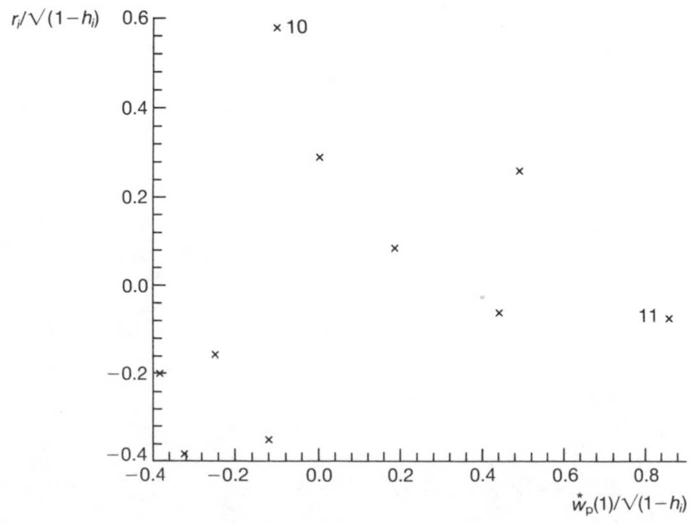


**Fig. 12.2** Example 14: Cook and Wang's 'data': standardized added variable plot of $w_p(1)$

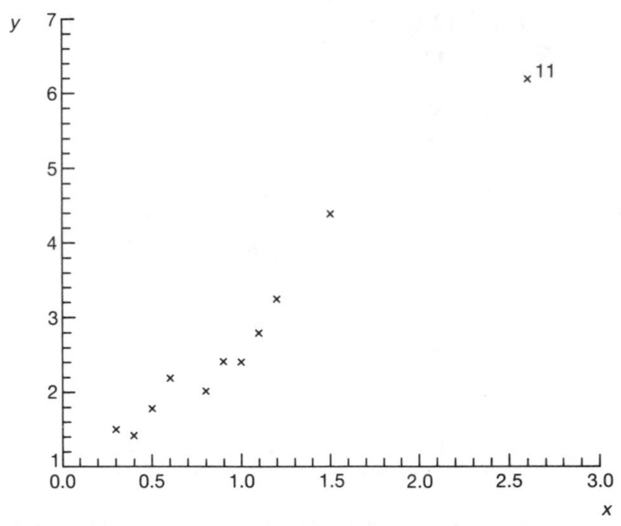


**Fig. 12.3** Example 14: Cook and Wang's 'data': scatter plot of $y$ against $x$

curve whereas observation 11 lies off the curve. This impression is strengthened by the plot of residuals $r$ against the fitted values $\hat{y}$, Fig. 12.4. The residuals from observation 1 to 10 form a roughly parabolic curve with observation 11 yielding a small residual rather apart from the other points. These plots indicate, to some extent, how the synthetic data were constructed: observations 1 to 10 follow a linear model in $x$ after a log transformation of $y$,

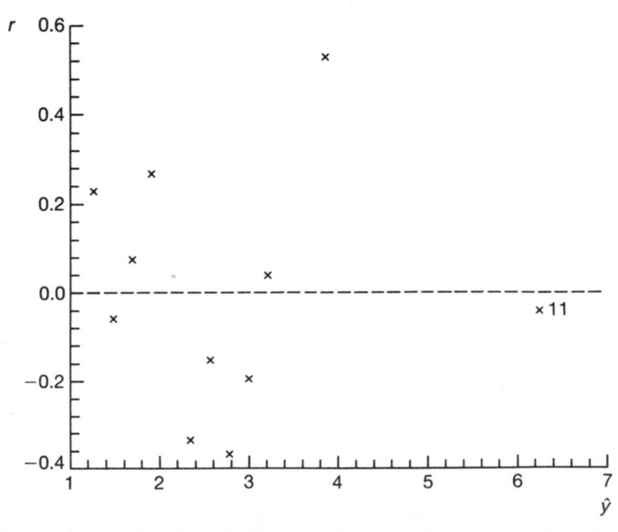


**Fig. 12.4** Example 14: Cook and Wang's 'data': residuals $r$ against fitted values $\hat{y}$

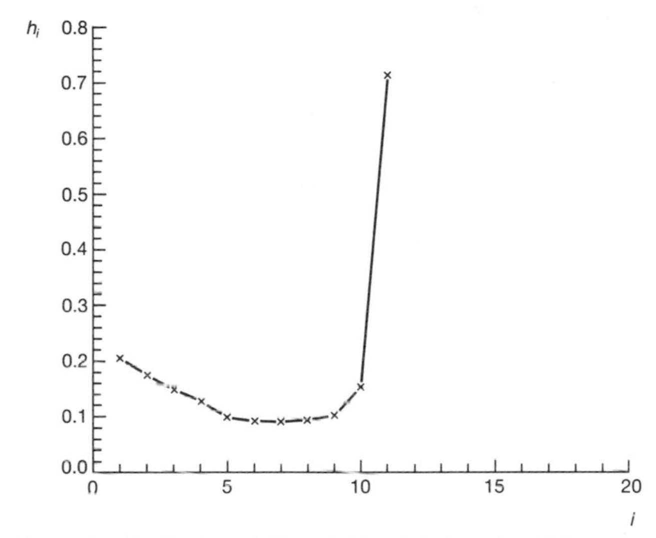


**Fig. 12.5** Example 14: Cook and Wang's 'data': index plot of leverage measure $h_i$

which model is not obeyed by observation 11. The values of $\hat{\lambda}$ and $T_p(1)$ after observation 11 has been deleted correctly lead to this model.

The scatter plot of the observations, Fig. 12.3, and the small residual for observation 11 in Fig. 12.4, suggest that this observation is at a leverage point. In fact, $h_{11} = 0.71$. The index plot of the $h_i$, Fig. 12.5, makes this point forcibly. From the theoretical development earlier in this section it is clear that this appreciable value of $h_i$ leads to the small values of both $r_{11}$ and $\overset{*}{w}_{11}$ shown in Fig. 12.1 and so to the failure of the added variable plot to detect the importance of observation 11 for the power transformation.

Although the graphical analysis of Figs. 12.3–5 suggests that observation 11 is anomalous, there is no reason, from the graphical analysis so far, to expect that observation 11 is important for the power transformation. This information is however obtained from diagnostic measures specific for the power transformation. Figure 12.6 is an index plot of three measures of the effect on $\hat{\lambda}$ of the deletion of individual observations. The easiest quantity to evaluate is (12.3.12), the quick estimate $\tilde{\lambda}_{(i)}$ evaluated at $\lambda = 1$. These values are shown joined by a continuous line. The dotted line gives, for comparison, the value of $\tilde{\lambda}$. This plot clearly shows that the effect of deleting observation 11 is to cause a change in the quick estimate of $\lambda$ from near one to near zero. The deletion of any of the other observations has only a slight effect. Also given in Fig. 12.6 are the values of the one-step estimate from $\hat{\lambda}$, that is $\hat{\lambda}^1_{(i)}$ (12.3.11) for deletion of a single observation, and the plot of the fully iterated estimate $\hat{\lambda}_{(i)}$. The three values are in sufficiently close agreement to give plots with a similar interpretation. An interesting feature is that, after observation 11 has been deleted, the three estimates are closer together than for deletion

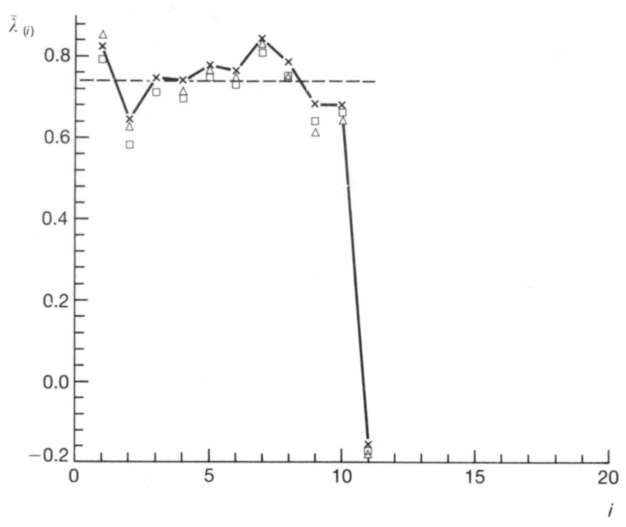


**Fig. 12.6** Example 14: Cook and Wang's 'data': index plot of deletion estimates of transformation parameter $\lambda$. × $\tilde{\lambda}_{(i)}$; □ $\hat{\lambda}^1_{(i)}$; △ $\hat{\lambda}_{(i)}$. The dotted line is $\tilde{\lambda}$

of any other point. One explanation is that the approximations to $\hat{\lambda}_{(i)}$ depend on a quadratic profile likelihood. Deletion of an outlying observation may well make the likelihood more quadratic.

Figure 12.6 provides a powerful diagnostic plot for the effect of observation

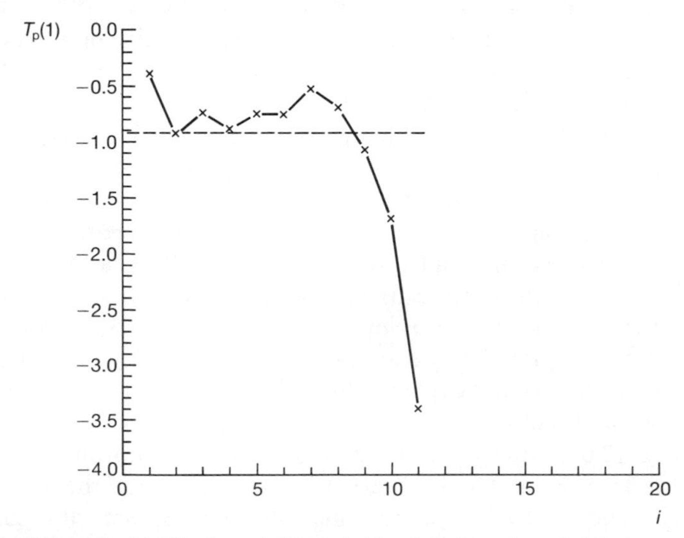


**Fig. 12.7** Example 14: Cook and Wang's 'data': index plot of deletion estimate of score statistic $T_p(1)$

11 on the transformation. It is however possible for an estimate to change appreciably, but not significantly, if its variance is large enough. But the plot itself provides some idea of the variability to be expected. Figure 12.7 is an index plot of $T_p(1)$, the score statistic for the hypothesis of no transformation as each observation in turn is deleted. This shows that deletion of observation 11 gives a significant value of the score statistic for transformation. The values plotted in the figure, however, are not suitable for routine diagnostic use as they have been calculated for each subsample of 10 observations. In particular, this calculation requires a new value of $\dot{y}$ for each statistic. As was mentioned above, one-step approximations to the score statistic could be derived, if necessary. But the index plot of $\tilde{\lambda}_{(i)}$ suffices to alert us to the anomalous and interesting properties of observation 11. Calculation of the value of the score statistic is a confirmation of something to which our attention has already been called. □

The previous example clearly shows how the techniques of this section can be used to provide diagnostic plots for power transformations. The example also shows how the added variable plots used from Chapter 6 onwards can fail due to the presence of an influential observation at a leverage point. In the three further examples of this section we exemplify the use of these new plots.

**Example 10 (concluded) Brownlee's stack loss data 2** The analysis of the stack loss data in Chapter 6 suggested the important properties of observation 21. The index plot of deletion estimates of $\lambda$ (Fig. 12.8) confirms

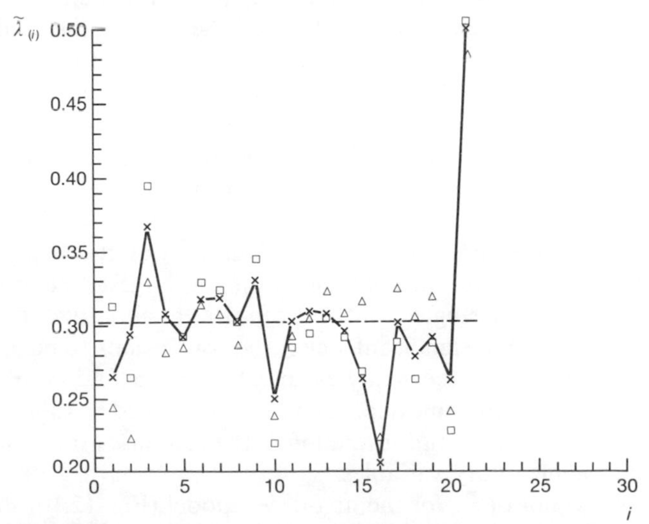


**Fig. 12.8** Example 10: Brownlee's stack loss data: index plot of deletion estimates of transformation parameter $\lambda$. × $\tilde{\lambda}_{(i)}$; □ $\hat{\lambda}^1_{(i)}$; △ $\hat{\lambda}_{(i)}$. The dotted line is $\tilde{\lambda}$

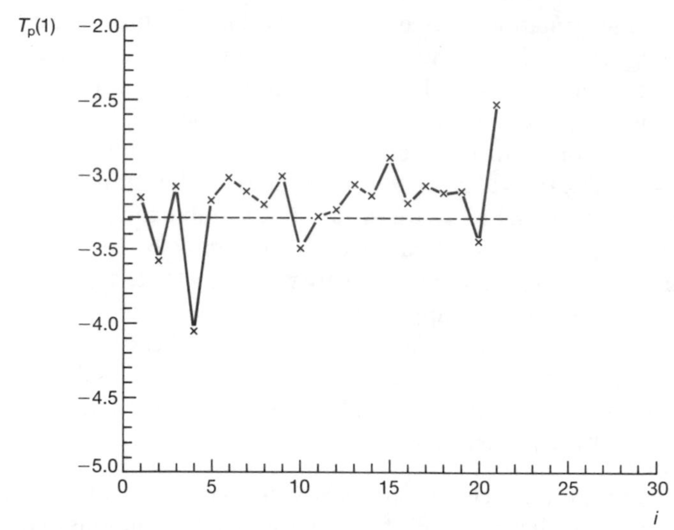


**Fig. 12.9** Example 10: Brownlee's stack loss data: index plot of deletion estimate of score statistic $T_p(1)$

this result for the first-order model in three explanatory variables. For all the data $\tilde{\lambda} = 0.302$ and $\hat{\lambda} = 0.297$. Deletion of individual observations yields values of $\tilde{\lambda}_{(i)}$ which oscillate around these values, except for the deletion of observation 21 for which $\tilde{\lambda}_{(21)} = 0.499$. The values of the other two estimates, $\hat{\lambda}^1_{(i)}$ and $\hat{\lambda}_{(i)}$, tell a similar tale. The three estimates are closer together after observation 21 has been deleted than they are in the majority of other cases. As in the previous example, deletion of the most influential observation causes the estimates to agree. The index plot of the deletion estimates of $T_p(1)$ (Fig. 12.9) shows that deletion of observation 21 reduces the evidence for a transformation, but not to the extent that the test becomes non-significant. As we have already seen, a log model with second-order terms fits the data well, but is also highly dependent upon observation 21. □

**Example 6 (concluded) Minitab tree data 7** For the tree data from the Minitab handbook the measurements on the largest tree, observation 31, have already been singled out as important, although deletion of this observation does not significantly alter the conclusions to be drawn from the data. As a result of the analyses in Chapters 6 and 8 several possible transformations were indicated, coupled with a set of possible linear models. Here we consider only transformation of the response for a first-order model in the two explanatory variables.

The index plot of $\tilde{\lambda}_{(i)}$ for the first-order model (Fig. 12.10) shows that, for deletion of any observation except 31, the values of $\tilde{\lambda}_{(i)}$ fluctuate around $\tilde{\lambda}$, which has the value 0.394. Deletion of observation 31 leads to the estimate

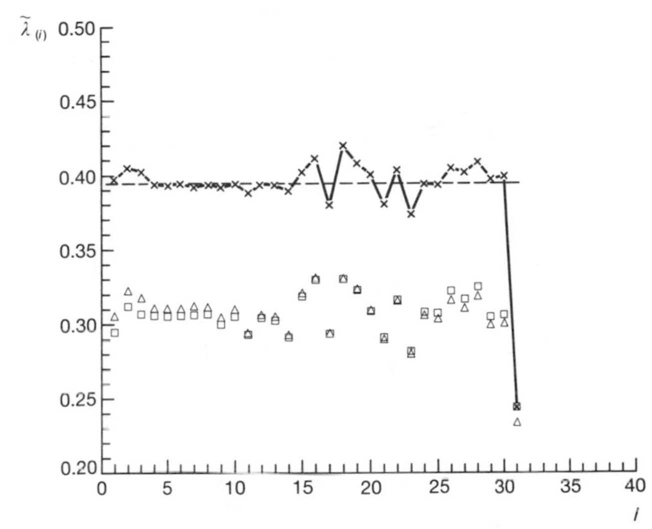


**Fig. 12.10** Example 6. Minitab tree data: index plot of deletion estimates of transformation parameter $\lambda$. $\times$ $\tilde{\lambda}_{(i)}$; $\square$ $\hat{\lambda}^{1}_{(i)}$; $\triangle$ $\hat{\lambda}_{(i)}$. The dotted line is $\tilde{\lambda}$

$\tilde{\lambda}_{(31)} = 0.243$. In this example the values of $\tilde{\lambda}_{(i)}$ are a poor approximation to $\hat{\lambda}_{(i)}$, as is $\tilde{\lambda}$ to the value of 0.307 for $\hat{\lambda}$. The one exception is for observation 31 when the two quick estimates and $\hat{\lambda}_{(i)}$ are all close together. In this example the index plot of $\tilde{\lambda}_{(i)}$ may over-emphasize the importance of observation 31. But the value of the score statistic, $-7.41$ for $T_{\mathrm{p}}(1)$ which becomes $-6.84$ after observation 31 has been deleted, makes it clear that a transformation should be considered, with or without observation 31. □

In Fig. 12.10 there was appreciable disagreement between $\hat{\lambda}$ and $\tilde{\lambda}$, but, even so, the index plot of $\tilde{\lambda}_{(i)}$ provided a useful diagnostic guide to the influence of individual observations on the transformation. We now consider the properties of the plot in an example where this disagreement is found in more extreme form.

**Example 8 (concluded) John's $3^{4-1}$ experiment 3** Of the examples analysed in this book, the one showing greatest disagreement between $\hat{\lambda}$ and $\tilde{\lambda}$ is Example 8, John's $3^{4-1}$ experiment. Because the data come from a designed experiment, the effects of leverage, which are the subject of this section, are not present. But the example is useful for the light it sheds on the properties of the quick estimates of the transformation parameter.

For all 27 observations $\hat{\lambda} = 8.27$ and $\tilde{\lambda} = 22.55$ with the score statistic $T_{\mathrm{p}}(1) = 2.98$. As we have already seen, this evidence for a power transformation depends entirely upon observation 11, a conclusion which is supported by the values of $\tilde{\lambda}_{(i)}$ in the index plot of Fig. 12.11. For all observations save 11

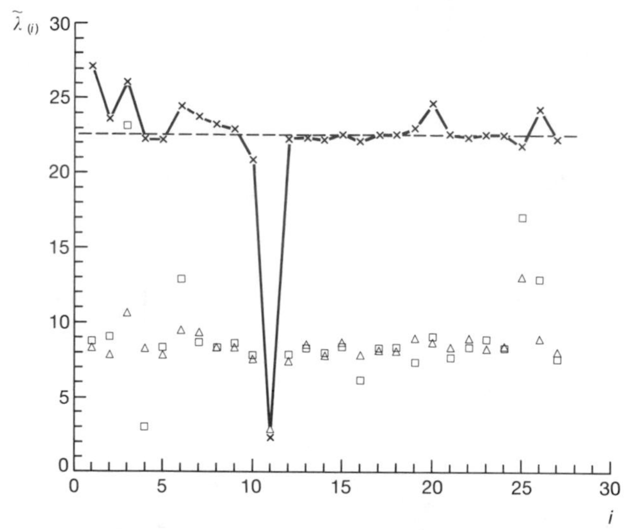


**Fig. 12.11** Example 8: John's $3^{4-1}$ experiment: index plot of deletion estimates of transformation parameter $\lambda$. X $\tilde{\lambda}_{(i)}$; □ $\hat{\lambda}^1_{(i)}$; △ $\hat{\lambda}_{(i)}$. The dotted line is $\tilde{\lambda}$

these values fluctuate around 22. When observation 11 is deleted the value is 2.27. The values are far from those for $\hat{\lambda}_{(i)}$ and the quick estimate $\hat{\lambda}^1_{(i)}$ which mostly lie between 7 and 9. An exception is the value for the deletion of observation 11, for which $\hat{\lambda}_{(11)} = 2.75$ close to the value of $\tilde{\lambda}_{(11)}$. The value of the one-step estimation from $\hat{\lambda}$ is $-6.72$, a poor approximation to the fully iterated value. In this example the lack of agreement between $\hat{\lambda}$ and $\tilde{\lambda}$ does not prevent the index plot of Fig. 12.11 from being a useful diagnostic tool. This conclusion agrees with those from the other examples, although the value of $\hat{\lambda}^1_{(11)}$ is anomalous.

The added variable plot of Fig. 6.11 provided an adequate method of assessing the influence of observation 11. The purpose of the alternative analysis here is to exemplify use of the index plot of $\tilde{\lambda}_{(i)}$. As a final use of such plots, Fig. 12.12 is an index plot showing the effect of deletion on the values of $T_p(1)$. This once more clearly shows that the evidence for the transformation depends only on observation 11. When observation 11 is deleted, the value of the score statistic is reduced to 0.57. □

## 12.4 Other analyses

Several of the sets of data which have served as examples in this book have been repeatedly analysed in the statistical literature. In this section no attempt will be made to present an exhaustive list of all earlier appearances of all 14 examples. Rather we consider analyses of the two examples which were

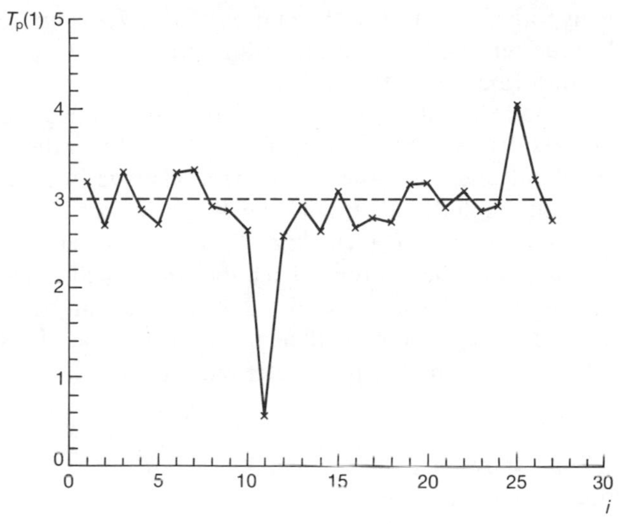


**Fig. 12.12** Example 8: John's $3^{4-1}$ experiment: index plot of deletion estimate of score statistic $T_p(1)$

given chapter length treatment by Daniel and Wood (1971). The emphasis is on points of methodological interest for diagnostic regression, rather than on giving a full account of the various analyses.

Of these two sets of data, Prater's gasoline data, Example 5, has attracted fewer analyses. Part of the attraction of these data for the purpose of this book lay in the misprint which was revealed by the half-normal plot of the deletion residuals $r_i^*$ shown in Fig. 4.21. The investigation of the uncorrupted data by Hader and Grandage (1958) consists of analysis of variance tables and related tests derived from fitting a first-order model in the four explanatory variables. A rather fuller, but basically similar, approach is followed in Chapter 6 of Wetherill (1981) where Table 6.4 provides analysis of variance tables for a variety of fitted models. Neither analysis employs graphical methods.

The discussion in Chapter 4 shows that these authors overlook the nested structure which was revealed by Daniel and Wood (1971, Chapter 8). In the analysis of Chapter 4 this structure was suggested by plots of the response against explanatory variables and confirmed by the plot of $x_2$ against $x_1$ (Fig. 4.26). Similar plots, but of residuals, alerted Daniel and Wood to the existence of the structure, which was made explicit by ordering the data on decreasing $x_1$. Ordering on $x_2$ or $x_3$, although not on $x_4$, would have done as well. Once the nested structure had been recognized, our analysis continued in Chapter 7 by considering a number of transformations of the response. These analyses, summarized in Tables 7.2–7.5, yield score statistics of about

$-3.5$ for the hypothesis of no transformation. Table 7.4 is an exception because, for the Guerrero and Johnson transformation, $\lambda = 1$ does not correspond to no transformation of the response.

Daniel and Wood do not consider transformations. Since the score tests give significant results and use of a transformation reduces the residual sum of squares by about one third, there is some advantage in using a suitable transformation, although the advantage may be greater for data with a larger range of response values. Daniel and Wood, however, make several important extensions to the analysis of the untransformed data.

The nested model (4.3.1) which was fitted to the data can be elaborated to include an individual regression coefficient for each crude oil as well as an individual intercept. This highly parameterized model is

$$E(Y_{ij}) = \alpha_i + \beta_i x_{4j} \qquad (i = 1, \ldots, 10; j = 1, \ldots, 32). \qquad (12.4.1)$$

Daniel and Wood show that this elaboration is not necessary and that the common slope of (4.3.1) is acceptable.

As a result of the analysis using (4.3.1) each crude is modelled with an individual intercept $\alpha_i$. This is useful provided the ten crudes are the only ones that will ever be met. But if the ten are a sample from a population of crude oils, it is potentially useful to know how the response depends on the measured properties $x_1, x_2$ and $x_3$. Daniel and Wood obtain this information from the predicted yield for each of the ten crudes at an end point, that is an $x_4$ value, of 332 °F. These derived responses are then used in a regression analysis with $x_1$ to $x_3$ as the explanatory variables. The best equation is selected using Mallows $C_p$ (Mallows 1973). They conclude with an analysis of variance in which the error is partitioned into components for between and within crude variation. As a result it is possible to make more precise statements about the differences to be expected between observations within the same and different crude oils.

A plot of one explanatory variable against another was valuable in detecting the nested nature of the gasoline data. For the stack loss data, which Daniel and Wood analyse in their Chapter 5, the plot of $x_2$ against $x_1$, Fig. 6.37 was likewise helpful, this time in displaying the special properties of observation 21. Several, but not all, of the published analyses of the stack loss data make use of graphical techniques. Of these, three are robust analyses.

Andrews (1974) employs normal plots of residuals from robust fits to identify observations which are not adequately explained by the first-order model. Denby and Mallows (1977) supplement Andrews' analysis with plots of the changes in residuals and parameter estimates as the trimming coefficient $c$ in (12.1.3) is altered. The plots of Chambers and Heathcote (1981), derived from estimation using the empirical characteristic function of the data, indicate the desirability of a transformation of the response.

Brownlee's original analysis, mentioned in Chapter 1, consisted of fitting the first-order model in the three explanatory variables to the untransformed response, dropping $x_3$ from the model and repeating the fitting process. This analysis is extended, while keeping in much the same framework, by Draper and Smith (1966, Exercise 6D). The much fuller analysis of Daniel and Wood (1971, Chapter 5) leads to dropping observations 1, 3, 4 and 21. This theme recurs not only in the robust analyses but also, for example, in Aitkin and Wilson (1980) where the EM algorithm is used to fit models in which the probability of an observation being an outlier is a parameter to be estimated. Cook (1979) uses methods similar to those of Chapter 4 to investigate the presence of multiple outliers.

Several of these authors justify their analyses as reproducing the results of Daniel and Wood by reasonably automatic methods. We conclude by summarizing their analysis (Daniel and Wood 1971, Chapter 5), stressing points of methodological interest.

Daniel and Wood start from the first-order model in three explanatory variables with untransformed response. As a result of fitting two further models suggested by the inadequacies of the first-order model, a systematic analysis is performed in which a $2^3$ factorial of possible equations is explored. The factors are the inclusion or exclusion of observation 21, $y$ or $\log y$ as response and $x_3$ in or out of the model. Their conclusion from the results of this factorial experiment are that $y$ should be used as a response, to drop $x_3$ from the model and to delete observation 21. In assessing transformation of the response Daniel and Wood work with $\log y$ rather than with $\dot{y} \log y$. They are therefore unable to make a direct comparison between logged and untransformed models on the basis of residual sums of squares. In fact, as we saw in Fig. 12.9, deletion of observation 21 does not remove all evidence of the need for a transformation.

Once $x_3$ has been excluded from the model, Daniel and Wood notice that the data contain near replicates which provide an approximate estimate of error against which to judge the fit of models. They then proceed to interpret the observations as part of a time series. Brownlee (1965, p. 454) says nothing about this, merely presenting the data as the results of 21 days operation of a chemical plant. If the data are treated as a record of operation on 21 successive days, there appears to be a time lag in the response to large changes in $x_1$. This interpretation suggests to Daniel and Wood that observations 1, 3, 4 and 21 should be omitted. This is, of course, the conclusion which has been reached by a variety of routes in many of the analyses mentioned above.

Once the four observations have been omitted, a further factorial experiment is used to build a second-order model in a systematic manner. All models include $x_1$ and $x_2$. Terms which are included or excluded are $x_1^2$, $x_2^2$, $x_1x_2$ and $x_3$. The log transformation of the response is also included as a final factor. The result of fitting these further 32 models is to drop $x_3$, include $x_1^2$

and to remain with the untransformed response, conclusions which only slightly extend those of the earlier factorial which was, however, run on all 21 observations. For the remaining 17 observations a plot of $x_1$ against $x_2$ shows how remote observation 2 is. This plot substantiates the results of Cook on the importance of observation 2 to the 17 observation second-order model and recalls the importance of observation 21 in Fig. 6.37, on which observation 2 is also shown.

An important methodological aspect of this analysis is the use of factorial experiments to explore possibilities in a systematic way. Daniel and Wood admit, with hindsight, that residual plots after fitting only two models would have led them to focus on observations 1, 3, 4 and 21. But an interesting aspect of the analysis they record is that it provides an example of the steps of an investigation, rather than being a summary with warts removed.

In their Chapter 7 Daniel and Wood return briefly to the stack loss data, using methods of error estimation from near neighbours, and extend the time series interpretation to allow for correlation between successive responses. Their final comment on these data will serve as my valediction to some of the sets of data which have been so extensively analysed in this book and elsewhere. 'We have overdone this study, hoping that readers will understand our purposes and that they will regard only charitably our final reduction of the problem almost to absurdity.'

## 12.5 Beyond diagnostic analyses

The diagnostic quantities derived in this book have tended to emerge in a rather informal and *ad hoc* manner. This is particularly true of the regression diagnostics of the first five chapters, which were often presented as intuitively reasonable rather than being derived from a general theory. In this last section an attempt will be made to amend this omission by considering what principles might be used to guide the development of new diagnostics. As a guide we shall draw on the experience of earlier chapters. The book concludes with a short discussion of the general framework in which diagnostic measures are used and the way in which that framework can be expected to alter due to increasingly sophisticated computer software.

The traditional methods of model checking described in Chapter 1 relied principally on a variety of plots of least squares residuals. These were used as omnibus tests for departures from the fitted model. The argument, from Chapter 3 onwards, is that specific tests should be designed to answer precisely formulated questions about the adequacy of the model. The general approach is that adequacy should be checked against a particular departure which, ideally, should be expressed in a parametric form. In the case of the addition of an extra carrier to a regression model, the parametric form is obvious. However, if the extension is to a power transformation of the response, possible transformations such as the square root, the logarithmic and the reciprocal have to be incorporated in a parametric family before appropriate tests can be developed. For the resulting diagnostic procedure to be useful it should not require appreciable computation above that needed to fit the tentative model. Score tests for parameterized departures are ideal in this respect. As we have seen, many diagnostic tests reduce to consideration of a single extra carrier in the regression model. The added variable plots of Chapter 5, the score tests for one-parameter families of transformations and the derivation of deletion residuals in Section 6.4 can all be considered as examples of the use of the score test.

The usefulness of the score test is enhanced by a suitable form of graphical display. We have seen numerous examples of both added variable plots and of index plots. Usually these suffice to call attention to interesting and important features of the data. But to avoid the dangers of over-interpretation some calibration of the diagnostics may be desirable. It is simpler if this calibration has a theoretical basis, such as the known null distribution of a test statistic. In the absence of such knowledge, or to assist the interpretation of a number of correlated tests, simulation can be employed with samples generated from the fitted model. One example is the introduction in Section 4.2 of simulation envelopes for $r_i^*$ and $C_i$. Although these were calculated only for a normal distribution of errors, extensions to other diagnostic measures and to other distributions are interesting possibilities.

A further desirable property of a diagnostic procedure is that if any departure from the model is indicated, a remedy should be suggested. An example is the added variable plot of the constructed variable for a power transformation. The presence of a significant slope in the plot indicates the desirability of transforming the response. But, in addition, the numerical value of the slope provides a quick estimate of the transformation parameter and so yields an indication of what a suitable transformation might be.

These and other general aspects of regression diagnostics are considered at greater length by Weisberg (1983) who divides his comments into two parts, the second of which is concerned with influence. One way of assessing influence was introduced informally in Chapter 1, where we looked at the effect of the deletion of a single observation on the conclusions drawn from the data. In succeeding chapters the deletion of observations one at a time was repeatedly used to obtain information on how particular aspects of the analysis depend on individual observations. For those problems in which the score statistic reduces to the regression coefficient of an additional carrier, the added variable plot is one way of displaying the information on influence. In other problems, index plots of the effect of deletion on either parameter estimates or score statistics are useful. One example is afforded by the index plots of Section 12.3. Sometimes a metric may be needed in which to combine several diagnostics. Thus changes in the individual parameters of a regression model are often not of direct interest. Combination of these changes in Cook's statistic, or one of its modifications, leads to plots containing a manageable amount of information from which incisive conclusions can be drawn.

In summary, the procedure is to find a suitable diagnostic measure which will often be a score test for a parameterized departure of interest. The effect of individual observations on this measure is then displayed graphically. In this procedure there is, of course, nothing specific to regression models and least squares, predominant though these are in this book. Box (1980) provides a discussion of model building and criticism from a Bayesian point of view.

This book has described many diagnostic measures in detail, sketched some further measures and hinted at others. Each has been for a distinct task in the process of model building and checking. But little has been said about the relationship of these procedures to any overall strategy. We now briefly consider such broader issues in the context of changing approaches to statistical analysis.

As was suggested in Chapter 1, the impact of the computer has been to broaden the scope of statistical analysis. It has also been to reduce the direct involvement of the expert statistician in the statistical analysis of data. In the days when the arithmetic needed for a regression analysis was performed laboriously on electromechanical calculators, the task was left to statisticians who accordingly had direct contact both with the data and with the analysis leading to a fitted model. With the advent of computer programs for regression analysis the computational burden was lifted, but problems remained for the user. Amongst these, non-statistical problems often predominated, such a manipulation of the data and the idiosyncrasies of individual computers. Statistical problems of interpretation tended to be pushed into the background.

The emergence of well-documented statistical packages has eliminated most problems associated with the computer. With the incorporation in packages of the diagnostic methods described in this book, scientists and engineers are in a position to respond to the statistical problems of model building and checking. But there remains the problem of how to combine the scientist's knowledge of his subject and data with the statistician's experience of the analysis of a wide variety of data. The diagnostic tools provide tactical information. How can the overall strategy of model building be guided?

This is clearly a large question which, in its general form, has aspects which go beyond statistics to psychological and philosophical problems of how we understand and model the world. But even within the statistical context, it is hard to see how to answer the question. An example of the kind of difficulty which can arise is provided by the analysis of the chimpanzee data in Chapter 11. What strategy would lead to the formulation of the lognormal model as one possibility with the gamma model as an alternative? Yet, as we have seen, both are plausible models. For the far narrower problem in which only one family of models is entertained, the tools of this book provide a way in which fitted models can be checked through plots of influence measures and residuals. But which checks should be performed? Even for the very restricted class of simple regression models the possibilities of model elaboration mentioned above include the addition of a further carrier, transformation of the response and deletion of one or more observations. Although the methods of this book provide evidence for each of these possibilities in turn, they provide no guidance as to which checks should be performed and in what order. Yet the order is important for, as we have seen several times, these three extensions are often nearly equivalent when the effect is measured by statistical tests. They are often very different, however, in their implications for modelling the physical world.

Some comments on the tactics and strategy of building and checking linear and nonlinear regression models are given by Cox (1977). The context is of choices made by human intervention. An important challenge is how the computer can be used in the process of model building. The techniques of artificial inteligence and, in particular, those of expert knowledge based systems, may be helpful in developing programs which learn, in a dynamic manner, how to choose between alternative elaborations and simplifications of a model. The program may thus lead to fitting a model with little human intervention. For this to become a reality the information provided by the data has to be supplemented by input both from the scientist whose data is being analysed and from the statistician. However such programs turn out, the diagnostic methods described in this book, or methods very similar to them, will be central.

# Appendix
# The inverse of a partitioned matrix

Suppose the carriers in a linear model are partitioned into two sets when the model can be written as

$$E(Y) = X\beta = X_1\beta_1 + X_2\beta_2. \tag{A1}$$

Then the covariance matrices of the least squares estimates $\hat{\beta}_1$ and $\hat{\beta}_2$ can be found from the formula for the inverse of a partitioned matrix. Let

$$(X^{\mathrm{T}}X) = \begin{pmatrix} X_1^{\mathrm{T}}X_1 & X_1^{\mathrm{T}}X_2 \\ X_2^{\mathrm{T}}X_1 & X_2^{\mathrm{T}}X_2 \end{pmatrix} = \begin{pmatrix} B & C \\ C^{\mathrm{T}} & D \end{pmatrix}$$

where, therefore, $B$ and $D$ are symmetrical. It is easy to verify that, if all necessary inverses exist, then

$$(X^{\mathrm{T}}X)^{-1} = \begin{pmatrix} B & C \\ C^{\mathrm{T}} & D \end{pmatrix}^{-1} = \begin{pmatrix} P & Q \\ Q^{\mathrm{T}} & U \end{pmatrix} = \begin{pmatrix} B^{-1} + FE^{-1}F^{\mathrm{T}} & -FE^{-1} \\ -E^{-1}F^{\mathrm{T}} & E^{-1} \end{pmatrix}$$

with

$$E = D - C^{\mathrm{T}}B^{-1}C \qquad \text{and} \qquad F = B^{-1}C. \tag{A2}$$

One consequence is that

$$\operatorname{var}(\hat{\beta}_2) = \sigma^2\{X_2^{\mathrm{T}}X_2 - X_2^{\mathrm{T}}X_1(X_1^{\mathrm{T}}X_1)^{-1}X_1^{\mathrm{T}}X_2\}^{-1}. \tag{A3}$$

In the special case of the addition of one extra carrier to a model, (A1) can be written in the notation of Chapter 5 as

$$E(Y) = X\beta + w\gamma$$

when, from (A3),

$$\operatorname{var}(\hat{\gamma}) = \sigma^2\{w^{\mathrm{T}}w - w^{\mathrm{T}}X(X^{\mathrm{T}}X)^{-1}X^{\mathrm{T}}w\}^{-1}. \tag{A4}$$

It can sometimes be helpful to rewrite (A4) in the various forms introduced in Section 5.2 as

$$\begin{aligned} \operatorname{var}(\hat{\gamma}) &= \sigma^2/w^{\mathrm{T}}(I - H)w \\ &= \sigma^2/w^{\mathrm{T}}Aw \\ &= \sigma^2/\overset{*}{w}{}^{\mathrm{T}}\overset{*}{w}. \end{aligned}$$

# References


Aitkin, M. (1982). Discussion of Dr Atkinson's paper. *J. R. Statist. Soc.* B **44**, 26.

Aitkin, M. and Wilson, G. T. (1980). Mixture models, outliers and the EM algorithm. *Technometrics* **22**, 325–31.

Akaike, H. (1973). Information theory and an extension of the maximum likelihood principle. In *Second International Symposium on Information Theory* (ed. B. N. Petrov and F. Czaki), pp. 267–81. Budapest: Akademiai Kiado.

Andrews, D. F. (1971). A note on the selection of data transformations. *Biometrika* **58**, 249–54.

——— (1974). A robust method for multiple linear regression. *Technometrics* **16**, 523–31.

Andrews, D. F. and Pregibon, D. (1978). Finding the outliers that matter. *J. R. Statist. Soc.* B **40**, 473–9.

Aranda-Ordaz, F. J. (1981). On two families of transformations to additivity for binary response data. *Biometrika* **68**, 357–63.

Atkinson, A. C. (1970). A method for discriminating between models (with discussion). *J. R. Statist. Soc.* B **32**, 323–53.

——— (1973). Testing transformations to normality. *J. R. Statist. Soc.* B **35**, 473–9.

——— (1981). Two graphical displays for outlying and influential observations in regression. *Biometrika* **68**, 13–20.

——— (1982a). Regression diagnostics, transformations and constructed variables (with discussion). *J. R. Statist. Soc.* B **44**, 1–36.

——— (1982b). Robust and diagnostic regression analyses. *Commun. Statist.* A **11**, 2559–71.

——— (1983). Diagnostic regression analysis and shifted power transformations. *Technometrics* **25**, 23–33.

Atkinson, A. C. and Pearce, M. C. (1976). The computer generation of beta, gamma and normal random variables. *J. R. Statist. Soc.* A **139**, 431–61.

Baker, R. J. and Nelder, J. A. (1978). *The GLIM System Release 3 Manual.* Oxford: Numerical Algorithms Group.

Barndorff-Nielsen, O. E. and Cox, D. R. (1984). Bartlett adjustments to the likelihood ratio statistic and the distribution of the maximum likelihood estimator. *J. R. Statist. Soc.* B **46**, 483–95.

Barnett, V. and Lewis, T. (1978). *Outliers in Statistical Data.* Chichester: Wiley.

Baskerville, J. C. and Toogood, J. H. (1982). Guided regression modeling for prediction and exploration of structure with many explanatory variables. *Technometrics* **24**, 9–17.

Bates, D. and Watts, D. (1980). Relative curvature measures of nonlinearity (with discussion). *J. R. Statist. Soc.* B **42**, 1–25.

Beale, E. M. L. (1960). Confidence regions in non-linear estimation (with discussion). *J. R. Statist. Soc.* B **22**, 41–88.

Belsley, D. A., Kuh, E. and Welsch, R. E. (1980). *Regression Diagnostics.* New York: Wiley.

Bickel, P. and Doksum, K. (1981). An analysis of transformations revisited. *J. Amer. Statist. Assoc.* **76,** 296–311.

Bliss, C. I. (1970). *Statistics in Biology,* Vol. 2. New York: McGraw-Hill.

Box, G. E. P. (1980). Sampling and Bayes inference in scientific modeling and robustness (with discussion). *J. R. Statist. Soc.* A **143**, 383–430.

Box, G. E. P. and Cox, D. R. (1964). An analysis of transformations (with discussion). *J. R. Statist. Soc.* B **26**, 211–46.

——— (1982). An analysis of transformations revisited, rebutted. *J. Amer. Statist. Assoc.* **77**, 209–10.

Box, G. E. P., Hunter, J. S. and Hunter, W. G. (1978). *Statistics for Experimenters.* New York: Wiley.

Box, G. E. P. and Tidwell, P. W. (1962). Transformations of the independent variables. *Technometrics* **4**, 531–50.

Bradley, J. V. (1982). The insidious L-shaped distribution. *Bull. Psychonomic Soc.* **20**, 85–8.

Brown, B. W. and Hollander, M. (1977). *Statistics: A Biomedical Introduction.* New York: Wiley.

Brownlee, K. A. (1960). *Statistical Theory and Methodology in Science and Engineering.* New York: Wiley.

——— (1965). *Statistical Theory and Methodology in Science and Engineering* (2nd edn.). New York: Wiley.

Carroll, R. J. (1980). A robust method for testing transformations to achieve approximate normality. *J. R. Statist. Soc.* B **42**, 71–8.

——— (1982). Two examples of transformations when there are possible outliers. *Appl. Statist.* **31**, 149–52.

Carroll, R. J. and Ruppert, D. (1981). On prediction and the power transformation family. *Biometrika* **68**, 609–15.

——— (1985). Transformations in regression: a robust analysis. *Technometrics* **27**, 1–12.

Chambers, J. M., Cleveland, W. S., Kleiner, B. and Tukey, P. A. (1983). *Graphical Methods for Data Analysis.* Belmont, California: Wadsworth.

Chambers, R. L. and Heathcote, C. R. (1981). On the estimation of slope and the identification of outliers in linear regression. *Biometrika* **68**, 21–33.

Cheng, R. C. H. and Amin, N. A. K. (1981). Maximum likelihood estimation of parameters in the inverse Gaussian distribution, with unknown origin. *Technometrics* **23**, 257–63.

——— (1983). Estimating parameters in continuous univariate distributions with a shifted origin. *J. R. Statist. Soc.* B **45**, 394–403.

Cook, R. D. (1977). Detection of influential observations in linear regression. *Technometrics* **19**, 15–18.

——— (1979). Influential observations in linear regression. *J. Amer. Statist. Assoc.* **74**, 169–74.

——— (1982). Discussion of Dr. Atkinson's paper. *J. R. Statist. Soc.* B **44**, 28.

Cook, R. D. and Wang, P. C. (1983). Transformations and influential cases in regression. *Technometrics* **25**, 337–43.

Cook, R. D. and Weisberg, S. (1980). Characterizations of an empirical influence function for detecting influential cases in regression. *Technometrics* **22**, 495–508.

——— (1982). *Residuals and Influence in Regression.* New York and London: Chapman and Hall.

——— (1983). Diagnostics for heteroscedasticity in regression. *Biometrika* **70**, 1–10.

Cox, D. R. (1961). Tests of separate families of hypotheses. *Proc. 4th Berkeley Symp.* **1**, 105–23.

——— (1962). Further results on tests of separate families of hypotheses. *J. R. Statist. Soc.* B **24**, 406–23.

——— (1970). *Analysis of Binary Data.* London: Chapman and Hall.

——— (1977). Nonlinear models, residuals and transformations. *Math. Operationsforsch. Statist., Ser. Statistics* **8**, 3–22.

Cox, D. R. and Hinkley, D. V. (1974). *Theoretical Statistics.* London: Chapman and Hall.

Cox, D. R. and McCullagh, P. (1982). Some aspects of analysis of covariance (with discussion). *Biometrics* **38**, 541–61.

Daniel, C. (1959). Use of half-normal plots for interpreting factorial two-level experiments. *Technometrics* **1**, 311–41.

Daniel, C. and Wood, F. S. (1971). *Fitting Equations to Data.* New York: John Wiley.

Dempster, A. P. and Gasko-Green, M. (1981). New tools for residual analysis. *Ann. Statist.* **9**, 945–59.

Denby, L. and Mallows, C. L. (1977). Two diagnostic displays for robust regression analysis. *Technometrics* **19**, 1–13.

Dobson, A. J. (1983). *An Introduction to Statistical Modelling.* London: Chapman and Hall.

Draper, N. R. and Smith, H. (1966). *Applied Regression Analysis.* New York: Wiley.

——— (1981). *Applied Regression Analysis* (2nd edn.). New York: Wiley.

Efron, B. and Hinkley, D. V. (1978). Assessing the accuracy of the maximum likelihood estimator: Observed versus expected Fisher information. *Biometrika* **65**, 457–87.

Ezekiel, M. and Fox, K. A. (1959). *Methods of Correlation and Regression Analysis* (3rd edn.). New York: Wiley.

Farewell, V. T. and Prentice, R. L. (1977). A study of distributional shape in life testing. *Technometrics* **19**, 69–75.

Finney, D. J. (1977). Dimensions of statistics. *Appl. Statist.* **26**, 285–9.

Fishman, G. S. (1978). *Principles of Discrete Event Simulation.* New York: Wiley-Interscience.

Furnival, G. M. and Wilson, R. W. (1974). Regression by leaps and bounds. *Technometrics* **16**, 499–511.

Gentleman, J. F. and Wilk, M. B. (1975a). Detecting outliers in a two-way table: 1. Statistical behavior of residuals. *Technometrics* **17**, 1–14.

——— (1975b). Detecting outliers. II. Supplementing the direct analysis of residuals. *Biometrics.* **31**, 387–410.

Gnanadesikan, R. (1977). *Methods for Statistical Data Analysis of Multivariate Observations.* New York: Wiley.

Green, P. J. (1984). Iteratively reweighted least squares for maximum likelihood estimation, and some robust and resistant alternatives (with discussion). *J. R. Statist. Soc.* B **46**, 149–92.

Griffiths, D. A. (1980). Interval estimation for the three-parameter lognormal distribution via the likelihood function. *Appl. Statist.* **29**, 58–68.

Guerrero, V. M. and Johnson, R. A. (1982). Use of the Box–Cox transformation with binary response models. *Biometrika* **69**, 309–14.

Hader, R. J. and Grandage, A. H. E. (1958). Simple and multiple regression analyses. In *Experimental Designs in Industry* (ed. V. Chew), pp. 108–37. New York: Wiley.

Harris, P. and Peers, H. W. (1980). The local power of the efficient scores test statistic. *Biometrika* **67**, 525–9.

Hawkins, D. M. (1980). *Identification of Outliers*. London: Chapman and Hall.

Hawkins, D. M., Bradu, D. and Kass, G. V. (1984). Location of several outliers in multiple-regression data using elemental sets. *Technometrics* **26**, 197–208.

Hoaglin, D. C., Mosteller, F. and Tukey, J. W. (1983). *Understanding Robust and Exploratory Data Analysis*. New York: Wiley.

Hoaglin, D. C. and Welsch, R. E. (1978). The hat matrix in regression and ANOVA. *Amer. Statist.* **32**, 17–22.

Huber, P. (1981). *Robust Statistics*. New York: Wiley.

——— (1983). Minimax aspects of bounded-influence regression with discussion. *J. Amer. Statist. Assoc.* **78,** 66–80.

John, J. A. (1978). Outliers in factorial experiments. *Appl. Statist.* **27**, 111–19.

John, J. A. and Draper, N. R. (1980). An alternative family of transformations. *Appl. Statist.* **29**, 190–7.

Johnson, W. and Geisser, S. (1983). A predictive view of the detection and characterization of influential observations in regression analysis. *J. Amer. Statist. Assoc.* **78**, 137–44.

Joiner, B. L. (1981). Lurking variables: some examples. *Amer. Statist.* **35**, 227–33.

Jørgensen, B. (1983). Maximum likelihood estimation and large-sample inference for generalized linear and nonlinear regression models. *Biometrika* **70**, 19–28.

Kalbfleisch, J. D. and Prentice, R. L. (1980). *The Statistical Analysis of Failure Time Data*. New York: Wiley.

Kennedy, W. J. and Gentle, J. E. (1980). *Statistical Computing*. New York: Marcel Dekker.

Knuth, D. E. (1973). *The Art of Computer Programming* (Vol. 3): *Sorting and Searching*. Reading, Mass.: Addison-Wesley.

——— (1981). *The Art of Computer Programming* (Vol. 2): *Seminumerical Algorithms* (2nd edn). Reading, Mass.: Addison-Wesley.

Königer, W. (1983). Remark AS R47. A remark on AS 177. Expected normal order statistics (exact and approximate). *Appl. Statist.* **32**, 223–4.

Krasker, W. S. and Welsch, R. E. (1982). Efficient bounded-influence regression estimation. *J. Amer. Statist. Assoc.* **77**, 595–604.

Larsen, W. A. and McCleary, S. A. (1972). The use of partial residual plots in regression analysis. *Technometrics* **14**, 781–90.

Mallows, C. L. (1973). Some comments on $C_p$. *Technometrics* **15**, 661–75.

——— (1982). Discussion of Dr. Atkinson's paper. *J. R. Statist. Soc.* B **44**, 29.

McCullagh, P. (1980). A comparison of transformations of chimpanzee learning data. *GLIM Newsletter* No. 2, 14–18.

——— (1983). Quasi-likelihood functions. *Ann. Statist.* **11**, 59–67.

McCullagh, P. and Nelder, J. A. (1983). *Generalized Linear Models*. London: Chapman and Hall.

Marsaglia, G. and Bray, J. A. (1964). A convenient method for generating normal variables. *S.I.A.M. Rev.* **6**, 260–4.

Milliken, G. A. and Graybill, F. A. (1970). Extensions of the general linear hypothesis. *J. Amer. Statist. Assoc.* **65**, 797–807.

Mosteller, F. and Tukey, J. W. (1977). *Data Analysis and Regression.* Reading, Mass.: Addison-Wesley.

Nelder, J. A. and Wedderburn, R. W. M. (1972). Generalized linear models. *J. R. Statist. Soc.* A **135**, 370–84.

Obenchain, R. L. (1977). Letter to the editor. *Technometrics* **19**, 348–9.

Pearson, E. S. and Hartley, H. O. (1966). *Biometrika Tables for Statisticians* (Vol. 1). Cambridge: University Press.

Pereira, B. de B. (1977). A note on the consistency and on the finite sample comparisons of some tests of separate families of hypotheses. *Biometrika* **64**, 109–13.

Peto, R. and Lee, P. (1973). Weibull distributions for continuous carcinogenesis experiments. *Biometrics* **29**, 457–70.

Pregibon, D. (1979). Data analytic methods for generalized linear models. Unpublished Ph.D. dissertation, University of Toronto.

——— (1980). Goodness of link tests for generalized linear models. *Appl. Statist.* **29**, 15–24.

——— (1981). Logistic regression diagnostics. *Ann. Statist.* **9**, 705–24.

——— (1982). Score tests in GLIM. In *GLIM 82: Proceedings of the International Conference on Generalized Linear Models* (ed. R. Gilchrist), pp. 87–97. New York: Springer (Lecture Notes in Statistics No. 14).

Prentice, R. L. (1974). A log gamma model and its maximum likelihood estimation. *Biometrika* **61**, 539–44.

Rao, C. R. (1973). *Linear Statistical Inference and its Applications* (2nd edn.). New York: Wiley.

Ripley, B. D. (1977). Modelling spatial patterns (with discussion). *J. R. Statist. Soc.* B **39**, 172–212.

——— (1981). *Spatial Statistics.* New York: Wiley.

——— (1983). Computer generation of random variables: a tutorial. *Int. Statist. Rev.* **51**, 301–19.

Rojas, B. A. (1973). On Tukey's test of additivity. *Biometrics* **29**, 45–52.

Royston, J. P. (1982). Algorithm AS 177. Expected normal order statistics (exact and approximate). *Appl. Statist.* **31**, 161–5.

Ruppert, D. and Carroll, R. J. (1980). Trimmed least squares estimation in the linear model. *J. Amer. Statist. Assoc.* **75**, 828–38.

Ryan, T. A., Joiner, B. L. and Ryan, B. F. (1976). *Minitab Student Handbook.* North Scituate, Mass.: Duxbury Press.

Schlesselman, J. (1971). Power families: a note on the Box and Cox transformation. *J. R. Statist. Soc.* B **33**, 307–11.

Schrader, R. M. and Hettmansperger, T. P. (1980). Robust analysis of variance based upon a likelihood ratio criterion. *Biometrika* **67**, 93–101.

Seber, G. A. F. (1977). *Linear Regression Analysis.* New York: Wiley.

Shen, S. M. (1982). A method for discriminating between models describing compositional data. *Biometrika* **69**, 587–95.

Sibson, R. (1974). $D_A$-optimality and duality. In *Progress in Statistics*, Vol. 2. (Proc. 9th European Meeting of Statisticians, Budapest, 1972. Ed. J. Gani, K. Sarkadi and I. Vincze). Amsterdam: North-Holland.

Silvey, S. D. (1980). *Optimal Design.* London: Chapman and Hall.

Smith, R. L. (1985). Maximum likelihood estimation in a class of non-regular cases. *Biometrika* **72**, 67–90.
Snedecor, G. W. and Cochran, W. G. (1967). *Statistical Methods* (6th edn.). Ames, Iowa: Iowa State University Press.
Sprent, P. (1982). Discussion of Dr. Atkinson's paper. *J. R. Statist. Soc.* B **44**, 22–4.
Tufte, E. R. (1983). *The Visual Display of Quantitative Information.* Cheshire, Conn.: Graphics Press.
Tukey, J. W. (1949). One degree of freedom for non-additivity. *Biometrics* **5**, 232–42.
Velleman, P. F. and Welsch, R. E. (1981). Efficient computing of regression diagnostics. *Amer. Statist.* **35**, 234–42.
Vinod, H. D. and Ullah, A. (1981). *Recent Advances in Regression Methods.* New York: Marcel Dekker.
Voorn, W. J. (1981). A class of variate transformations causing unbounded likelihood. *J. Amer. Statist. Assoc.* **76**, 709–12.
Weisberg, S. (1980). *Applied Linear Regression.* New York: Wiley.
——— (1981). A statistic for allocating $C_p$ to cases. *Technometrics* **23**, 27–31.
——— (1983). Some principles for regression diagnostics and influence analysis. *Technometrics* **25**, 240–4.
Wetherill, G. B. (1981). *Intermediate Statistical Methods.* London: Chapman and Hall.
Williams, D. A. (1982). Discussion of Dr. Atkinson's paper. *J. R. Statist. Soc.* B **44**, 33.
Yates, F. (1972). A Monte-Carlo trial on the behaviour of the non-additivity test with nonnormal data. *Biometrika* **59**, 253–61.

# Author index

# Subject index